Numerical Methods
for Engineers

NUMERICAL METHODS FOR ENGINEERS

BILAL M. AYYUB

Department of Civil Engineering
University of Maryland at College Park

RICHARD H. McCUEN

Department of Civil Engineering
University of Maryland at College Park

 PRENTICE HALL, Upper Saddle River, New Jersey 07458

Library of Congress Cataloging-in-Publication Data

Ayyub, Bilal M.
    Numerical methods for engineers / Bilal M. Ayyub, Richard H. McCuen.
      p.   cm.
    Includes bibliographical references and index.
    ISBN 0-13-337361-4
    1. Engineering mathematics.   2. Numerical analysis.   I. McCuen,
Richard H.   II. Title.
TA335.A99 1996
620'.001'5194—dc20
                                  95–20132
                                  CIP

Acquisition editor: *Linda Ratts Engelman*
Editorial/production supervision: *Raeia Maes*
Cover designer: *Bruce Kenselaar*
Manufacturing buyer: *Donna Sullivan*

 © 1996 by Prentice-Hall, Inc.
Simon & Schuster/A Viacom Company
Upper Saddle River, New Jersey 07458

Printed in the United States of America

10 9 8 7 6 5

ISBN 0-13-337361-4

PRENTICE-HALL INTERNATIONAL (UK) LIMITED, *London*
PRENTICE-HALL OF AUSTRALIA PTY. LIMITED, *Sidney*
PRENTICE-HALL CANADA INC., *Toronto*
PRENTICE-HALL HISPANOAMERICANA, S.A., *Mexico*
PRENTICE-HALL OF INDIA PRIVATE LIMITED, *New Delhi*
PRENTICE-HALL OF JAPAN, INC., *Tokyo*
SIMON & SCHUSTER ASIA PTE. LTD., *Singapore*
EDITORA PRENTICE-HALL DO BRASIL, LTDA., *Rio de Janeiro*

Dedicated to

my wife, Deena, and our three children, Omar, Rami, and Samar,
and
my mother and father for supporting me in all my endeavors

Bilal Ayyub

Contents

Preface

In preparing this book, we strove to achieve a set of educational objectives that include (1) introducing numerical methods to engineering students and practicing engineers, (2) emphasizing the practical aspects of the use of these methods, and (3) establishing the limitations, advantages, and disadvantages of the methods. Although the book was developed with emphasis on engineering and technological problems, the numerical methods can also be used to solve problems in other fields of science.

Due to the rapid advancement of computer technology and its availability to practicing engineers, we have demonstrated the power of numerical methods in the context of computer use. Studies show that engineering practice will continue to tackle more complex design problems, thus necessitating increased reliance on numerical methods because of the lack of analytical solution methodologies. Therefore, this book is intended to better prepare future engineers, as well as assist practicing engineers, in understanding the fundamentals of numerical methods, especially their application, limitations, and potentials.

STRUCTURE, FORMAT, AND MAIN FEATURES

We have developed this book with a dual use in mind, as both a self-learning guidebook and as a required textbook for a course. In either case, the text has been designed to achieve important educational objectives.

The 10 chapters of the book cover the following subjects: (1) introduction to the text; (2) matrix analysis; (3) introduction to numerical methods; (4) the identification of the roots of equations; (5) the solution of linear simultaneous equations; (6) the interpolation of values of a dependent variable for a given set of discrete measurements; (7) approximation of the differential or integral of an unknown function for a given set of discrete measurements from the function; (8) solutions to differential equations; (9) describing data graphically and computing important characteristics of sample measurements and making basic statistical computations, and (10) fitting a curve or a model to data. The book was designed for an introductory course in numerical analysis with emphasis on applications. In developing the book, a set of educational outcomes as detailed in Chapter 1 motivated the structure and content of its text. Ultimately, serious readers will find the material very useful in engineering problem solving and decision making.

Each chapter of the book contains many computational examples, as well as a section on applications that contain additional engineering examples. Also, each chapter includes a set of exercise problems that cover the materials of the chapter. The problems were designed carefully to meet the needs of instructors in assigning homework and the readers in practicing the fundamental concepts.

The book can be covered in one semester or two semesters in conjunction with a programming language. The chapter sequence can be followed as a recommended sequence. However, if needed, instructors can choose a subset of the chapters of the book for courses that do not permit a complete coverage of the materials or a coverage that cannot follow the presented order. After completing Chapters 1, 2, 3, the readers will have sufficient background to follow and understand the materials in Chapters 4 through 8 in any sequence. However, before reading Chapter 10, Chapter 9 should be studied. The accompanying figure illustrates the logical sequence of these chapters in terms of their interdependencies:

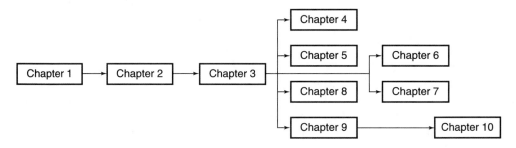

ACKNOWLEDGMENTS

This book was developed over several years and draws on our experiences in teaching courses on numerical methods and statistics. Drafts of most sections of the book were tested in several courses at the University of Maryland for about two years before its publication. This testing period has proved to be a very valuable tool in establishing its contents and final format and structure.

We would like to acknowledge all the students who participated in the testing of the book for their comments, input, and suggestions. The students who took courses on computational methods in civil engineering in the semesters of 1992 and 1994 contributed to this endeavor. Their feedback was very helpful and greatly contributed to the final product. Also, we would like to acknowledge the instructors who used the book during this testing period, Drs. P. Johnson and R. Muhanna. The assistance of Mr. Ru-Jen Chao, Dr. Maguid Hassan, and Mr. Ibrahim Assakkaf in critically reviewing the problems at the end of the chapters and developing some of the example problems and problem solutions is gratefully acknowledged. The reviewers' comments that were provided to us by the publisher were used to improve the book to meet the needs of instructors and enhance the education process. We acknowledge the publisher and the reviewers for these comments.

We would like to invite users of this book, including both readers and instructors, to send us any comments on the book by E-mail at ba15@umail.umd.edu. These comments will be used in developing future editions of the book.

Bilal M. Ayyub
Richard H. McCuen

About the Authors

Bilal M. Ayyub, professor of civil engineering at the University of Maryland at College Park, completed his B.S. degree in civil engineering in 1980 and completed both the M.S. (1981) and Ph.D. (1983) in civil engineering at the Georgia Institute of Technology. Dr. Ayyub has an extensive background in risk-based analysis and design, simulation, and marine structural reliability. He is engaged in research work involving structural reliability, marine structures, uncertainty modeling and analysis, and mathematical modeling using the theories of probability, statistics, and fuzzy sets. He has completed various research projects that were funded by the National Science Foundation, the U.S. Coast Guard, the U.S. Navy, the Department of Defense, the U.S. Army Corps of Engineers, the Maryland State Highway Administration, the American Society of Mechanical Engineers, and several engineering companies. Currently, he is the chairman of the American Society of Civil Engineers (ASCE) Committee on the Reliability of Offshore Structures. He also was the general chairman of the first and second International Symposia on Uncertainty Modeling and Analysis that were held in 1990 and 1993. He is the author of about 200 publications in journals and conference proceedings and reports. Dr. Ayyub is the double co-recipient of the American Society of Naval Engineers (ASNE) Jimmie Hamilton Award for the best paper in the *Naval Engineers Journal* in 1985 and 1992. Also, he was co-recipient of the ASCE Outstanding Research Oriented Paper in the *Journal of Water Resources Planning and Management* for 1987 and the ASCE Edmund Friedman Award in 1989.

Richard H. McCuen, professor of civil engineering at the University of Maryland at College Park, received degrees from Carnegie Mellon University and the Georgia Institute of Technology. He is the author of eleven books and over 175 professional papers. He received the Icko Iben Award from the American Water Resource Association and was co-recipient of the 1988 Outstanding Research Award from the American Society of Civil Engineers Water Resources Planning and Management Division. Topics in statistical hydrology and stormwater management are his primary research interest.

NUMERICAL METHODS
FOR ENGINEERS

1

Introduction

1.1 NUMERICAL ANALYSIS IN ENGINEERING

The advancement of computer technology and its availability to practicing engineers has affected engineering education in many aspects. To prepare engineering students and practicing engineers for current and future challenges, an introduction to and sometimes emphasis on computer-based methods are needed. Currently, practical engineering problems demand the use of computers to obtain solutions in a timely manner and with acceptable accuracy.

The solutions of engineering problems can be obtained using analytical methods or numerical methods. The complexity of many practical engineering problems makes it necessary to use numerical methods; that is, very often analytical solutions cannot be obtained. For example, engineering students learn analytical differentiation and integration of mathematical functions within the first two years of their college education. However, analytical differentiation and integration provide a closed-form derivative and integral, respectively, only for simple functions. As the problems take on greater reality and the complexity of these functions increases, our ability to use analytical differentiation or integration diminishes until the point where analytical solutions are not possible. Also, it is common in engineering to require the use of functions that cannot be expressed in closed form for which derivatives or integrals are needed. In such cases, analytical methods of differentiation and integration cannot be used. Therefore, there is a need for another class of solution methods. Numerical methods belong to this second class of solution tools that can be used where analytical methods are not capable of or practical for providing solutions.

The objective of this book is to introduce numerical methods to engineering students and practicing engineers. The practical aspects of the use of these methods are emphasized throughout the book, with sections in each chapter where practical applications are provided. Although the book was developed with emphasis on engineering and technological problems, numerical methods can be used to solve problems in many other fields. Numerical methods are provided with varying levels of detail and emphasis. A critical presentation of these methods is provided to enhance the readers' understanding.

The numerical methods can be solved by hand or with a computer. The importance of numerical methods in engineering can be effectively demonstrated in cases dealing with complex problems for which analytical solutions cannot be obtained or hand calculations cannot be made. In this book, we use common engineering problems to demonstrate the computational procedures. The examples were intentionally selected with traceable solutions so that readers can reproduce them. It is helpful, but not necessary, if the readers are familiar with the fundamentals of a computer language. We present algorithms in the book as pseudocodes for selected methods. Also, programming considerations for selected methods are provided.

The use of any computational method, analytical or numerical, without the proper understanding of the limitations and shortcomings can have serious consequences. Before using them, users should become well versed with the numerical methods in terms of both the computational details and their limitations, shortcomings, and strengths. Numerical methods should not be used as black boxes with input and output or as numerical recipes.

1.1.1 Decision Making in Engineering

Engineering is a profession that continually places its members in decision-making positions. To make the best decisions, engineers must be aware of and fully understand alternative solution procedures. This need has become more important in professional practice as the problems that engineers must address have become more important to society and as the design methods have become more complex. The term *best* decisions is multifaceted. Engineers need accurate solutions that are both unbiased and have high precision. The solution must be cost effective and have minimal environmental consequences. The adopted solution should improve the community and be esthetically appealing. These are just some of the criteria that engineers use in making design decisions.

When engineers are given a design problem, they typically approach the problem using a systematic procedure. One formulation of the process is as follows: (1) identify the problem, (2) state the objectives, (3) develop alternative solutions, (4) evaluate the alternatives, and (5) implement the best alternative.

Engineering decision problems can be classified into single- and multiple-objective problems. Example objectives are minimizing the total expected cost, maximizing safety, maximizing the total expected utility value, and maximizing the total expected profit. Decision analysis requires the definition of these objectives. For cases of multiple objectives, the objectives need to be stated in the same units, and weights that reflect the importance of the objectives and that can be used to combine the objectives need to be assigned. Then the problem can be formulated in a suitable form for attaining the decision objectives. The development of alternative solutions comes next, which is an especially critical stage of the process. The team of engineers must have a sound technical background and a broad understanding of alternative design methods. To properly evaluate each alternative, a complete understanding of the technical basis of the design method is required. Then alternatives need to be evaluated, and selection can then be made.

Computers have increased the number of alternative solution procedures that are available to engineers. Whereas the engineers of past generations were limited to graphical, analytical, and simple empirical methods, engineers can now consider numerical and simulation methods in their design work. The increased complexity of the numerical and simulation methods is believed to improve the accuracy of the solutions, and since these methods are easily implemented with a computer, the increased effort over the less detailed methods is minimal. Thus the increase in design effort is more than offset by the expected improvement in accuracy. However, to achieve this benefit, the engineer must fully understand the more complex methods so that they can be properly applied.

1.1.2 Expected Educational Outcomes

While a textbook can be the principal medium for someone to individually gain an understanding of a body of knowledge, books such as this one are typically used as part of a course on numerical methods. This book has been designed to be used in either capacity, as a self-learning guidebook or as a required textbook for a course. In either case, the text has been designed to produce the following educational outcomes:

1. The reader of the book should be able to numerically solve problems in (a) the identification of roots of equations; (b) the solution of linear simultaneous equations; (c) the interpolation of values of a dependent variable given a set of discrete measurements; (d) approximating the differential or integral of an unknown function given a set of discrete measurements from the function; (e) finding solutions to differential equations; (f) describing data graphically and computing important characteristics of sample measurements; (g) making basic statistical computations, and (h) fitting a curve or a model to data. Also, the reader will learn the fundamentals of matrix analysis.

2. The reader should be able to select from alternative methods the one method that is most appropriate for a specific problem and state reasons for the selection.

3. The reader should be able to formulate algorithms to solve problems numerically.

4. The reader should understand the limitations of each numerical method, especially the conditions under which they may fail to converge to a solution.

5. The reader should recognize engineering applications of the various numerical methods.

In developing the topics presented in the book, these expected educational outcomes motivated the structure and content of the text. Ultimately, the serious reader will find the material very useful in engineering problem solving and decision making.

1.2 ANALYTICAL VERSUS NUMERICAL ANALYSIS

In the previous section, problem-solving methods were separated into two classes, analytical and numerical. These two approaches are distinguished on the basis of their algorithms. Analytical calculus forms the basis for analytical problem solving, while finite-difference arithmetic forms the basis of numerical methods. As an example, finding the minimum or maximum of a simple function is a common need in engineering. For the analytical approach, the derivative is computed analytically and set equal to zero, and the resulting equation is solved for the value of the unknown. If the minimum of the function $y = x^2 - 3x + 2$ was of interest, it is easily obtained analytically by computing the derivative as $2x - 3$. Therefore, the minimum solution occurs at $x = 1.5$, which is obtained by setting the derivative to zero and solving for x. One numerical solution is to iterate over a range of x values at a constant increment Δx and select the value of x for which y is smallest. For example, if we define the interval from 1 to 2 and use an increment of 0.2, the following tabulation indicates that the minimum value of y falls in the interval $1.4 < x < 1.6$:

x	1.0	1.2	1.4	1.6	1.8	2.0
y	0	−0.16	−0.24	−0.24	−0.16	0

To improve the accuracy of the solution, the search interval of x could be established as $1.4 < x < 1.6$, the increment decreased to 0.04, and the search repeated. For this example, the analytical approach is clearly superior. But what if the function is complex and an analytical derivative cannot be found? Then a numerical solution is necessary.

Both classes have advantages and disadvantages. Analytical techniques provide a direct solution and will result in an exact solution, if one exists. Analytical methods usually require less time to find a solution. However, this occurs partly because analytical methods are practical only for functions that have a simple, closed-form mathematical structure. Also, the analytical solution procedure becomes considerably more complex when constraints are involved.

Numerical analysis can be used with functions with a moderately complex structure, and it is easy to include constraints on the unknowns in the solution. However, numerical methods most often require a considerable number of iterations in order to approach the true solution. Furthermore, the solution usually is not exact, and it is also necessary to provide initial estimates of the unknowns. Since the solution is not exact, error analysis and error estimation are often necessary.

1.3 TAYLOR SERIES EXPANSION

A Taylor series is commonly used as a basis of approximation in numerical analysis. The objective of this section is to provide a review of the Taylor series.

A Taylor series is the sum of functions based on continually increasing derivatives.

For a function $f(x)$ that depends on only one independent variable x, the value of the function at point $x_0 + h$ can be approximated by the following Taylor series:

$$f(x_0 + h) = f(x_0) + hf^{(1)}(x_0) + \frac{h^2}{2!}f^{(2)}(x_0) + \frac{h^3}{3!}f^{(3)}(x_0)$$

$$+ \cdots + \frac{(h^n)}{n!}f^n(x_0) + R_{n+1}$$

(1-1a)

in which x_0 is some base value (or starting value) of the independent variable x; h is the distance between x_0 and the point x at which the value of the function is needed; that is, $h = x - x_0$; $n!$ is the factorial of n [i.e., $n! = n(n - 1)(n - 2) \ldots 1$]; R_{n+1} is the remainder of the Taylor series expansion; and the superscript (n) on the function $f^{(n)}$ indicates the nth derivative of the function $f(x)$. The derivatives of $f(x)$ in Eq. 1-1a are evaluated at x_0. Equation 1-1a can also be written in the following form:

$$f(x_0 + h) = \sum_{k=0}^{\infty} \frac{h^k}{k!}f^{(k)}(x_0)$$

(1-1b)

where $0! = 1$. Equations 1-1 are based on the assumption that continuous derivatives exist in an interval that includes the points x and x_0.

Equations 1-1 can be used with a finite number of terms to produce approximations for the evaluated function at $(x_0 + h)$. The order of the approximation is defined by the order of the highest derivative that is included in the approximation. For example, the first-order approximation is defined by the terms of the Taylor series expansion up to the first derivative as follows:

$$f(x_0 + h) \approx f(x_0) + hf^{(1)}(x_0)$$

(1-2a)

Similarly, the second- and third-order approximations are given, respectively, by

$$f(x_0 + h) \approx f(x_0) + hf^{(1)}(x_0) + \frac{h^2}{2!}f^{(2)}(x_0)$$

(1-2b)

and

$$f(x_0 + h) \approx f(x_0) + hf^{(1)}(x_0) + \frac{h^2}{2!}f^{(2)}(x_0) + \frac{h^3}{3!}f^{(3)}(x_0)$$

(1-2c)

The accuracy of the approximation improves as the order of the approximation is increased.

The importance of the individual terms of the Taylor series depends on the nature of the function and the distance h. The higher-order terms become more important as the nonlinearity of the function increases and the difference $x - x_0$ increases. If the function is linear, then only the term with the first derivative is necessary. As the separation distance h increases, the importance of the nonlinear terms usually increases. Thus the error increases as fewer terms of the Taylor series are included.

Example 1-1. Nonlinear Polynomials

Taylor series are used to evaluate a function that is of practical interest, but with an unknown expression. If the expression of the function were known, then the Taylor series approximation would not be necessary. In this example, the true function will be assumed to be known in order to demonstrate the computational aspects of Taylor series and the sizes of the remainder terms in the series. For this purpose, the following fourth-order polynomial is used:

$$f(x) = \frac{1}{8}x^4 - \frac{1}{6}x^3 + \frac{1}{2}x^2 - \frac{1}{2}x + 2 \tag{1-3}$$

The function has the following derivatives:

$$f^{(1)}(x) = 0.5x^3 - 0.5x^2 + x - 0.5 \tag{1-4}$$

$$f^{(2)}(x) = 1.5x^2 - x + 1 \tag{1-5}$$

$$f^{(3)}(x) = 3x - 1 \tag{1-6}$$

$$f^{(4)}(x) = 3 \tag{1-7}$$

$$f^{(5)}(x) = 0 \tag{1-8}$$

For an assumed base point (x_0) of 1.0, the value of the function and its derivatives are

$$f(x) = 1.95833 \tag{1-9a}$$

$$f^{(1)}(x) = 0.5 \tag{1-9b}$$

$$f^{(2)}(x) = 1.5 \tag{1-9c}$$

$$f^{(3)}(x) = 2 \tag{1-9d}$$

$$f^{(4)}(x) = 3 \tag{1-9e}$$

$$f^{(5)}(x) = 0 \tag{1-9f}$$

Using a separation distance h of 0.5, the values of the individual terms of the Taylor series are shown in Table 1-1a. For $x = 1.5$, the linear term is dominant. For x from 2 to 3, the second term is dominant, with less emphasis given to the linear term and more emphasis given to the cubic term. For $x = 3.5$, the cubic term is the most important, while for $x = 4$, the fourth-order term is most important. These results illustrate the interaction between the importance of the terms and both the nonlinearity of the function and the separation distance. The cumulative values for the terms of the Taylor series are given in Table 1-1b, and the differences between the cumulative values and the true values are given in Table 1-1c. The dif-

TABLE 1-1 (a) Values of the Individual Terms of the Taylor Series Expansion of Eq. 1-1 for the Quadratic Function of Eq. 1-3; (b) Cumulative Function of the Taylor Series Expansion; (c) Difference between the Cumulative Function and the True Value of the Function

Part	x	h	$hf^{(1)}(x_0)$	$\dfrac{h^2}{2!}f^{(2)}(x_0)$	$\dfrac{h^3}{3!}f^{(3)}(x_0)$	$\dfrac{h^4}{4!}f^{(4)}(x_0)$
a. Individual	1.5	0.5	0.25	0.1875	0.04167	0.0078125
terms	2.0	1.0	0.50	0.7500	0.33333	0.1250000
	2.5	1.5	0.75	1.6875	1.12500	0.6328125
	3.0	2.0	1.00	3.0000	2.66667	2.0000000
	3.5	2.5	1.25	4.6875	5.20833	4.8828125
	4.0	3.0	1.50	6.7500	9.00000	10.1250000
b. Cumulative	1.5		2.20833	2.39583	2.43750	2.44531
function	2.0		2.45833	3.20833	3.54167	3.66667
	2.5		2.70833	4.39538	5.52083	6.15365
	3.0		2.95833	5.95833	8.62500	10.62500
	3.5		3.20833	7.89583	13.10417	17.98698
	4.0		3.45833	10.20833	19.20833	29.33333
c. Errors	1.5		−0.23698	−0.04948	−0.00781	0.00000
	2.0		−1.20833	−0.45833	−0.12500	0.00000
	2.5		−3.44531	−1.75781	−0.63281	0.00000
	3.0		−7.66667	−4.66667	−2.00000	0.00000
	3.5		−14.77865	−10.09115	−4.88281	0.00000
	4.0		−25.87500	−19.12500	−10.12500	0.00000

ferences show the need to consider the higher-order terms as the separation distance increases. This is clearly evident from Fig. 1-1, which shows the cumulative values from Table 1-1b as a function of separation distance h.

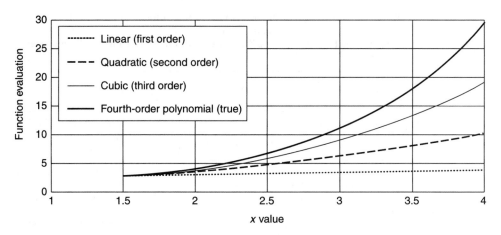

Figure 1-1   Taylor Series Approximation of a Fourth-order Polynomial

1.4 APPLICATIONS

1.4.1 Taylor Series Expansion of the Square Root

The square-root function is used in this section to illustrate the Taylor series expansion. The square-root function can be expressed as

$$f(x) = \sqrt{x} \tag{1-10}$$

To evaluate the Taylor series, the derivatives of the function are developed as follows:

$$f^{(1)}(x) = \frac{1}{2}x^{-0.5} \tag{1-11a}$$

$$f^{(2)}(x) = -\frac{1}{4}x^{-1.5} \tag{1-11b}$$

$$f^{(3)}(x) = \frac{3}{8}x^{-2.5} \tag{1-11c}$$

Higher-order derivatives can be developed similarly. For a base point $x_0 = 1$ and $h = 0.001$, the four terms of the Taylor series produce the following estimate for the square root of 1.001:

$$f(1.001) = \sqrt{1.001} \approx \sqrt{1} + 0.5(0.001)(1)^{-0.5} - \frac{1}{4(2!)}(0.001)^2(1)^{-1.5} \tag{1-12a}$$

$$+ \frac{3}{8(3!)}(0.001)^3(1)^{-2.5}$$

or

$$f(1.001) \approx 1 + 0.5 \times 10^{-3} - 0.125 \times 10^{-6} + 0.625 \times 10^{-10} = 1.0004999 \qquad (1\text{-}12b)$$

which equals the true value to the number of decimal points shown. Because the interval h is so small, the linear approximation would have been accurate to five decimal points.

1.4.2 Example Taylor Series

In this section, example Taylor series are provided. These series might be familiar to the readers from previous courses in mathematics. For example, the exponential evaluation to the base e of x can be expressed by the following series:

$$e^x = 1 + x + \frac{x^2}{2!} + \frac{x^3}{3!} + \cdots + \frac{x^k}{k!} + \cdots = \sum_{k=0}^{\infty} \frac{x^k}{k!} \qquad (1\text{-}13)$$

This equation is valid only for x that is finite. Similarly, the natural logarithm of x can be expressed using a Taylor series as

$$\ln(x) = (x - 1) - \frac{(x-1)^2}{2!} + \frac{(x-1)^3}{3!} - \cdots = \sum_{k=0}^{\infty} (-1)^{k-1} \frac{(x-k)^k}{k!}$$

$$for \; 0 < x \le 2 \qquad (1\text{-}14)$$

The sine and cosine functions can also be expressed using Taylor series as follows, respectively:

$$\sin(x) = x - \frac{x^3}{3!} + \frac{x^5}{5!} - \cdots = \sum_{k=0}^{\infty} (-1)^k \frac{x^{2k+1}}{(2k+1)!} \qquad (1\text{-}15)$$

$$\cos(x) = 1 - \frac{x^2}{2!} + \frac{x^4}{4!} - \cdots = \sum_{k=0}^{\infty} (-1)^k \frac{x^{2k}}{(2k)!} \qquad (1\text{-}16)$$

Equations 1-15 and 1-16 are valid only for x values that are finite. The reciprocal of $(1 - x)$, where $|x| < 1$, can be expressed by the following Taylor series:

$$\frac{1}{1-x} = 1 + x + x^2 + x^3 + \cdots = \sum_{k=0}^{\infty} x^k \qquad (1\text{-}17)$$

These example Taylor series can be used to evaluate their corresponding functions at any point x using a base value of $x_0 = 0$ and increment h.

1.5 PROBLEMS

1-1. Engineering problems can be solved using analytical or numerical methods. Describe the differences between these two classes of methods. What are the limitations, advantages, and disadvantages of the two classes?

1-2. Using the Taylor series expansion for $\cos(x)$ provided in Section 1.4.2, evaluate the series starting at $x_0 = 0$ radian with $h = 0.1$ to $x = 1$ radian. Use one term, two terms, and three terms of the series. Evaluate the error in each value, and discuss the effect of incrementally adding the terms to the series. Compare your results with the true solution.

1-3. Using the Taylor series for $\sin(x)$ of Eq. 1-15, evaluate the series starting at $x_0 = 0$ radian with $h = 0.1$ to $x = 1$ radian. Use one term, two terms, and three terms of the series. Evaluate the error in each value, and discuss the effect of incrementally adding the terms to the series. Compare your results with the true solution.

1-4. Using the Taylor series for e^x provided by Eq. 1-13, evaluate the series starting at $x_0 = 0$ with $h = 0.1$ to $x = 1$. Use one term, two terms, and three terms of the series. Evaluate the error in each value, and discuss the effect of incrementally adding the terms to the series. Compare your results with the true solution.

1-5. Develop a Taylor series expansion of the following polynomial:

$$f(x) = x^3 - 3x^2 + 5x + 10$$

Use $x = 2$ as the base (or starting) point and h as the increment. Evaluate the series for $h = 0.1, 0.2, 0.3, 0.4, 0.5, 0.6, 0.7, 0.8, 0.9,$ and 1.0. Discuss the accuracy of the method as the terms of the Taylor series are added incrementally.

1-6. Develop a Taylor series expansion of the following polynomial:

$$f(x) = x^5 - 5x^4 + x^2 + 6$$

Use $x = 2$ as the base (or starting) point and h as the increment. Evaluate the series for $h = 0.1, 0.2, 0.3, 0.4, 0.5, 0.6, 0.7, 0.8, 0.9,$ and 1.0. Discuss the accuracy of the method as the terms of the Taylor series are added incrementally.

1-7. Develop a Taylor series expansion of the following function:

$$f(x) = x^2 - 5x^{0.5} + 6$$

Use $x = 2$ as the base (or starting) point and h as the increment. Evaluate the series for $h = 0.1, 0.2, 0.3, 0.4, 0.5, 0.6, 0.7, 0.8, 0.9,$ and 1.0. Discuss the accuracy of the method as the terms of the Taylor series are added incrementally.

1-8. Write a computer program in FORTRAN to evaluate the Taylor series expansion for Problem 1-2.

1-9. Write a computer program in FORTRAN to evaluate the Taylor series expansion for Problem 1-4.

1-10. Write a computer program in FORTRAN to evaluate the Taylor series expansion for Problem 1-6.

1-11. Write a computer program in FORTRAN to evaluate the Taylor series expansion for Problem 1-7.

2

Matrices

2.1 INTRODUCTION

2.1.1 Definition of a Matrix

What is a matrix? Simply stated, it is a rectangular array of numbers. Each column has the same number of values r, where r is the number of rows. Similarly, each row has the same number of values c, where c is the number of columns. It is not necessary for the number of rows to equal the number of columns. A rectangular matrix with three rows and two columns is shown in Fig. 2-1a. A capital letter will be used to denote the name of the array; the values contained in the matrix are called elements and are denoted as a_{ij}, where the lowercase letter a is used to indicate the element of matrix $\mathbf{A}$, i refers to the row, and j refers to the column. Figure 2-1b shows a matrix with r rows and c columns; thus, element b_{57} refers to the value in fifth row and seventh column of matrix $\mathbf{B}$.

(a) 3×2 matrix $\qquad \mathbf{A} = \begin{bmatrix} a_{11} & a_{12} \\ a_{21} & a_{22} \\ a_{31} & a_{32} \end{bmatrix}$

(b) $r \times c$ matrix $\qquad \mathbf{B} = \begin{bmatrix} b_{11} & b_{12} & \cdots & b_{1c} \\ b_{21} & b_{22} & \cdots & b_{2c} \\ \vdots & \vdots & \vdots & \vdots \\ b_{r1} & b_{r2} & \cdots & b_{rc} \end{bmatrix}$

Figure 2-1 Example Matrices

2.1.2 Formation of a Matrix

What produces a matrix? Matrices are useful for representing data sets. Figure 2-2 shows a data matrix of concentrations of the following three water-quality indicators from five wells: hardness (H), alkalinity (A), and pH. The matrix can be represented with the well number either as the row variable (Fig. 2-2a) or the column variable (Fig. 2-2b). The form of the matrix depends on style or use. It is easier to compare the values of pH by reading down the columns (Fig. 2-2a) than across (Fig. 2-2b), so Fig. 2-2a is probably the preferred form. The form may also be dictated by computational requirements, which are illustrated later. The variables can be properties that vary with time, space, and individuals or on the basis of any characteristic. For example, if measurements of average temperature, wind speed, humidity, and evaporation were made daily for a month, then the 30 by 4, denoted as 30×4, matrix would have the time of measurement as the row variable and the climatic variable as the column variable.

Matrices can also be the result of matrix computations. For example, if we were interested in the correlation between the water-quality variables of Fig. 2-2, where correlation is an indicator of the strength of linear association between variables, matrix computations on the data matrix of Fig. 2-2a would produce the following correlation matrix:

$$
\begin{array}{c}
 \\
\text{H} \\
\text{A} \\
\text{pH}
\end{array}
\begin{array}{ccc}
\text{H} & \text{A} & \text{pH} \\
\left[\begin{array}{ccc}
1.00 & -0.23 & -0.27 \\
-0.23 & 1.00 & 0.64 \\
-0.27 & 0.64 & 1.00
\end{array} \right]
\end{array}
\tag{2-1}
$$

			H	A	pH
		1	140	35	7.7
		2	195	12	7.1
(a) 5×3 matrix	Well	3	283	53	6.4
		4	132	188	8.3
		5	60	55	6.5

				Well		
(b) 3×5 matrix		1	2	3	4	5
	H	140	195	283	132	60
	A	35	12	53	188	55
	pH	7.7	7.1	6.4	8.3	6.5

Figure 2-2 Data Matrix of Water-Quality Concentrations of Hardness (H mg/L), Alkalinity (A, mg/L), and pH from Five Wells

This matrix shows that the measurements of hardness and alkalinity have an inverse relationship (-0.23) that is not strong; pH and alkalinity have a positive moderate relationship (0.64).

Matrices can also result from the physical structure of a problem. Figure 2-3 shows a simple structure that is subject to a distributed load w and a point load P. The two joints rotate through angles θ_1 and θ_2 and are displaced distance d, with the sign of the angles assuming counterclockwise movement as positive and the displacement d shown. For given values of w, P, length (L), moment of inertia (I), and modulus of elasticity (E), the rotations and displacement are determined by the solution of the following system of three equations:

$$4\theta_1 + \theta_2 - \frac{3}{L}d = \frac{wL^3}{24E_1} \tag{2-2a}$$

$$\theta_1 + 4\theta_2 - \frac{3}{L}d = \frac{-wL^3}{24E_1} \tag{2-2b}$$

$$\theta_1 + \theta_2 - \frac{4}{L}d = \frac{PL^2}{6EI} \tag{2-2c}$$

For the three unknowns (θ_1, θ_2, and d), the three equations can be represented by the following 3×4 matrix:

$$\begin{bmatrix} 4 & 1 & \frac{-3}{L} & \frac{wL^3}{24EI} \\ 1 & 4 & \frac{-3}{L} & \frac{-wL^3}{24EI} \\ 1 & 1 & \frac{-4}{L} & \frac{PL^2}{6EI} \end{bmatrix} \tag{2-3}$$

This matrix is a numerical shorthand for Eqs. 2-2, and the matrix representation is useful for numerical calculations.

Figure 2-3 Development of a Matrix for a Portal Frame

2.1.3 Types of Matrices

A *square matrix* is a special case in which the number of rows equals the number of columns; the correlation matrix of Eq. 2-1 is an example. The set of elements a_{ij} for which $i = j$ is called the *principal diagonal*. For the correlation matrix of Eq. 2-1, all the values on the principal diagonal equal 1.

An *upper triangular* matrix has values on the principal diagonal and above, but contains zero values for the matrix elements below the principal diagonal. All the elements above the principal diagonal of a *lower triangular* matrix are zero. For a *strictly upper triangular* matrix or *strictly lower triangular* matrix, the elements of the principal diagonal would also be zero.

A *diagonal matrix* is a square matrix with all the elements equal to zero except for the elements on the principal diagonal. The following is an $n \times n$ diagonal matrix that contains n eigenvalues (λ_i, $i = 1, 2, \ldots, n$):

$$
\begin{bmatrix}
\lambda_1 & 0 & 0 & \cdots & 0 \\
0 & \lambda_2 & 0 & \cdots & 0 \\
0 & 0 & \lambda_3 & \cdots & 0 \\
\vdots & \vdots & \vdots & \ddots & \vdots \\
0 & 0 & 0 & \cdots & \lambda_n
\end{bmatrix}
\tag{2-4}
$$

A *banded matrix* is a square matrix with elements of zero except for the principal diagonal and values in the positions adjacent to the principal diagonal. For example, a *tridiagonal matrix* is the special case of a banded matrix that has zeros except in the principal diagonal and the two adjacent diagonals:

$$
\begin{bmatrix}
a_{11} & a_{12} & 0 & 0 & 0 \\
a_{21} & a_{22} & a_{23} & 0 & 0 \\
0 & a_{32} & a_{33} & a_{34} & 0 \\
0 & 0 & a_{43} & a_{44} & a_{45} \\
0 & 0 & 0 & a_{54} & a_{55}
\end{bmatrix}
\tag{2-5}
$$

A *unit matrix* (**I**) is a commonly used diagonal matrix; **I** has 1's on the principal diagonal, with zeros for all other elements.

All the elements of a *null matrix* are zero. It is usually denoted as **0**. It does not have to be a square matrix.

A *symmetric matrix* is a square matrix in which $a_{ij} = a_{ji}$; the correlation matrix of Eq. 2-1 is symmetric.

A *skew-symmetric matrix* is a square matrix with all values on the principal diagonal equal to zero and with off-diagonal values such that $a_{ij} = -a_{ji}$.

The *transpose* of matrix **A** is the matrix that results if the rows of **A** are written as the columns of the new matrix denoted as $\mathbf{A}^T$. In terms of the elements, $a_{ij}^T = a_{ji}$. If matrix **A** has r rows and c columns, then $\mathbf{A}^T$ has c rows and r columns. Note that $(\mathbf{A}^T)^T = \mathbf{A}$. The matrix of Fig. 2-2a is the transpose of the matrix of Fig. 2-2b.

2.2 MATRIX OPERATIONS

The primary arithmetic operations are addition, subtraction, multiplication, and division. Matrix algebra has operations called matrix addition, subtraction, and multiplication; there is no operation called matrix division. Instead, there is a matrix operation called matrix inversion. Each of these matrix operations is discussed in this section.

2.2.1 Matrix Equality

Two matrices are said to be equal, that is, $\mathbf{A} = \mathbf{B}$, if every element in matrix $\mathbf{A}$ is equal to the corresponding element in matrix $\mathbf{B}$; that is, $a_{ij} = b_{ij}$ for all i and j. When performing matrix computations on a computer, it is often necessary to store the original contents of a matrix for later use prior to performing computations on the matrix. In such cases, matrix $\mathbf{A}$ can be formed immediately after matrix $\mathbf{B}$ is inputted; this would involve two logic loops (e.g., logic loops such as DO loops), one for the rows and one for the columns. At that point, matrix $\mathbf{A}$ equals matrix $\mathbf{B}$.

2.2.2 Matrix Addition and Subtraction

To perform matrix addition, that is, $\mathbf{C} = \mathbf{A} + \mathbf{B}$, the two or more matrices must have the same number of rows and the same number of columns. The elements of $\mathbf{C}$ are then the sum of the corresponding elements of the matrices being summed; that is,

$$c_{ij} = a_{ij} + b_{ij} \tag{2-6}$$

To perform matrix subtraction, $\mathbf{C} = \mathbf{A} - \mathbf{B}$, the matrices must have the same number of rows and the same number of columns. The elements of $\mathbf{C}$ are the differences between the corresponding elements of the matrices being subtracted; that is,

$$c_{ij} = a_{ij} - b_{ij} \tag{2-7}$$

Whereas matrix addition is not directional (commutative), that is,

$$\mathbf{A} + \mathbf{B} = \mathbf{B} + \mathbf{A} \tag{2-8}$$

matrix subtraction depends on the direction of the operation (noncommutative):

$$\mathbf{A} - \mathbf{B} \neq \mathbf{B} - \mathbf{A} \tag{2-9}$$

Example 2-1. Matrix Addition and Subtraction

For the following two matrices, $\mathbf{A}$ and $\mathbf{B}$,

$$\mathbf{A} = \begin{bmatrix} 2 & 4 & 6 \\ 1 & 3 & 5 \\ 9 & 10 & 11 \end{bmatrix} \tag{2-10}$$

$$\mathbf{B} = \begin{bmatrix} 0 & 2 & 3 \\ 7 & 8 & 4 \\ 2 & 1 & 1 \end{bmatrix} \tag{2-11}$$

the resulting matrix (**C**) of the addition of **A** and **B** is

$$C = A + B = \begin{bmatrix} 2 & 6 & 9 \\ 8 & 11 & 9 \\ 11 & 11 & 12 \end{bmatrix} \tag{2-12}$$

The subtraction of **B** from **A** is the following matrix **D**:

$$D = A - B = \begin{bmatrix} 2 & 2 & 3 \\ -6 & -5 & 1 \\ 7 & 9 & 10 \end{bmatrix} \tag{2-13}$$

It should be noted that when adding or subtracting matrices the matrices must be of the same size. In this example, the matrices **A**, **B**, **C**, and **D** have the same 3×3 size.

2.2.3 Matrix Multiplication: An Introductory Example

Matrix multiplication is slightly more complex than matrix addition and subtraction, so it is introduced with an example. A contractor is laying a pipe and has three primary components of cost: excavation, pipe, and inlets. These costs are $5, $2, and $3 per unit, respectively, and can be represented by the one-dimensional cost matrix **C**:

$$C = \begin{bmatrix} 5 \\ 2 \\ 3 \end{bmatrix} \tag{2-14}$$

On one project, the contractor uses 4, 6, and 1 units of the three components, which is represented by the one-dimensional matrix **S**:

$$S = [4 \quad 6 \quad 1] \tag{2-15}$$

The total project cost **T** can be represented as the product of **S** and **C**:

$$T = S \cdot C = [4 \quad 6 \quad 1] \cdot \begin{bmatrix} 5 \\ 2 \\ 3 \end{bmatrix} \tag{2-16a}$$

$$= 4(5) + 6(2) + 1(3) = [\$35] \tag{2-16b}$$

If the contractor has a second project on which 3, 8, and 4 units of the three components are used, this second one-dimensional matrix **S** could be multiplied with **C** to get the total cost for the second project. However, the problem can be simplified by writing **S** as a two-dimensional matrix before multiplying with **C**. The multiplication of **S** and **C** will yield the total cost for each project:

$$T = S \cdot C = \begin{bmatrix} 4 & 6 & 1 \\ 3 & 8 & 4 \end{bmatrix} \cdot \begin{bmatrix} 5 \\ 2 \\ 3 \end{bmatrix} \tag{2-17a}$$

$$= \begin{bmatrix} 4(5) + 6(2) + 1(3) \\ 3(5) + 8(2) + 4(3) \end{bmatrix} = \begin{bmatrix} \$35 \\ \$43 \end{bmatrix} \tag{2-17b}$$

If the contractor is interested in computing the labor requirements as well as the cost, the **C** matrix can be expanded into a two-dimensional matrix of two columns. The first column contains the cost per unit, and the second column contains the labor requirements per unit. For this example, we can assume that there are labor requirements of 7, 3, and 9 for the three components; that is, 7 units of labor are required per unit of excavation, 3 units of labor are required to install 1 unit of pipe, and 9 units of labor are required to install each inlet. Multiplying **S** and **C** yields both the total cost and the total labor requirements for each project:

$$\mathbf{T} = \mathbf{S} \cdot \mathbf{C} = \begin{bmatrix} 4 & 6 & 1 \\ 3 & 8 & 4 \end{bmatrix} \cdot \begin{bmatrix} 5 & 7 \\ 2 & 3 \\ 3 & 9 \end{bmatrix} \tag{2-18a}$$

$$= \begin{bmatrix} 4(5) + 6(2) + 1(3) & 4(7) + 6(3) + 1(9) \\ 3(5) + 8(2) + 4(3) & 3(7) + 8(3) + 4(9) \end{bmatrix} = \begin{bmatrix} \$35 & 55 \\ \$43 & 81 \end{bmatrix} \tag{2-18b}$$

The **T** matrix, which has two rows and two columns, shows that project 1 costs $35 and requires 55 units of labor and project 2 costs $43 and requires 81 units of labor. Row i corresponds to project i, while column j corresponds to project criterion j ($j = 1$ for cost and $j = 2$ for labor).

This problem could be expanded to include as many projects as the contractor has and to requirements other than cost and labor. The **S** matrix would have three columns and as many rows as there are projects. The **C** matrix would have three rows and one column for each project criterion. The resulting **T** matrix would have one row for each project and one column for each project criterion.

2.2.4 Matrix Multiplication: General Rules

Using **A**, **B**, and **C** to denote three matrices for the matrix product $\mathbf{C} = \mathbf{A} \cdot \mathbf{B}$, the following are the rules of matrix multiplication:

1. The number of columns in the first matrix **A** must equal the numbers of rows in the second matrix **B**.
2. The number of rows in the product matrix **C** equals the number of rows in the first matrix **A**.
3. The number of columns in the product matrix **C** equals the number of columns in the second matrix **B**.
4. The element of matrix **C** in row i and column j, c_{ij}, is equal to the sum of the products $a_{ik} b_{kj}$:

$$c_{ij} = \sum_{k=l}^{m} a_{ik} b_{kj} \qquad (2\text{-}19)$$

where m is the number of columns in **A** (which is the number of rows in **B**).

5. Matrix multiplication is not commutative; that is,

$$\mathbf{A} \cdot \mathbf{B} \neq \mathbf{B} \cdot \mathbf{A} \qquad (2\text{-}20)$$

6. Matrix multiplication is associative; that is,

$$(\mathbf{A} \cdot \mathbf{B}) \cdot \mathbf{C} = \mathbf{A} \cdot (\mathbf{B} \cdot \mathbf{C}) \qquad (2\text{-}21)$$

The example of Eq. 2-18 can be used to illustrate the first four rules. In this case, **S** is a 2×3 matrix and **C** is 3×2; therefore, **T** is 2×2, and each summation involves three products, with each element of **T** computed using Eq. 2-19.

The terms premultiplication and postmultiplication are used in expressing matrix multiplication operations. For the two matrices **A** and **B**, the following apply:

1. Premultiplication of **B** by **A** means $\mathbf{A} \cdot \mathbf{B}$.
2. Premultiplication of **A** by **B** means $\mathbf{B} \cdot \mathbf{A}$.
3. Postmultiplication of **A** by **B** means $\mathbf{A} \cdot \mathbf{B}$.
4. Postmultiplication of **B** by **A** means $\mathbf{B} \cdot \mathbf{A}$.

Example 2-2. Matrix Multiplication

The matrix **E** in Example 2-1 can be computed from the following two matrices, **A** and **B**:

$$\mathbf{A} = \begin{bmatrix} 2 & 4 & 6 \\ 1 & 3 & 5 \\ 9 & 10 & 11 \end{bmatrix}$$

and

$$\mathbf{B} = \begin{bmatrix} 0 & 2 & 3 \\ 7 & 8 & 4 \\ 2 & 1 & 1 \end{bmatrix}$$

The matrix **E** that results from the product of **A** and **B** is

$$\mathbf{E} = \mathbf{A} \cdot \mathbf{B} = \begin{bmatrix} 40 & 42 & 28 \\ 31 & 31 & 20 \\ 92 & 109 & 78 \end{bmatrix} \qquad (2\text{-}22)$$

The product of **B** and **A** is

$$\mathbf{F} = \mathbf{B} \cdot \mathbf{A} = \begin{bmatrix} 29 & 36 & 43 \\ 58 & 92 & 126 \\ 14 & 21 & 28 \end{bmatrix} \qquad (2\text{-}23)$$

Therefore, matrix multiplication is not commutative; that is, $\mathbf{A} \cdot \mathbf{B} \neq \mathbf{B} \cdot \mathbf{A}$.

2.2.5 Matrix Multiplication by a Scalar

The multiplication of a matrix $\mathbf{A}$ by a scalar s has the effect of multiplying each element in the matrix (a_{ij}) by the scalar. The resulting elements of a matrix $\mathbf{B}$ can be expressed as

$$b_{ij} = sa_{ij} \tag{2-24}$$

Example 2-3. Matrix Multiplication by a Scalar

For the matrices $\mathbf{A}$ discussed in Example 2-1,

$$\mathbf{A} = \begin{bmatrix} 2 & 4 & 6 \\ 1 & 3 & 5 \\ 9 & 10 & 11 \end{bmatrix}$$

the resulting matrix $\mathbf{F}$ of the product of $\mathbf{A}$ by a scalar s that has a value of 10 is

$$\mathbf{F} = s\mathbf{A} = \begin{bmatrix} 20 & 40 & 60 \\ 10 & 30 & 50 \\ 90 & 100 & 110 \end{bmatrix} \tag{2-25}$$

2.2.6 Matrix Inversion

Matrix inversion, which applies to square matrices of n rows and n columns, is another matrix operation and is defined by

$$\mathbf{A}^{-1} \cdot \mathbf{A} = \mathbf{I} \tag{2-26}$$

in which $\mathbf{A}^{-1}$ is the inverse of $\mathbf{A}$, and $\mathbf{I}$ is the unit matrix. The inverse can be determined by forming n^2 simultaneous equations and solving for the n^2 unknowns. For the case of $n = 2$, the four equations are (letting the elements of $\mathbf{A}^{-1}$ be indicated by c_{ij})

$$\begin{aligned} c_{11}a_{11} + c_{12}a_{21} &= 1 \\ c_{11}a_{12} + c_{12}a_{22} &= 0 \\ c_{21}a_{11} + c_{22}a_{21} &= 0 \\ c_{21}a_{12} + c_{22}a_{22} &= 1 \end{aligned} \tag{2-27}$$

Solving Eqs. 2-27 yields the values of c_{ij}, which form the inverse. Equations 2-27 can be solved as two pairs of two simultaneous equations.

Example 2-4. Matrix Inversion

For the matrix $\mathbf{A} = \begin{bmatrix} 2 & 3 \\ 5 & 7 \end{bmatrix}$, Eq. 2-26 appears as

$$\begin{bmatrix} c_{11} & c_{12} \\ c_{21} & c_{22} \end{bmatrix} \cdot \begin{bmatrix} 2 & 3 \\ 5 & 7 \end{bmatrix} = \begin{bmatrix} 1 & 0 \\ 0 & 1 \end{bmatrix} \tag{2-28a}$$

which yields the following simultaneous equations:

$$
\begin{aligned}
2c_{11} + 5c_{12} &&&= 1 \\
3c_{11} + 7c_{12} &&&= 0 \\
&2c_{21} + 5c_{22} &&= 0 \\
&3c_{21} + 7c_{22} &&= 1
\end{aligned}
\tag{2-28b}
$$

The solution of Eqs. 2-28 yields the inverse $\mathbf{A}^{-1} = \begin{bmatrix} -7 & 3 \\ 5 & -2 \end{bmatrix}$. Performing the multiplication of Eq. 2-26 with $\mathbf{A}^{-4}$ and $\mathbf{A}$ verifies that the solution of the simultaneous equations yields the inverse matrix.

2.2.7 Matrix Singularity

A matrix can be referred to as being singular or nonsingular. If the inverse of a matrix exists, then the matrix is said to be nonsingular. Conversely, if the inverse of the matrix does not exist, then the matrix is said to be singular. One implication of matrix singularity in solving a system of simultaneous equations is that a unique solution for the equations does not exist. This is most easily understood by illustrating it using a system of equations. Consider, for example, the following two linear simultaneous equations:

$$
2X_1 + 3X_2 = a \tag{2-29a}
$$

$$
4X_1 + 6X_2 = b \tag{2-29b}
$$

If $2a = b$, then there are an infinite number of solutions; for example, three possibilities are (1) $X_1 = 2$, $X_2 = 1$, $a = 7$, and $b = 14$; (2) $X_1 = -1$, $X_2 = 4$, $a = 10$, and $b = 20$; and (3) $X_1 = 0$, $X_2 = -2$, $a = -6$, and $b = -12$. If $2a \neq b$, then there is no feasible solution; for example, if $a = 2$ and $b = 3$, there are no values of X_1 and X_2 that can satisfy the equalities of the two equations. In both cases, the matrix $\mathbf{A}$ formed by the coefficients on the left side of the equations is said to be a singular matrix. The matrix is given by

$$
\mathbf{A} = \begin{bmatrix} 2 & 3 \\ 4 & 6 \end{bmatrix} \tag{2-30a}
$$

The inability to find a solution is easily shown using Eq. 2-26 as

$$
\begin{bmatrix} c_{11} & c_{12} \\ c_{21} & c_{22} \end{bmatrix} \cdot \begin{bmatrix} 2 & 3 \\ 4 & 6 \end{bmatrix} = \begin{bmatrix} 1 & 0 \\ 0 & 1 \end{bmatrix} \tag{2-30b}
$$

After forming the four simultaneous equations, it can easily be shown that the equations do not have a solution; therefore, the matrix $\mathbf{A}$ is singular.

2.2.8 Additional Topics in Matrix Algebra

The *trace* of a square matrix equals the sum of the diagonal elements:

$$
tr(\mathbf{A}) = \sum_{i=1}^{n} a_{ii} \tag{2-31}
$$

Matrix augmentation is the addition of a column or columns to the initial matrix. For example, the following shows matrix **A** and an augmented matrix:

$$\mathbf{A} = \begin{bmatrix} 2 & 3 & 1 \\ 1 & 4 & 1 \\ 2 & 3 & 4 \end{bmatrix} \qquad (2\text{-}32a)$$

$$\mathbf{A}_a = \left[\begin{array}{ccc|ccc} 2 & 3 & 1 & 1 & 0 & 0 \\ 1 & 4 & 1 & 0 & 1 & 0 \\ 2 & 3 & 4 & 0 & 0 & 1 \end{array} \right] \qquad (2\text{-}32b)$$

A matrix can be partitioned by separating it into smaller matrices. For example, matrix **A** can be partitioned into four other matrices:

$$\mathbf{A} = \left[\begin{array}{c|c} A_{11} & A_{12} \\ \hline A_{21} & A_{22} \end{array} \right] \qquad (2\text{-}33)$$

where, for example,

$$\mathbf{A}_{11} = \begin{bmatrix} 2 & 3 \\ 1 & 4 \end{bmatrix} \qquad (2\text{-}34a)$$

$$\mathbf{A}_{12} = \begin{bmatrix} 1 \\ 1 \end{bmatrix} \qquad (2\text{-}34b)$$

$$\mathbf{A}_{21} = \begin{bmatrix} 2 & 3 \end{bmatrix} \qquad (2\text{-}34c)$$

$$\mathbf{A}_{22} = \begin{bmatrix} 4 \end{bmatrix} \qquad (2\text{-}34d)$$

Example 2-5. Matrix Partition

A correlation matrix is often partitioned into a matrix of correlation coefficients between one set of variables called the predictor variables and a vector of predictor-criterion correlation coefficients. For example, a correlation analysis of an engineering problem resulted in the following correlation matrix **R** for three variables X_1, X_2, and Y:

$$\mathbf{R} = \begin{array}{c} \\ X_1 \\ X_2 \\ Y \end{array} \begin{array}{c} \begin{array}{ccc} X_1 & X_2 & Y \end{array} \\ \begin{bmatrix} 1 & 0.2 & 0.9 \\ 0.2 & 1 & 0.8 \\ 0.9 & 0.8 & 1 \end{bmatrix} \end{array} \qquad (2\text{-}35)$$

The variables X_1 and X_2 are considered to be the predictor (independent) variables, and Y is the criterion (dependent) variable. The correlation matrix can be partitioned as follows:

$$\mathbf{R} = \begin{array}{c} \\ X_1 \\ X_2 \\ Y \end{array} \begin{array}{c} \begin{array}{ccc} X_1 & X_2 & Y \end{array} \\ \left[\begin{array}{cc|c} 1 & 0.2 & 0.9 \\ 0.2 & 1 & 0.8 \\ \hline 0.9 & 0.8 & 1 \end{array} \right] \end{array} \qquad (2\text{-}36)$$

The matrix

$$\mathbf{R}_{11} = \begin{bmatrix} 1 & 0.2 \\ 0.2 & 1 \end{bmatrix} \tag{2-37}$$

is called the intercorrelation matrix for the predictor variables. The reason for partitioning the matrix in this case is because the intercorrelation coefficients are assessed differently than the predictor-criterion correlation coefficients.

2.3 VECTORS

2.3.1 Definitions

In the example used to introduce matrix multiplication, the **C** and **S** matrices were initially presented as one dimensional, with **C** having one column and **S** having one row. Such matrices are called vectors. A vector of one row is called a *row vector*. A vector of one column is called a *column vector*. An element of a vector can be indicated with a single subscript, so a_3 is the third element of the vector **A**, and b_j is the jth element of the vector **B**.

The name vector indicates that a one-dimensional matrix with n elements can be represented as a vector in n dimensions, with one end of the vector at the origin and the other end at the point defined by the values of the elements of the vector. For example, the vector [2, 3] would be a line from the origin (0, 0) to the point located two units along the first axis and three units along the second axis. Figure 2-4 provides another example; in this case the vectors have two ordinates, measured on the axes X_1 and X_2. Vector **A** is indicated by (2, 1), and vector **B** is indicated by $\left(1, -\frac{2}{3}\right)$. In each case, the first element of the vector corresponds to X_1 while the second element corresponds to X_2.

A *null vector* is a vector that has zeros for all its elements. It corresponds to the null matrix.

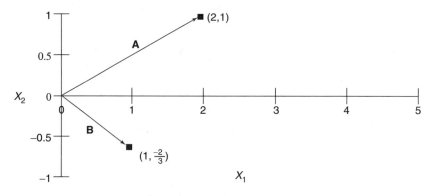

Figure 2-4   Vectors

2.3.2 Vector Operations

The matrix operations of addition, subtraction, and multiplication can be applied to vectors. Two row (column) vectors can be added or subtracted. A row vector with n elements can be postmultiplied by a column vector with n elements to equal a scalar value. In Eq. 2-16, row vector $\mathbf{S}$ is postmultiplied by the column vector $\mathbf{C}$ to produce a scalar value. The number of rows in the column vector must equal the number of columns in the row vector. This is easily understood if the row vector is viewed as a $1 \times n$ matrix and the column vector is viewed as an $n \times 1$ matrix. Postmultiplication of a $(1 \times n)$ by a $(n \times 1)$ yields a (1×1) matrix, which is a scalar. It also shows why the number of columns and rows in the vector must be equal. A column vector with n elements can be multiplied by a row vector with m elements to produce an $n \times m$ matrix. If the $n \times 1$ column vector is postmultiplied by the $1 \times m$ row vector, then the resulting matrix is $n \times m$. Thus, in this case, the number of rows and number of columns do not have to be equal. It is important to note that the order of the multiplication is not arbitrary; it depends on the concept that underlies the problem that was the basis for the development of vectors.

2.3.3 Orthogonal and Normalized Vectors

Two vectors are said to be *orthogonal* if their product equals zero. For the vector product $\mathbf{A} \cdot \mathbf{B}$, $\mathbf{A}$ is a row vector and $\mathbf{B}$ is a column vector; the resulting vector product is a scalar value. If two vectors that are orthogonal are plotted in the *n-dimensional* space, the vectors will be perpendicular to each other. For example, the vector product of $\mathbf{A}$ and $\mathbf{B}$ in Fig. 2-5 is zero:

$$\begin{bmatrix} 2 & 3 \end{bmatrix} \cdot \begin{bmatrix} 1 \\ -\frac{2}{3} \end{bmatrix} = [0] \tag{2-38}$$

Since the vector product is zero, the vectors are perpendicular to each other.

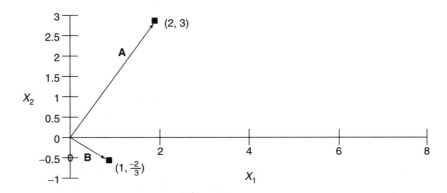

Figure 2-5   Orthogonal Vectors

The *length of a vector* **V** equals the square root of the sum of the squares of its elements; that is,

$$\text{vector length} = \left(\sum_{i=1}^{n} v_i^2 \right)^{0.5} \tag{2-39}$$

where $v_1, v_2, \ldots, v_n$ are the n elements of the vector **V**. The length of a vector is a scalar quantity.

A vector can be *normalized* by dividing each element of the vector by its length. Therefore, a normalized vector has a length that is equal to 1. Two vectors that are both normalized and orthogonal to each other are said to be *orthonormal* vectors.

Example 2-6. Vectors

For the two vectors

$$\mathbf{V}_1 = [2 \ -3 \ 5] \tag{2-40a}$$

and

$$\mathbf{V}_2 = \begin{bmatrix} -1 \\ 1 \\ 1 \end{bmatrix} \tag{2-40b}$$

the vector lengths are

$$\text{length of } \mathbf{V}_1 = \sqrt{(2)^2 + (-3)^2 + (5)^2} = \sqrt{38} = 6.164 \tag{2-41a}$$

$$\text{length of } \mathbf{V}_2 = \sqrt{(-1)^2 + (1)^2 + (1)^2} = \sqrt{3} = 1.732 \tag{2-41b}$$

The normalized vectors are

$$\mathbf{V}_{1n} = \begin{bmatrix} \dfrac{2}{\sqrt{38}} & \dfrac{-3}{\sqrt{38}} & \dfrac{5}{\sqrt{38}} \end{bmatrix} \tag{2-42a}$$

and

$$\mathbf{V}_{2n} = \begin{bmatrix} \dfrac{-1}{\sqrt{3}} & \dfrac{1}{\sqrt{3}} & \dfrac{1}{\sqrt{3}} \end{bmatrix} \tag{2-42b}$$

Since the vector product $\mathbf{V}_1 \cdot \mathbf{V}_2$ equals zero, $\mathbf{V}_1$ and $\mathbf{V}_2$ are orthogonal, whereas the vectors $\mathbf{V}_{1n}$ and $\mathbf{V}_{2n}$ are orthonormal.

2.4 DETERMINANTS

The *determinant* of a square matrix is a unique scalar number that represents the value of a matrix. A determinant of a matrix is denoted by $|\mathbf{A}|$. By definition, the determinant of a 2×2 matrix is computed by

$$\begin{vmatrix} a & b \\ c & d \end{vmatrix} = ad - bc \qquad (2\text{-}43)$$

The right side of Eq. 2-43 is referred to as the *expansion* of the determinant shown on the left side.

For a 3×3 matrix, the value of the determinant is found by the following expansion:

$$\begin{vmatrix} a_{11} & a_{12} & a_{13} \\ a_{21} & a_{22} & a_{23} \\ a_{31} & a_{32} & a_{33} \end{vmatrix} = a_{11} \begin{vmatrix} a_{22} & a_{23} \\ a_{32} & a_{33} \end{vmatrix} - a_{12} \begin{vmatrix} a_{21} & a_{23} \\ a_{31} & a_{33} \end{vmatrix} + a_{13} \begin{vmatrix} a_{21} & a_{22} \\ a_{31} & a_{32} \end{vmatrix} \qquad (2\text{-}44)$$

The 2×2 determinants shown in the expansion of Eq. 2-44 are called *minors*. The minor of an element is formed by eliminating the row and column in which the element appears. For example, the minor of element a_{ij} is the matrix that results when row i and column j are eliminated; the minor for element a_{ij} of matrix **A** is denoted $\mathbf{A}_{ij}$. For an $n \times n$ matrix, any minor will be an $(n - 1) \times (n - 1)$ matrix.

The determinant of an $n \times n$ matrix is given by

$$|\mathbf{A}| = a_{11} |\mathbf{A}_{11}| - a_{12} |\mathbf{A}_{12}| + a_{13} |\mathbf{A}_{13}| - \cdots + (-1)^{(n + 1)} a_{1n} |\mathbf{A}_{1n}| \qquad (2\text{-}45)$$

where the matrix $\mathbf{A}_{ij}$ is the minor of element a_{ij}. The sign of the term is plus if the sum of the row number and column number in which the element lies is even and is minus if odd. The product of $(-1)^{i+j}$ and the minor of a_{ij} is called the *cofactor* of a_{ij}. It can be stated that the determinant in Eq. 2-45 was performed using the first row of the matrix. In general, any row can be used, but the signs of the different terms need to be revised. The general expression for calculating the determinant of an $n \times n$ matrix using the ith row is

$$|\mathbf{A}| = (-1)^{i+1} a_{i1} |\mathbf{A}_{i1}| - (-1)^{i+1} a_{i2} |\mathbf{A}_{i2}| + (-1)^{i+1} a_{i3} |\mathbf{A}_{i3}| - \cdots \qquad (2\text{-}46)$$
$$+ (-1)^{i+1} (-1)^{n+1} a_{in} |\mathbf{A}_{in}|$$

The following are properties of a determinant:

1. If the elements of any two rows (columns) are equal, the determinant equals zero. For example, the following matrix has a zero determinant because the first column is equal to the third column:

$$\mathbf{A} = \begin{bmatrix} 1 & 2 & 1 \\ 2 & 14 & 2 \\ 3 & 5 & 3 \end{bmatrix}, \quad \text{where } |\mathbf{A}| = 0$$

2. If the values in any row (column) are proportional to the corresponding values in another row (column), the determinant equals zero. For example, the following matrix has a zero determinant because the first column is proportional to the third column:

$$\mathbf{A} = \begin{bmatrix} 1 & 2 & 2 \\ 2 & 14 & 4 \\ 3 & 5 & 6 \end{bmatrix}, \quad \text{where } |\mathbf{A}| = 0$$

3. If all the elements in any row (column) equal zero, the determinant equals zero.

4. If all the elements of any row (column) are multiplied by a constant c, the value of the determinant is multiplied by c. The following matrix illustrates this property:

$$\mathbf{A} = \begin{bmatrix} 6 & 4 \\ 4 & 5 \end{bmatrix} = \begin{bmatrix} 2(3) & 2(2) \\ 4 & 5 \end{bmatrix}, \quad \text{where } |\mathbf{A}| = 2[3(5) - 2(2)] = 14$$

5. The value of the determinant is not changed by adding any row (column) multiplied by a constant c to another row (column). The following matrix illustrates this property:

$$\mathbf{A} = \begin{bmatrix} 3 & 2 \\ 4 & 5 \end{bmatrix}, \quad \text{where } |\mathbf{A}| = 3(5) - 2(4) = 7$$

Multiplying the second row by (-1) and adding it to the first row produces the following revised matrix $\mathbf{B}$:

$$\mathbf{B} = \begin{bmatrix} -1 & -3 \\ 4 & 5 \end{bmatrix}, \quad \text{where } |\mathbf{B}| = -1(5) + 3(4) = 7$$

6. If any two rows (columns) are interchanged, the sign of the determinant is changed. The determinants are computed for a matrix, and another with interchanged columns to illustrate this property as

$$\begin{vmatrix} 3 & 2 \\ 4 & 5 \end{vmatrix} = 3(5) - 2(4) = 7 \quad \text{and} \quad \begin{vmatrix} 2 & 3 \\ 5 & 4 \end{vmatrix} = 2(4) - 3(5) = -7$$

7. Multiplying an $n \times n$ matrix by a constant c results in multiplying the value of the determinant by c^n. The following matrix illustrates this property:

$$\mathbf{A} = 2^2 \begin{bmatrix} 3 & 2 \\ 4 & 5 \end{bmatrix}, \quad \text{where } |\mathbf{A}| = 2^2[3(5) - 2(4)] = 28$$

or

$$\mathbf{A} = \begin{bmatrix} 2(3) & 2(2) \\ 2(4) & 2(5) \end{bmatrix}, \quad \text{where } |\mathbf{A}| = 6(10) - 4(8) = 28$$

8. The determinant of a matrix equals that of its transpose; that is, $|\mathbf{A}| = |\mathbf{A}^T|$. The following determinants illustrate this property:

$$\begin{vmatrix} 3 & 2 \\ 4 & 5 \end{vmatrix} = 3(5) - 2(4) = 7 \quad \text{and} \quad \begin{vmatrix} 3 & 4 \\ 2 & 5 \end{vmatrix} = 3(5) - 4(2) = 7$$

9. If a matrix $\mathbf{A}$ is placed in diagonal form using property 5, then the product of the elements on the diagonal equals the determinant of $\mathbf{A}$. Consider the following matrix:

$$\mathbf{A} = \begin{bmatrix} 3 & 2 \\ 4 & 5 \end{bmatrix}, \quad \text{with } |\mathbf{A}| = 3(5) - 2(4) = 7$$

Multiplying the first row by $-\frac{4}{3}$ and adding it to the second row produces a matrix with a zero element in the second row and first column, as follows:

$$\begin{bmatrix} 3 & 2 \\ 0 & \frac{7}{3} \end{bmatrix}$$

Then, multiplying the second row by $-\frac{6}{7}$ and adding it to the first row results in the following diagonal matrix:

$$\begin{bmatrix} 3 & 0 \\ 0 & \frac{7}{3} \end{bmatrix}$$

Therefore, the determinant of **A** is

$$|\mathbf{A}| = \begin{vmatrix} 3 & 0 \\ 0 & \frac{7}{3} \end{vmatrix} = 3\left(\frac{3}{7}\right) = 7$$

10. If a matrix **A** has a zero determinant, then **A** is a singular matrix; that is, the inverse of **A** does not exist.

Example 2-7. Matrix Determinant

The matrix **A** discussed in Example 2-1 can be used to illustrate the calculation of the determinant. The matrix **A** is given by

$$\mathbf{A} = \begin{bmatrix} 2 & 4 & 6 \\ 1 & 3 & 5 \\ 9 & 10 & 11 \end{bmatrix}$$

The determinant is computed using the elements of the first row and their minors as follows:

$$|\mathbf{A}| = a_{11}|\mathbf{A}_{11}| - a_{12}|\mathbf{A}_{12}| + a_{13}|\mathbf{A}_{13}| \tag{2-47a}$$

or

$$|\mathbf{A}| = \begin{vmatrix} 2 & 4 & 6 \\ 1 & 3 & 5 \\ 9 & 10 & 11 \end{vmatrix} = 2\begin{vmatrix} 3 & 5 \\ 10 & 11 \end{vmatrix} - 4\begin{vmatrix} 1 & 5 \\ 9 & 11 \end{vmatrix} + 6\begin{vmatrix} 1 & 3 \\ 9 & 10 \end{vmatrix} \tag{2-47b}$$

$$|\mathbf{A}| = 2[3(11) - 5(10)] - 4[1(11) - 5(9)] + 6[1(10) - 3(9)] = 0 \tag{2-47c}$$

The determinant could also have been computed using the elements of the second column and their minors as follows:

$$|\mathbf{A}| = -a_{12}|\mathbf{A}_{12}| + a_{22}|\mathbf{A}_{22}| - a_{32}|\mathbf{A}_{32}| \tag{2-47d}$$

or

$$|\mathbf{A}| = \begin{vmatrix} 2 & 4 & 6 \\ 1 & 3 & 5 \\ 9 & 10 & 11 \end{vmatrix} = -4\begin{vmatrix} 1 & 5 \\ 9 & 11 \end{vmatrix} + 3\begin{vmatrix} 2 & 6 \\ 9 & 11 \end{vmatrix} - 10\begin{vmatrix} 2 & 6 \\ 1 & 5 \end{vmatrix} \tag{2-27e}$$

$$|\mathbf{A}| = -4[11 - 45] + 3[22 - 54] - 10[10 - 6] = 0 \tag{2-27f}$$

Therefore, **A** is a singular matrix; that is, the inverse of **A** does not exist.

2.5 RANK OF A MATRIX

A matrix of r rows and c columns is said to be of *order r by c*. If it is a square matrix, r by r, then the matrix is of order r.

The rank of a matrix is another term used to classify matrices. There are several concepts used to explain the rank of a matrix, three of which are presented here.

First, if we examine all possible square submatrices (by partitioning) of an r by c matrix and evaluate those having nonzero determinants, then the rank equals the order of the highest-order nonsingular submatrix. For example, a 2×3 matrix has a rank of 2 if one of its 2×2 submatrices has a nonzero determinant.

The second definition uses vector terminology as its basis. Specifically, the rank of a matrix is the minimum number of dimensions of an orthogonal coordinate system that is needed to graph the rows or columns of the matrix as vectors, permitting the axis system to be rotated to minimize the necessary number of dimensions.

The third definition is closely related to the first two definitions. The rank of a matrix is the number of vectors (rows or columns) in the matrix that is linearly independent. The rows (or columns) of a matrix are called linearly independent if none of them can be produced by adding any row (column) multiplied by a constant to another row (column) or other rows (or columns).

Example 2-8. Rank of Matrix

The three definitions of the rank of a matrix can be illustrated with a single example. Consider the following matrix:

$$2 \times 3 \text{ order matrix}, \quad \mathbf{R} = \begin{bmatrix} 1 & 2 & 4 \\ 2 & 4 & 8 \end{bmatrix} \tag{2-48}$$

Three square submatrices can be formed from this as follows:

$$\mathbf{R}_1 = \begin{bmatrix} 1 & 2 \\ 2 & 4 \end{bmatrix} \tag{2-49a}$$

$$\mathbf{R}_2 = \begin{bmatrix} 1 & 4 \\ 2 & 8 \end{bmatrix} \tag{2-49b}$$

$$\mathbf{R}_3 = \begin{bmatrix} 2 & 4 \\ 4 & 8 \end{bmatrix} \tag{2-49c}$$

Each of these has a determinant of zero, so the rank must be less than 2. Thus the first definition indicates that $\mathbf{R}$ has a rank of 1.

If matrix $\mathbf{R}$ is viewed as three vectors in a two-dimensional space (see Fig. 2-6), it is evident that the vectors overlap each other; the angle between them is zero. If the axis system is rotated so that it passes through the first vector (1, 2), it is evident that the axis will coincide with the other two vectors. Thus only one axis is needed to represent the three vectors, and so matrix $\mathbf{R}$ is of rank 1.

The elements of the three vectors are related by the following equations:

$$v_{i2} = 2v_{i1}, \quad \text{for } i = 1 \text{ and } 2 \tag{2-50a}$$

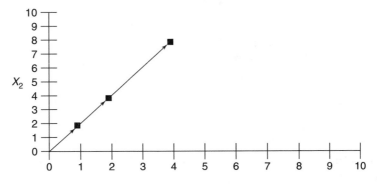

Figure 2-6   Rank of a Matrix

$$v_{i3} = 4v_{i1}, \qquad \text{for } i = 1 \text{ and } 2 \tag{2-50b}$$

$$v_{i3} = 2v_{i2}, \qquad \text{for } i = 1 \text{ and } 2 \tag{2-50c}$$

where the first subscript refers to the element number of the column vector indicated by the second subscript. Since the three equations are linear, the three vectors are linearly dependent; so there is only one linearly independent vector, which means that matrix **A** is of rank 1.

Example 2.9 Rank of a Matrix

The matrix of Example 2-7, which is of the third order (that is, a 3×3 square matrix), is used to illustrate the computation of the rank. The matrix is

$$\mathbf{A} = \begin{bmatrix} 2 & 4 & 6 \\ 1 & 3 & 5 \\ 9 & 10 & 11 \end{bmatrix}$$

Since the determinant of **A** is zero, the rank is not 3. The following submatrix has a nonzero determinant:

$$\begin{vmatrix} 2 & 4 \\ 1 & 3 \end{vmatrix} = 2(3) - 4(1) = 2$$

Therefore, the rank is 2. It should be noted that one submatrix of order 2 with a nonzero determinant is needed to obtain a rank of 2.

2.6 APPLICATIONS

2.6.1 Reactions of a Beam Due to Loads

A beam is supporting a concentrated load (P), a distributed load (q), and a moment (M) as shown in Fig. 2-7. The span of the beam is $S = 10$. There are three types of loads: dead

Figure 2-7 Loaded Beam

load (D), live load (L), and wind load (W). The magnitudes of these loads are given by the following matrix:

$$\text{loads} = \begin{array}{c} \\ P \\ q \\ M \end{array} \begin{array}{ccc} D & L & W \\ \left[\begin{matrix} 10 & 12 & -2 \\ 1 & 2 & 1 \\ 2 & 4 & 3 \end{matrix}\right] \end{array} \tag{2-51}$$

The loads cause structural reactions where the beam is attached to the support. Two reactions are of interest: R_1 = force reaction at the fixed end, and R_2 = moment reaction at the fixed end. The computations of the two reactions depend on the load type. For an applied force (P) by itself, the resulting reactions are

$$R_1 = P \tag{2-52a}$$

$$R_2 = PS \tag{2-52b}$$

For an applied uniform load (q) by itself, the resulting reactions are

$$R_1 = qS \tag{2-52c}$$

$$R_2 = \frac{qS^2}{2} \tag{2-52d}$$

For an applied moment (M) at the free end by itself, the resulting reactions are

$$R_1 = 0 \tag{2-52e}$$

$$R_1 = M \tag{2-52f}$$

The beam reactions can be expressed as

$$\begin{bmatrix} R_1 \\ R_2 \end{bmatrix} = \begin{bmatrix} 1 & S & 0 \\ S & \frac{S^2}{2} & 1 \end{bmatrix} \cdot \begin{bmatrix} P \\ q \\ M \end{bmatrix} \tag{2-53a}$$

For $S = 10$, Eq. 2-53a can be written as

$$\begin{bmatrix} R_1 \\ R_2 \end{bmatrix} = \begin{bmatrix} 1 & 10 & 0 \\ 10 & 50 & 1 \end{bmatrix} \cdot \begin{bmatrix} P \\ q \\ M \end{bmatrix} \tag{2-53b}$$

This means that R_1 and R_2 are computed as follows:

$$R_1 = P + 10q + 0M = P + 10q \tag{2-54a}$$

$$R_2 = 10P + 50q + M \tag{2-54b}$$

The values for P, q, and M due to dead load D are given in Eq. 2-51. The reactions due to the dead load $\mathbf{R}_D$ can be computed using Eqs. 2-52 and are given by the following matrix:

$$\mathbf{R}_D = \begin{matrix} & \begin{matrix} R_1 & R_2 \end{matrix} \\ \begin{matrix} P \\ q \\ M \end{matrix} & \begin{bmatrix} 10 & 100 \\ 10 & 50 \\ 0 & 2 \end{bmatrix} \end{matrix} \tag{2-55}$$

Similarly, the reactions due to the live load $\mathbf{R}_L$ are given by the following matrix:

$$\mathbf{R}_L = \begin{matrix} & \begin{matrix} R_1 & R_2 \end{matrix} \\ \begin{matrix} P \\ q \\ M \end{matrix} & \begin{bmatrix} 12 & 120 \\ 20 & 100 \\ 0 & 4 \end{bmatrix} \end{matrix} \tag{2-56}$$

Similarly, the reations due to the wind load $\mathbf{R}_W$ are given by the following matrix:

$$\mathbf{R}_W = \begin{matrix} & \begin{matrix} R_1 & R_2 \end{matrix} \\ \begin{matrix} P \\ q \\ M \end{matrix} & \begin{bmatrix} -2 & -20 \\ 10 & 50 \\ 0 & 3 \end{bmatrix} \end{matrix} \tag{2-57}$$

In the matrices $\mathbf{R}_D$, $\mathbf{R}_L$, and $\mathbf{R}_W$, the rows represent the types of load and the columns represent the types of reaction at the support. The total reactions $\mathbf{R}_T$ are given by

$$\mathbf{R}_T = \mathbf{R}_D + \mathbf{R}_L + \mathbf{R}_W = \begin{matrix} & \begin{matrix} R_1 & R_2 \end{matrix} \\ \begin{matrix} P \\ q \\ M \end{matrix} & \begin{bmatrix} 20 & 200 \\ 40 & 200 \\ 0 & 9 \end{bmatrix} \end{matrix} \tag{2-58}$$

Load combinations can be evaluated, such as $D + L$ and $D + W$. The resulting reactions are

$$\mathbf{R}_{D+L} = \mathbf{R}_D + \mathbf{R}_L = \begin{matrix} & \begin{matrix} R_1 & R_2 \end{matrix} \\ \begin{matrix} P \\ q \\ M \end{matrix} & \begin{bmatrix} 22 & 220 \\ 30 & 150 \\ 0 & 6 \end{bmatrix} \end{matrix} \tag{2-59a}$$

and

$$\mathbf{R}_{D+W} = \mathbf{R}_D + \mathbf{R}_W = \begin{array}{c} \\ P \\ q \\ M \end{array} \overset{\displaystyle R_1 \quad R_2}{\begin{bmatrix} 8 & 80 \\ 20 & 100 \\ 0 & 5 \end{bmatrix}} \tag{2-59b}$$

The combined reaction due to the total effects of P, q, and M can be computed by determining the column summations in Eqs. 2-55 to 2-58. The combined reactions $\mathbf{R}_C$ can be expressed using the following vectors:

$$\begin{array}{cc} R_1 & R_2 \end{array}$$
$$\mathbf{R}_{CD} = [20 \quad 152] \qquad \text{for dead load} \tag{2-60a}$$

$$\mathbf{R}_{CL} = [32 \quad 224] \qquad \text{for live load} \tag{2-60b}$$

$$\mathbf{R}_{CW} = [8 \qquad 33] \qquad \text{for wind load} \tag{2-60c}$$

and

$$\mathbf{R}_{CT} = [60 \quad 409] \quad \text{for total loads} \tag{2-60d}$$

Load combinations can be evaluated, such as $D + L$ and $D + W$. The resulting combined reactions are

$$\begin{array}{cc} R_1 & R_2 \end{array}$$
$$\mathbf{R}_{C(D+L)} = [52 \quad 376] \qquad \text{for dead and live loads} \tag{2-61a}$$

and

$$\mathbf{R}_{C(D+W)} = [28 \quad 185] \qquad \text{for dead and wind loads} \tag{2-61b}$$

2.6.2 Correlation Analysis of Water-quality Data

A 5×3 matrix is given in Fig. 2-2a, with water-quality concentrations given for three water-quality characterisitcs; it will be denoted as matrix $\mathbf{X}$. The correlations between the three constituents for the five wells are given in the matrix of Eq. 2-1. The correlation matrix is computed using several matrix operations, as follows:

1. Compute a row vector that contains the means for each water-quality constituent; it will be denoted as $\mathbf{X}_m$:

$$\mathbf{X}_m = [162 \quad 68.6 \quad 7.2] \tag{2-62a}$$

The mean value is computed for each column as

$$x_{mj} = \frac{1}{5} \sum_{i=1}^{5} x_{ij} \tag{2-62b}$$

2. Compute a row vector, denoted as **S**, that contains the standard deviations of each of the three constituents:

$$\mathbf{S} = [82.94 \ 68.95 \ 0.806] \qquad (2\text{-}62c)$$

The standard deviation is computed for each column as

$$s_j = \sqrt{\frac{1}{5-1} \sum_{i=1}^{5} (x_{ij} - x_{mj})^2} \qquad (2\text{-}62d)$$

3. A rectangular matrix, denoted as **Z**, is computed by transforming the matrix **X** using both $\mathbf{X}_m$ and **S**. The elements z_{ij} are computed by

$$z_{ij} = \frac{x_{ij} - x_{mj}}{s_j}, \quad \text{for } i = 1, 2, \ldots 5, \text{ and } j = 1, 2, 3 \qquad (2\text{-}62e)$$

Thus, for the values in Eqs. 2-62a and 2-62c, **Z** is

$$\mathbf{Z} = \begin{bmatrix} -0.265 & -0.487 & 0.620 \\ 0.398 & -0.821 & -0.124 \\ 1.459 & -0.226 & -0.993 \\ -0.362 & 1.732 & 1.365 \\ -1.230 & -0.197 & -0.868 \end{bmatrix} \qquad (2\text{-}62f)$$

where, for example, element z_{23} is computed by

$$z_{23} = \frac{z_{23} - x_{m3}}{s_3} = \frac{7.1 - 7.2}{0.806} = -0.124$$

4. The correlation matrix, **R**, is computed by the matrix product divided by a scalar n_r, which is the number of rows, as follows:

$$\mathbf{R} = \frac{1}{n_r - 1} \mathbf{Z}^T \mathbf{Z} \qquad (2\text{-}62g)$$

The resulting matrix is

$$\mathbf{R} = \frac{1}{5-1} \begin{bmatrix} -0.265 & 0.398 & 1.459 & -0.362 & -1.230 \\ -0.487 & -0.821 & -0.226 & 1.732 & -0.197 \\ 0.620 & -0.124 & -0.993 & 1.365 & -0.868 \end{bmatrix} \cdot$$

$$\begin{bmatrix} -0.265 & -0.487 & 0.620 \\ 0.398 & -0.821 & -0.124 \\ 1.459 & -0.226 & -0.993 \\ -0.362 & 1.732 & 1.365 \\ -1.230 & -0.197 & -0.868 \end{bmatrix} \qquad (2\text{-}62h)$$

or

$$\mathbf{R} = \begin{bmatrix} 1.000 & -0.228 & -0.272 \\ -0.228 & 1.000 & 0.640 \\ -0.272 & 0.640 & 1.000 \end{bmatrix} \tag{2-62i}$$

This example illustrates matrix operations that involve vectors, a scalar (n_r in step 4), a transpose, and matrix multiplication.

2.6.3 Regression Analysis for Predicting Bus Ridership

The objective of this example is to demonstrate that matrices have a wide variety of applications in engineering, especially in developing prediction models using analytical tools such as regression analysis. Regression analysis is a method used to fit a linear equation to a set of data. This example uses only the matrix-related part of this analysis.

For this application, assume that a prediction equation is necessary to estimate bus ridership (Y) as a function of the population in an area (X_1) and the cost of a one-way trip (X_2):

$$\hat{Y} = b_0 + b_1 X_1 + b_2 X_2 \tag{2-63}$$

where $\hat{Y}$ is an estimator of Y. The coefficient b_0, b_1, and b_2 are unknown regression coefficients. Values for the coefficients are usually computed by transforming Eq. 2-63 to the following form:

$$\hat{Z} = t_1 Z_1 + t_2 Z_2 \tag{2-64}$$

where Z, Z_1, and Z_2 are linear transformations of Y, X_1, and X_2, respectively. The t values are linear transformations of the b values. A solution to the unknown t values in Eq. 2-64 is usually found by using the matrix of measured values of Y, X_1, and X_2 to compute the correlation matrix, which in this case would be a 3×3 matrix because there are three variables: Y, X_1, and X_2.

Assume for the bus ridership problem that the correlation matrix ($\mathbf{R}$) is

$$\mathbf{R} = \begin{array}{c} \\ X_1 \\ X_2 \\ Y \end{array} \begin{array}{c} X_1 \quad\;\; X_2 \quad\;\;\; Y \\ \begin{bmatrix} 1.00 & -0.30 & 0.64 \\ -0.30 & 1.00 & -0.42 \\ 0.64 & -0.42 & 1.00 \end{bmatrix} \end{array} \tag{2-65}$$

A solution of the problem is found using the system of equations represented by the following matrix expression:

$$\mathbf{R}_{11}\mathbf{t} = \mathbf{R}_{12} \tag{2-66}$$

where $\mathbf{R}_{11}$ is the 2×2 matrix of the intercorrelation between X_1 and X_2, $\mathbf{t}$ is the column vector of t values, and $\mathbf{R}_{12}$ is the column vector of predictor-criterion correlation. To find the solution to Eq. 2-66, it is necessary to make the following matrix-algebra formulation:

1. Premultiply both sides of Eq. 2-66 by R_{11}^{-1}:

$$R_{11}^{-1} R_{11} t = R_{11}^{-1} R_{12} \qquad (2\text{-}67)$$

2. Based on Eq. 2-26, Eq. 2-67 becomes

$$It = R_{11}^{-1} R_{12} \qquad (2\text{-}68)$$

3. The product of a vector and the unit matrix (**I**) equals the vector; that is, $It = t$; therefore, Eq. 2-68 becomes

$$t = R_{11}^{-1} R_{12} \qquad (2\text{-}69)$$

In Eq. 2-69, the known quantities are on the right side and the unknowns are on the left side. To find the t values, it is necessary to find the inverse.

For the intercorrelation matrix of Eq. 2-65, Eq. 2-26 can be used to compute the inverse as

$$\begin{bmatrix} c_{11} & c_{12} \\ c_{21} & c_{22} \end{bmatrix} \cdot \begin{bmatrix} 1.00 & -0.30 \\ -0.30 & 1.00 \end{bmatrix} = \begin{bmatrix} 1 & 0 \\ 0 & 1 \end{bmatrix} \qquad (2\text{-}70)$$

Solving Eq. 2-70 for the c_{ij} values yields the following inverse matrix:

$$R_{11}^{-1} = \begin{bmatrix} 1.10 & 0.33 \\ 0.33 & 1.10 \end{bmatrix} \qquad (2\text{-}71)$$

Substituting into Eq. 2-69 and using the vector of predictor-criterion correlation yields

$$\begin{bmatrix} t_1 \\ t_2 \end{bmatrix} = \begin{bmatrix} 1.10 & 0.33 \\ 0.33 & 1.10 \end{bmatrix} \cdot \begin{bmatrix} 0.64 \\ -0.42 \end{bmatrix} = \begin{bmatrix} 0.57 \\ -0.25 \end{bmatrix} \qquad (2\text{-}72)$$

Thus Eq. 2-64 becomes

$$\hat{Z} = 0.57 Z_1 - 0.25 Z_2 \qquad (2\text{-}73)$$

Equation 2-63 can be found by making the necessary linear transformations used in regression analysis.

2.7 PROBLEMS

2-1. Using weather summaries from the daily newspapers for the last week, create a 7×2 matrix of the daily maximum and minimum temperatures. Compute the mean for each day, and present the daily means as a column vector. Compute the means of the daily maximums and minimums and present the values as a row vector with two elements.

2-2. Create a matrix using scores for all tests that you had in a course as well as the m scores for two or more other students; make sure that all scores are on a 0 to 100 scale. Create a column vector that has the same number of elements m as there are tests, with the elements of the vector being the weights assigned by the instructor, with the weights standardized so that they sum to 1.0. For a matrix of n students and m tests, postmultiply the test score matrix by the column vector of weights. Discuss the resulting vector.

2-3. A mechanical engineer is studying the efficiency of a two-step process for producing a one-piece machine part. First, the part is cut from the raw material and then the rough-cut part is precision machined. The engineer observes that the time required for machining is longer if the machinist takes very little time to rough-cut the material. Similarly, if the machinist spends more time cutting the material, less time is required to machine the part. Create a rectangular matrix in which the two rows represent the cutting time (short, long) and three columns represent the machining time (short, moderate, long). Values in the matrix represent the probability p_{ij} that it takes j time units to machine a part given that i time units were taken to cut the part; thus the matrix elements in the rows must sum to 1.

2-4. An agricultural engineer is establishing a policy on irrigation. The policy statement will indicate the number of units of irrigated water that will be applied to crops depending on recent rainfall and rainfall that is expected in the future. The probabilities of rainfall rates expected in the future (low, average, high) are correlated with past rainfall rates (low, average, high). Square matrices can be created using the three rows to represent the rainfall in the most recent time period and the three columns to represent the rainfall in the next time period. The matrix elements are the probabilities p_{ij} that rainfall for the next time period is at level j given that rainfall for the most recent time period is at level i. The matrix elements in the rows must sum to 1. Create two matrices, one in which low rainfalls tend to follow high rainfalls and a second in which low rainfalls follow low rainfalls.

2-5. Show the transpose of each of the following matrices:

(a) $\mathbf{A} = \begin{bmatrix} 2 & 6 \\ -3 & 4 \end{bmatrix}$

(b) $\mathbf{B} = \begin{bmatrix} -1 & 9 & 6 \\ 3 & 0 & -7 \end{bmatrix}$

(c) $\mathbf{C} = \begin{bmatrix} 0.2 & -0.5 \\ -0.8 & 0.1 \\ -0.3 & 0.7 \end{bmatrix}$

2-6. Using the following matrices, show that matrix addition is not directional, but that matrix subtraction is directional:

$$\mathbf{A} = \begin{bmatrix} 3 & -2 \\ 0 & 4 \end{bmatrix}, \quad \mathbf{B} = \begin{bmatrix} -2 & 3 \\ 1 & 2 \end{bmatrix}$$

2-7. Given matrices $\mathbf{C}$ and $\mathbf{E}$, find matrix $\mathbf{D}$ such that $\mathbf{C} + \mathbf{D} = \mathbf{E}$. The matrices are given by

$$\mathbf{C} = \begin{bmatrix} 0.4 & -0.3 & 0.5 \\ 0.0 & 0.7 & -0.2 \end{bmatrix} \quad \text{and} \quad \mathbf{E} = \begin{bmatrix} 0.1 & -0.8 & 0.9 \\ -0.6 & 0.3 & -0.4 \end{bmatrix}$$

2-8. Using the matrices of Problem 2-6, show that matrix multiplication is not commutative.

2-9. Using the following matrices, show that matrix multiplication is not commutative:

$$\mathbf{F} = \begin{bmatrix} 0.1 & 0.7 \\ -0.3 & 0.4 \end{bmatrix} \quad \text{and} \quad \mathbf{G} = \begin{bmatrix} -0.2 & 0.8 \\ 0.5 & -0.6 \end{bmatrix}$$

2-10. For each of the following matrix pairs, compute the matrix product $\mathbf{C} = \mathbf{A} \cdot \mathbf{B}$.

(a) $\mathbf{A} = \begin{bmatrix} 0.1 & -0.3 & 0.2 \\ 0.6 & -0.1 & 0.4 \end{bmatrix}$, $\mathbf{B} = \begin{bmatrix} 1 \\ -4 \\ 2 \end{bmatrix}$

(b) $\mathbf{A} = \begin{bmatrix} 0.3 \\ 1.2 \\ -0.5 \end{bmatrix}$, $\mathbf{B} = [-0.2 \quad 0.7 \quad 0.5]$

(c) $\mathbf{A} = [-0.2 \quad 0.7 \quad 0.5]$, $\mathbf{B} = \begin{bmatrix} 0.3 \\ 1.2 \\ -0.5 \end{bmatrix}$

(d) $\mathbf{A} = \begin{bmatrix} 0.4 & 0.6 \\ 1.3 & -0.2 \\ -0.8 & 1.5 \end{bmatrix}$, $\mathbf{B} = \begin{bmatrix} 2 & 3 & -1 & 4 \\ 6 & -4 & 0 & 5 \end{bmatrix}$

2-11. Compute the inverse of matrix $\mathbf{B}$ of Problem 2-6.

2-12. Compute the inverse of the following matrix:

$$\begin{bmatrix} 3 & -1 & 2 \\ 1 & 4 & -3 \\ 2 & 5 & 3 \end{bmatrix}$$

2-13. Show that the inverse of the inverse of matrix $\mathbf{X}$ is $\mathbf{X}$.

2-14. Show which of the following matrices are singular:

(a) $\mathbf{A} = \begin{bmatrix} 0.1 & 0.3 \\ 0.4 & 1.2 \end{bmatrix}$

(b) $\mathbf{B} = \begin{bmatrix} -2 & 6 & 2 \\ 1 & -3 & 2 \\ 2 & -6 & -2 \end{bmatrix}$

2-15. What does the trace of a correlation matrix equal?

2-16. Compute the trace of matrix $\mathbf{B}$ in problem 2-14b.

2-17. Add the following vectors:

(a) $\mathbf{V}_1 = [2 \quad 6 \quad -3]$ and $\mathbf{V}_2 = [-3 \quad 1 \quad 2]$

(b) $\mathbf{V}_1 = \begin{bmatrix} 0.6 \\ 0.2 \\ 0.0 \\ 0.3 \end{bmatrix}$ and $\mathbf{V}_2 = \begin{bmatrix} -0.4 \\ 0.1 \\ -0.5 \\ 0.2 \end{bmatrix}$

2-18. Subtract vector $\mathbf{V}_1$ from vector $\mathbf{V}_2$ in Problem 2-17a and b.

2-19. Find the vector product $\mathbf{V}_1^T\mathbf{V}_2$ using the vectors of Problem 2-17a and b.

2-20. Find the vector product $\mathbf{V}_1\mathbf{V}_2^T$ using the vectors of Problem 2-17a and b.

2-21. If the following matrix is viewed as three column vectors, which of the vectors are normalized? Which pair, if any, of the column vectors is orthonormal?

$$\begin{bmatrix} 0.5883 & -0.5921 & -0.1830 \\ 0.7845 & -0.1502 & 0.3649 \\ -0.1961 & 0.7917 & 0.9129 \end{bmatrix}$$

2-22. The coordinates of a point in a three-dimensional axis system are the elements of a row vector

$$\mathbf{X} = [X_1 \quad X_2 \quad X_3]$$

For two points B and A in the axis system, the elements of a vector of the direction cosines of the line connecting the two points are (i.e., a vector from B to A)

$$\cos \theta_i = \frac{a_i - b_i}{L}$$

$$L = \sqrt{\sum_{i=1}^{3} (a_i - b_i)^2}$$

Compute the θ_i for the following two points (i.e., a vector from $\mathbf{S}$ to $\mathbf{T}$):

$$\mathbf{S} = [3 \quad 4 \quad -2]$$

$$\mathbf{T} = [-1 \quad 3 \quad 3]$$

2-23. Compute the value of x that is necessary for the following vector $\mathbf{A}$ to be normal:

$$\mathbf{A} = [0.7 \quad 0.4 \quad -0.2 \quad -0.3 \quad x]$$

2-24. Compute the value of b_{22} that yields a normalized column vector for the following matrix. Is the matrix orthonormal?

$$\mathbf{B} = \begin{bmatrix} 0.23 & 0.66 \\ -0.58 & b_{22} \\ 0.78 & 0.26 \end{bmatrix}$$

2-25. Compute the values of a_{12} and a_{32} that are necessary for matrix $\mathbf{A}$ to be orthonormal in the columns:

$$\mathbf{A} = \begin{bmatrix} 0.5 & a_{12} \\ 0.2 & -0.6 \\ -0.843 & a_{32} \end{bmatrix}$$

2-26. Compute the determinants of matrices $\mathbf{A}$ and $\mathbf{B}$ of Problem 2-6.

2-27. Compute the determinant of the matrix of Problem 2-12.

2-28. Compute the determinant of matrix $\mathbf{B}$ in Problem 2-14.

2-29. Determine the rank of the matrices in Problem 2-10.

2-30. Determine the rank of the following matrices:

$$
(a) \begin{bmatrix} 1 & 2 & 4 \\ 1.5 & 4 & 6 \end{bmatrix}
\qquad
(b) \begin{bmatrix} 3 & 2 & 1 \\ 1 & 3 & 2 \\ 2 & 1 & 3 \end{bmatrix}
\qquad
(c) \begin{bmatrix} 2.0 & 8.0 & 8.0 \\ 1.5 & 4.5 & 6.0 \\ 3.5 & 24.5 & 14.0 \\ 1.5 & 4.5 & 6.0 \end{bmatrix}
$$

2-31. A soil scientist collects the erosion rate Y from four test plots. The plots differ in percent slope (X_1) and soil type; the mean diameter of the soil particles (X_2) is used to represent the soil type. Compute the correlation matrix method of Section 2.6.3. The collected data are

$$
\begin{array}{ccc}
X_1 & X_2 & Y \\
\begin{bmatrix} 2 & 0.10 & 11 \\ 5 & 0.09 & 14 \\ 6 & 0.04 & 12 \end{bmatrix}
\end{array}
$$

3

Introduction to Numerical Methods

3.1. INTRODUCTION

Many operations in mathematical analysis can be identified as belonging to one of two types, analytical or numerical. For example, when we need to solve a second-order polynomial $aX^2 + bX + c = 0$, for the values of X that satisfy the equality, it is common to use the closed-form solution $X = \left(-b \pm \sqrt{b^2 - 4ac}\right)/2a$. This is an *analytical* solution. Alternatively, we could assume some value for the solution (X), and through a systematic method of trial and error, we could obtain essentially the same solution as the analytical approach. However, this *numerical* approach is much more tedious, and so when an analytical solution exists it is usually the preferred method. For many types of problems, such as a fifth-order polynomial, a closed-form or analytical solution does not exist. Then the iterative, or numerical, approach must be used.

In spite of the simplicity of most numerical methods, it is somewhat difficult to give a specific definition for the term *numerical method*. Instead of trying to define the term, we will discuss the characteristics, advantages, and disadvantages of numerical methods and then provide an example to illustrate these characteristics.

Numerical methods can be used to solve a variety of quantitative problems faced by engineers. For many engineering problems, an analytical solution is either impossible or the cost or effort of performing an analytical solution would be prohibitive. Furthermore, in an effort to achieve greater accuracy in engineering solutions, more complex analyses are being used. Very often, the complexity of the problem becomes such that an analytical solution is not possible. In cases where analytical solutions are not practical or the problem is very complex, numerical methods can often be used to find a solution.

3.1.1 Characteristics of Numerical Methods

Numerical methods have most of the following characteristics:

1. The solution procedure is iterative, with the accuracy of the estimated solution improving with each iteration.
2. The solution procedure provides only an approximation to the true, but unknown, solution.
3. An initial estimate of the solution may be required.
4. The solution procedure is conceptually simple, with algorithms representing the solution procedure that can be easily programmed on a digital computer.
5. The solution procedure may occasionally diverge from rather than converge to the true solution.

Each of these characteristics will be evident in the numerical methods introduced in many of the remaining chapters of this book. Example 3-1 is used to illustrate the concept of a numerical method.

Example 3-1. Square Root

Finding the square root of a number is a frequent task. On a hand calculator, we simply enter the number and then press the $\sqrt{x}$ key. On a digital computer, the function SQRT is used. For the purposes of illustrating numerical methods, let's assume that SQRT is not available, and we need to develop a method for estimating $\sqrt{x}$, where x is any positive real value. We can start by assuming that we have an initial estimate of $\sqrt{x}$, which we will denote as x_0, and x_0 is in error by an unknown amount Δx. If we know Δx, then we would have the following equality:

$$x_0 + \Delta x = \sqrt{x} \qquad (3\text{-}1)$$

If both sides of Eq. 3-1 are squared and we assume that $(\Delta x)^2$ is much smaller than Δx and can be neglected, we can solve for Δx:

$$\Delta x = \frac{x - x_0^2}{2x_0} \qquad (3\text{-}2)$$

The value of Δx computed with Eq. 3-2 can be added to x_0 to get a revised estimate of x. Thus the new estimate x_1 of the true solution is

$$x_1 = x_0 + \Delta x \qquad (3\text{-}3\text{a})$$

Generalizing the notation, the concept behind Eqs. 3-2 and 3-3a yields

$$x_{i+1} = x_i + \Delta x \qquad (3\text{-}3\text{b})$$

where x_i and x_{i+1} are the estimates of x on trials i and $(i+1)$, respectively, and

$$\Delta x = \frac{x - x_i^2}{2x_i} \qquad (3\text{-}3\text{c})$$

To illustrate the use of Eqs. 3-3, assume that $x = 150$. We know that $12^2 = 144$, so $x_0 = 12$ is a reasonable initial estimate. Therefore, the first iteration of Eq. 3-3c yields

$$\Delta x = \frac{x - x_0^2}{2x_0} = \frac{150 - 12^2}{2(12)} = 0.25 \tag{3-4a}$$

Thus, from Eq. 3-3b, x_1 equals 12.25. A second iteration of Eq. 3-3c can now be applied using the revised estimate x_1:

$$\Delta x = \frac{x - x_1^2}{2x_1} = \frac{150 - (12.25)^2}{2(12.25)} = -0.00255 \tag{3-4b}$$

Applying Eq. 3-3b again yields $x_2 = 12.24745$. A third iteration of Eq. 3-3c yields

$$\Delta x = \frac{x - x_2^2}{2x_2} = \frac{150 - (12.24745)^2}{2(12.24745)} = -0.12861 \times 10^{-5} \tag{3-4c}$$

The revised value x_3 equals 12.2474487, which is identical to x_2 to seven significant digits. While the true value is not usually known, the true solution of $\sqrt{150}$ equals 12.24744871. Thus x_2 was accurate to seven significant digits, and x_3 was accurate to nine digits.

The purpose of the above example is not to provide a numerical method for estimating $\sqrt{x}$; rather, the objective is to illustrate the characteristics of numerical methods. With respect to the first characteristic identified at the beginning of this section, the solution procedure is iterative in that Eqs. 3-3 were applied three times, with the accuracy improving with each iteration. Second, the solution procedure provided an approximation of the true value; the initial estimate of 12 was a first approximation, while x_1, x_2, and x_3 were subsequent approximations. Third, the solution procedure required an initial estimate; for our example problem, an initial estimate of 12 was used. Fourth, the solution procedure is conceptually simple and can be easily programmed; an algorithm is provided in Fig. 3-1. Fifth, the procedure would have converged even if a less accurate initial estimate, such as 10 or 20 were used. Other numerical methods may not converge if the initial estimate is not a reasonably good estimate.

```
        FUNCTION SQRTN(X, X0, TOL)
   c    X = VALUE FOR WHICH SQUARE ROOT IS NEEDED
   c    X0 = AN INPUT, INITIAL ESTIMATE OF SQUARE ROOT OF X
   c    X0 = FINAL ESTIMATE OF SQUARE ROOT OF X
   c    TOL = MAXIMUM ALLOWABLE (TOLERABLE) ERROR IN
   c    SQUARE ROOT OF X
   1    DELX=(X − X0**2)/(2.0*X0)
        X0=X0+DELX
        IF (ABS(DELX).GT.TOL)GO TO 1
        SQRTN=X0
        END
```

Figure 3-1 Subprogram for the Numerical Solution of $\sqrt{x}$

3.2 ACCURACY, PRECISION, AND BIAS

When dealing with solutions from numerical analyses, we need to address the issues of *accuracy, precision,* and *bias.* The term *accuracy* is commonly used, for example, as in reference to the accuracy of the media's report on a particular issue or to the accuracy of an opinion poll taken just before a presidential election. Of interest here is the meaning of the term as it applies to numerical solutions.

Consider the four targets of Fig. 3-2. Assuming that those shooting at the targets were aiming at the center, the person shooting at target *A* was successful. While not every bullet went through the exact center, the distances between the holes in the target and the center are small. The holes in target *B* are similarly clustered as in target *A,* but they show large deviation from the center. The large deviations or errors between the holes and the center of the target suggest a lack of exactness or correctness. While all the holes deviate significantly from the center, there is, however, a measure of consistency in the holes. In summary, the holes in target *B* show two important characteristics; they tend to agree with each other, but they deviate considerably from where the shooter was aiming.

The holes in target *C* are very different in character from the holes in either target *A* or *B.* First, most are not near the center, and second, they are not near each other. Thus, they lack both correctness and consistency. Because most of them are not near the center of the target, the shooter is not exact. Because of the wide scatter, there is a lack of consistency. In comparing the holes on targets *B* and *C,* they both lack exactness, but there is a measure of consistency for the holes in target *B* that is missing in target *C.*

A fourth distribution of holes is shown in target *D.* Like target *B,* all of the holes are to the left of the center; but, unlike target *B,* the holes lack consistency. Both targets *C* and *D* show considerable scatter, but with target *D* the scatter is concentrated to one side of the target.

A comparison of the four targets indicates that there are three important characteristics of data. The holes in targets *A* and *B* show a measure of consistency. In data analyses, this consistency is more formally referred to as precision. *Precision* is defined as the ability to give multiple estimates that are near to each other. In terms of the targets *A* and *B,* the shooters were precise. The shooters of targets *C* and *D* were imprecise since the holes show a lot of scatter.

Bias is a second important characteristic of data. The holes in targets *B* and D are

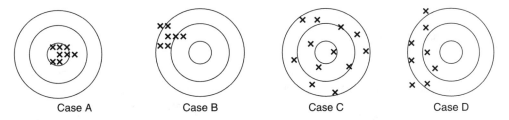

Case A   Case B   Case C   Case D

Figure 3-2   Accuracy, Precision, and Bias

TABLE 3-1 Summary of Bias, Precision, and Accuracy for Figure 3-2

Target	Bias	Precision	Accuracy
A	None (unbiased)	High	High
B	High	High	Low
C	None (unbiased)	Low	Low
D	Moderate	Low	Low

consistently to the left of center; that is, there is a systematic distortion of the holes with respect to the center of the target. If we found a point that could represent the center of the holes in any target, then the difference between the center of the holes and the center of the target would be a measure of the shooter's bias. More formally, *bias* is a systematic deviation of values from the true value. The holes in targets B and D show a systematic deviation to the left, while targets A and C are considered to be unbiased because there is no systematic deviation.

Precision and bias are elements of accuracy. Specifically, bias is a measure of systematic deviation and precision is a measure of random deviations. Inaccuracy can result from either a bias or a lack of precision. In referring to the targets, the terms correctness and exactness were used. More formally, *accuracy* is the degree to which the measurements deviate from the true value. If the deviations are small, then the method or process that led to the measurement is accurate.

With reference to the numerical method of Example 3-1, it provides accurate estimates because it closely approaches the true value regardless of the initial estimate. If the method consistently led to estimates that exceeded the true value by some amount δ, then the numerical method would have a positive bias of δ. If the estimates were consistently less than the true value by an amount δ, then the bias is negative. If the deviations were often greater than the true value and just as often less than the true value, then the method would be imprecise and inaccurate.

Table 3-1 provides a summary of the concepts of accuracy, precision, and bias as they characterize the holes in the targets of Fig. 3-2. Of course, terms like high, moderate, and low are somewhat subjective.

3.3 SIGNIFICANT FIGURES

Consider the problem of measuring the distance between two points using a ruler that has a scale with 1 mm between the finest divisions. If we record our measurements in centimeters and if we estimate fractions of a millimeter, then a distance recorded as 3.76 cm gives two precise digits (i.e., the 3 and the 7) and one estimated digit (i.e., the 6). If we define a significant digit to be any number that is relatively precise, then the measurement of 3.76 cm has three significant digits. Even though the last digit could be a 5 or a 7, it still provides some information about the length, and so it is assumed to be significant. If we recorded the number as 3.762, we would still have only three significant digits since the 2 is not precise. Only one imprecise digit can be considered as a significant digit.

The number of significant digits is used to reflect the accuracy of a number. For example, if the number 46.23 is exact to the four digits shown, it is said to have four significant digits. The error is then no more than 0.005.

The number of significant digits is set either by the scale measurement or the physical significance of the numbers. For example, a digital bathroom scale that shows weight to the nearest pound uses up to three significant digits. If the scale shows, for example, 159 pounds, then the individual assumes that his or her weight is within 0.5 pound of the observed value. In this case, the scale has set the number of significant digits. The number of significant digits in someone's height is set by the physical significance of the numbers. The ruler may provide for measurements to $\frac{1}{32}$ inch, but the number may be recorded to the nearest inch because more precise values are not important. In that case we would have two significant digits, since a measurement of 5 ft, $11\frac{3}{32}$ in. would be recorded as 71 in. In this case, the meaningfulness of the information content of the number set the number of significant digits.

The rule for identifying the number of significant digits is as follows: *The digits from 1 to 9 are always significant, with zero being significant when it is not being used to set the position of the decimal point.* For example, each of the following have three significant digits: 2,410, 2.41, and 0.00241. In the first example, the 0 is only used to set the decimal place. Confusion can be avoided by using scientific notation: 2.41×10^3. This means that 2.41×10^3 has three significant digits, where 2.410×10^3 has four significant digits. The numbers 13 and 13.00 differ in that the former is recorded at two significant digits, and the latter has four significant digits. In this case, the two zeros are significant.

The location of the decimal point does not influence the number of significant digits. The measurements of 3.76 cm could be written as 37.6 mm or 0.0376 m; each of these three values has three significant digits.

When performing computations, the following is a general rule on setting the number of significant digits in a computed value: *Any mathematical operation using an imprecise digit is imprecise.* When combined with the rule that only one imprecise digit can be considered significant, then the number of significant digits is set. Consider the following multiplication of two numbers (4.2<u>6</u> and 8.3<u>9</u>), each having three significant digits, with the last digit of each being imprecise:

4.2<u>6</u>	Starting number
8.3<u>9</u>	Starting number
0.<u>3</u>834	0.0<u>9</u> times 4.2<u>6</u>
1.27<u>8</u>	0.3 times 4.2<u>6</u>
34.0<u>8</u>	8 times 4.2<u>6</u>
35.<u>7</u>414	Total (product result)

The digits that depend on imprecise digits are underlined. In the final answer, only the first two digits (35) are not based on imprecise digits. Since one and only one imprecise digit can be considered as significant, then the result is recorded as 35.7.

Just because a set of numbers is only accurate to a specified number of digits does not mean that computations based on those numbers should be rounded and reported to that number of digits. Computed values should only be rounded at the end of the computation process. For intermediate steps, the computed numbers should be determined and reported to more digits than the number of significant digits in the original set of numbers. The reporting of coefficients of equations provides an illustration of the problem. Consider the equation with values of the coefficients reported to five significant digits:

$$\hat{Y} = 11.587 + 1.9860x \tag{3-5a}$$

where $\hat{Y}$ is the predicted value of the variable y in the regression model of Eq. 3-5a. For three significant digits the equation would be

$$\hat{Y} = 11.6 + 1.99x \tag{3-5b}$$

Using three, four, and five digits for the regression coefficients yields the predicted values $(\hat{Y}_3, \hat{Y}_4, \text{ and } \hat{Y}_5)$ for given values of x as shown in Table 3-2. Rounding the regression coefficients to three digits does not cause an error for $x = 1$, but errors of 0.1, 0.2, and 1 occur for $x = 20$, $x = 40$, and $x = 100$, respectively, assuming that the final values are rounded to three significant digits. Using the regression coefficients with four digits produces estimated values for all of the values of x that are without rounding error when evaluated at three significant digits. In this case, estimated values of Y can be rounded to three significant digits, but the regression coefficients should not be rounded to three digits; the rounding should be made at the end of the computation, not at intermediate calculations.

Example 3-2. Arithmetic Operations and Significant Digits

When performing mathematical operations as a part of the solution of a numerical analysis, it is important to consider the number of significant digits in the solution. If we are measuring the lengths of the base and height of a triangle with an instrument that provides values to the nearest *tenth of an inch,* then the measured lengths of $b=12.3$ in. and $h=17.2$ in. are ex-

TABLE 3-2   Rounding Numerical Calculations

x	$\hat{Y}_3$	$\hat{Y}_4$	$\hat{Y}_5$
1	13.59	13.576	13.573
20	51.40	51.310	51.307
40	91.20	91.030	91.027
100	210.60	210.19	210.187

pressed to three significant digits. If we use these values to compute the area of a triangle, then we should express the solution as a number having the smallest number of significant digits; thus the area would be

$$A = 0.5bh = 0.5(12.3)(17.2) = 106 \text{ in.}^2 \tag{3-6a}$$

Note that if we ignore the concept of significant digits the area would be 105.78 in.^{2}; however, using this value would imply that there are five significant digits. We only know that the true value is expected to lie between

$$0.5(12.25)(17.15) = 105.04375 \text{ cm}^2 \tag{3-6b}$$

and

$$0.5(12.35)(17.25) = 106.51875 \text{ cm}^2 \tag{3-6c}$$

It is also of interest to note that, whereas the 0.5 in Eq. 3-6a appears to only be expressed with one significant digit, it is considered to be an exact value.

It is common in engineering to deal with quantities that cannot be expressed in a finite number of significant digits, for example, e (for exponential), π, $\sqrt{3}$, and $\sqrt{11}$. Although these quantities have specific numerical meaning, their absolutely correct values can only be expressed in an infinite number of significant digits. Therefore, for a specified engineering application, an engineer needs to specify the number of significant digits that should be used to express these quantities. For example, $\pi = 3.14159265358979323\ldots$ can be expressed to three significant digits as 3.14. As a result, a round-off error is introduced, which is the true value minus the approximate value. The round-off error in this case is $-0.00159265358979323.\ldots$ This example illustrates the importance of significant digits and their role in introducing errors or accuracy in numerical solutions.

3.4 ANALYSIS OF NUMERICAL ERRORS

3.4.1 Error Types

An *error* in estimating or determining a quantity of interest can be defined as a deviation from its unknown true value. In general, errors can be classified based on their sources as non-numerical and numerical errors.

Non-numerical errors include (1) modeling errors, (2) blunders and mistakes, and (3) uncertainty in information and data. Prediction models used for assessing the parameters of an engineering problem are commonly based on a set of assumptions and limitations. Therefore, the models can give predicted values that might deviate from the corresponding true values of the problem. Unfortunately, the true values of the problem are unknown. Assumptions and limitations that are made for formulating the model reduce the analytical complexity to a reasonable level. The deviations of the predicted values from the true values are the modeling errors.

Blunders and gross mistakes can generally be attributed to human errors. They can take several forms, for example, arithmetic mistakes, omissions, incorrect use of equations or computational procedures, incorrect development of computational algorithms including computer programs, incorrect input, incorrect understanding of output, incorrect interpretation of computational algorithms and computer programs, and the use of prediction models beyond the range of their validity. This source of error is usually not considered in error analysis. However, it can be controlled using, for example, checking procedures, simplified algorithms, and simplified human–machine interfaces, such as control panels for equipment or input formats for computer programs.

Uncertainty in information can be attributed to two main sources, objective and subjective. The objective source includes inherent variability in some engineering parameters and the use of limited information; for example, a sample is used to draw conclusions about a problem. This source of uncertainty can be dealt with using probability and statistics, and it is not numerically based. The subjective uncertainty in information is encountered in cases where an engineer has to make a subjective assessment, for example, the condition of a corroded steel beam. In this case, the assessment might vary from one engineer to another, resulting in subjective uncertainty in information.

Numerical errors include (1) round-off errors, (2) truncation errors, (3) propagation errors, and (4) mathematical-approximation errors. Round-off errors were discussed in Section 3.3. They are due to expressing some quantities of interest in a limited number of significant digits, whereas the true values have a large number of significant digits. Round-off errors can be computed as the difference between a true value and its rounded value.

In numerical analysis, it is common to deal with mathematical expressions that consist of the addition of many terms, such as the first term of a Taylor series or the sum of the first five terms of an infinite series. The numerical contribution of these terms is often of diminishing value as more terms are added to the expression. To keep a computational effort to a reasonable level, it is common to truncate such mathematical expressions to a certain number of terms. The truncated terms contribute to the total error. However, terms that are truncated are considered to contribute insignificantly to error in the final estimate of the model. The difference between the evaluation of the complete (nontruncated) expression and the truncated expression is called the *truncation error.* An example of these mathematical expressions is a mathematical infinite series (e.g., the Taylor series). The Taylor series that is discussed in Chapter 1 contains an infinite number of terms. The first term is called the zero-order term. The second term is the first-order term because it is based on the first derivative. The third term uses the second derivative and is called the second-order term. The additional terms are referred to similarly. It is common to truncate the Taylor series to the first-order terms, or sometimes to the second-order terms. Therefore, truncation errors are introduced. The effects of these errors can be significant unless the numerical algorithm is properly formulated.

In Eqs. 3-6 and 3-7, it was noted that an arithmetic operation such as multiplication of two numbers that are expressed to three significant digits results in a product that can only be expressed to three significant digits. The product is uncertain to any larger number of significant digits. The product expressed to three significant digits in Eq. 3-6 is

106, but the true value is expected to lie between 105.04375 and 106.51875. Therefore, arithmetic operations result in a propagation of errors. Sometimes, by establishing a computational order for the arithmetic operations in an algorithm, the propagation errors can be reduced. For example, in evaluating the summation of several numerical values using a specified number of significant digits, it is desirable to rank order the numerical values in ascending order before performing the summation. The result of this summation can, therefore, have a reduced propagation error as compared to an arbitrary order for performing the summation.

Mathematical-approximation errors result from using approximate mathematical expressions for a relationship or solution. For example, numerical methods sometimes call for the use of a linear model for representing a nonlinear mathematical expression. The difference in prediction between the nonlinear and linear models is an error of this type.

In this chapter, only numerical errors are of interest and need to be evaluated, because their assessment provides a means of determining when a numerical iterative solution should be terminated. Also, the understanding of the errors in numerical methods can assist an engineer in selecting appropriate numerical methods for solving a problem.

3.4.2 Measurement and Truncation Errors

When dealing with engineering measurements, the accuracy and precision of an estimate may be just as important as the estimate itself. We can define the *error* (e) as the difference between the computed (x_c) and true (x_t) values of a number x:

$$e = x_c - x_t \tag{3-7}$$

The *relative true error* (e_r) is defined as the error e relative to the true value:

$$e_r = \frac{x_c - x_t}{x_t} = \frac{e}{x_t} \tag{3-8}$$

The relative true error could be expressed as a percentage by multiplying e_r of Eq. 3-8 by 100.

To illustrate the concept of error analysis, consider the atomic weight of oxygen, which is 15.9994 and is believed to be accurate to ± 0.0001. The relative error would then be

$$e_r = \frac{0.0001}{15.9994} = \pm 0.6 \times 10^{-5} \tag{3-9}$$

Even though the atomic weight is expressed to six significant digits, the relative error is only expressed to one significant digit because the error is only given to one significant digit.

Example 3-3. Truncation Error in Atomic Weight

In addition to measurement errors, such as the measurement of atomic weights, engineers may also be concerned with errors due to truncation. If we round the atomic weight of oxygen to 16, which is common, we introduce a truncation error, which can be computed using Eq. 3-7 as

$$e = 16 - 15.9994 = 0.0006 \tag{3-10}$$

Relative to the true value, this results in a relative error, which is computed with Eq. 3-8 as

$$e_r = \frac{0.0006}{15.9994} = 0.4 \times 10^{-4} \tag{3-11}$$

Thus the relative error due to truncation (Eq. 3-11) is seven times greater than the relative error due to measurement (Eq. 3-9).

3.4.3 Error Analysis in Numerical Solutions

As indicated in Section 3.1, numerical methods provide an approximation to a problem's solution, rather than the true value. Furthermore, numerical methods are iterative. Based on these two characteristics of numerical methods, it would be of interest to have an estimate of the error with each iteration so that the computations could be discontinued when sufficient accuracy has been achieved. The purpose of the variable TOL in Figure 3-1 was to provide a means of exiting from the loop where the error is within some user-specified tolerance.

In practice, the true value is not known, so Eqs. 3-7 and 3-8 cannot be used to compute the error. In such cases, the best estimate of the number x should be used. Unfortunately, the best estimate is the computed estimate, and so Eqs. 3-7 and 3-8 lose their meaning when directly applied to the accuracy of solutions from numerical methods. Instead, we will apply Eq. 3-7 iteratively; thus Eq. 3-7 is rewritten as

$$e_i = x_i - x_t \tag{3-12}$$

where e_i is the error in x at iteration i, and x_i is the computed value of x from iteration i; similarly, the error for iteration $(i+1)$ is

$$e_{i+1} = x_{i+1} - x_t \tag{3-13}$$

Therefore, the change in the error (called simply the *relative error*) can be computed from Eqs. 3-12 and 3-13 as

$$\Delta e_i = e_{i+1} - e_i = (x_{i+1} - x_t) - (x_i - x_t) = x_{i+1} - x_i \tag{3-14}$$

It can be shown that e_{i+1} is expected to be smaller than Δe_i; so if we continue to iterate until Δe is smaller than our tolerable error, we can be reasonably sure that x_{i+1} is sufficiently close to x_t.

To illustrate the calculation of the error using Eq. 3-14, assume that we wish to find the value of x that makes the following third-order polynomial an identity:

$$x^3 + ax^2 + bx + c = 0 \tag{3-15}$$

If we divide Eq. 3-15 by x and solve for x using the x^2 term, we get

$$x = \sqrt{-ax - b - \frac{c}{x}} \qquad (3\text{-}16)$$

We can find the positive value of x for Eq. 3-15 by selecting an initial estimate of x and solving the right side of Eq. 3-16 to provide a new estimate of x (i.e., the left side of Eq. 3-16). This new estimate can be inserted into the right side of Eq. 3-16 to find a better estimate. The iterations are continued until the change, Δe, which is given by Eq. 3-14, is less than some stated tolerance.

Example 3-4. Numerical Errors Analysis

The following polynomial is used to illustrate the numerical solution according to Eq. 3-16:

$$x^3 - 3x^2 - 6x + 8 = 0 \qquad (3\text{-}17)$$

Equation 3-17 can be expressed in the following form:

$$x = \sqrt{3x + 6 - \frac{8}{x}} \qquad (3\text{-}18)$$

If we choose $x_0 = 2$ as the initial estimate of the positive value of x, Eq. 3-18 becomes

$$x_1 = \sqrt{3x_0 + 6 - \frac{8}{x_0}} = 2.828427 \qquad (3\text{-}19)$$

Thus Eq. 3-14 yields an error of

$$e_1 = x_1 - x_0 = 0.828427 \qquad (3\text{-}20)$$

Using the value of x_1 from Eq. 3-19 as input to Eq. 3-18 yields a revised estimate of x_2 of 3.414214, which has an error from Eq. 3-14 of $e_2 = 0.585786$. The results of 10 iterations are given in Table 3-3. It is evident that the solution approaches $x = 4$, and substitution of $x = 4$ into Eq. 3-17 shows that it is a valid solution. On trial 1, the change Δe is actually smaller than the true error, which is also shown in Table 3-3, but after trial 2, Δe_i is always larger than or equal to the true error. If the tolerable error were set at 0.025, the iterative solution would be stopped after trial 7; for trial 6, the change $\Delta e = 0.030$ was larger than the tolerance and therefore one more iteration was necessary. However, the estimate on trial 6 is actually within the stated tolerance with respect to the true value, but since the true value was not known, trial 7 was necessary. Ten trials were necessary to achieve a level of accuracy within a tolerance of 0.002.

In numerical analysis, the number of significant digits that should be used in computation is a common question. Even though the numbers are given in Eq. 3-17 to only one significant digit, the values are reported to seven digits in Table 3-3; this occurs only because these were the values stored in the hand calculator used in developing Table 3-3. If fewer digits were used, the solution to this problem would have converged in the same number of trials. In practice, there is no reason to truncate the values after each iteration if the values can be stored to a large number of digits. It is only necessary to carry the number of digits that would be needed to check the tolerable error for the accuracy considered desirable.

TABLE 3-3 Error Analysis of the Solution of the Polynomial of Equation 3-17 with $x_t = 4$

Trial i	x_i	Δe_i	$x_t - x_i$
0	2.000000	—	2.000000
1	2.828427	0.828427	1.171573
2	3.414214	0.585786	0.585786
3	3.728203	0.313989	0.271797
4	3.877989	0.149787	0.122011
5	3.946016	0.068027	0.053984
6	3.976265	0.030249	0.023735
7	3.989594	0.013328	0.010406
8	3.995443	0.005849	0.004557
9	3.998005	0.002563	0.001995
10	3.999127	0.001122	0.000873

3.5 ADVANTAGES AND DISADVANTAGES OF NUMERICAL METHODS

As indicated in the introduction to this chapter, numerical methods are an alternative to analytical solutions. When it is possible, analytical solutions are usually preferred. With analytical problem solving, the solution is found directly, rather than iteratively as with numerical methods; but it is often not possible to obtain an analytical solution, and a numerical method of solution is necessary. Analytical methods of problem solving yield an exact solution, if one exists; a numerical solution only yields an approximation, although if a sufficient number of iterations are performed, there may be no difference between the numerical solution and the true value. Since analytical methods yield the solution directly, the computation time required is usually less than the time required for a numerical solution.

In addition to advantages, analytical problem solving has disadvantages. First, the solution procedures are only valid for simple problems and cannot be used for many types of problems that engineers need to solve. Second, with many engineering problems, the solution is subject to a number of constraints. In general, it is usually difficult to solve a problem with an analytical method when the problem involves constraints; in such cases, numerical methods have the advantage because constraints are relatively easy to handle.

In summary, numerical methods have the following advantages: (1) they are conceptually simple, (2) they are easily programmed on a digital computer, (3) constraints are easily handled, and (4) solutions are possible for complex problems. However, numerical methods also have disadvantages, including that (1) they are solved iteratively and, thus, more computation time is required, (2) an exact solution may not be found, and (3) initial estimates of the solution are often required.

3.6 APPLICATIONS

3.6.1 Pipe Design

Consider the case shown in Fig. 3-3 where a long pipe connects two reservoirs, with a pump located in the system to pump the water from the low-elevation reservoir to the high-elevation reservoir. The reservoirs are open to the atmosphere, and the water surfaces are denoted as points 1 and 3. The location of the pump is denoted as point 2. The elevation of any point i in the system above a datum is indicated as Z_i; thus Z_1 and Z_3 are the elevations of the water surfaces in the reservoirs, and Z_2 is the elevation of the pump.

The following equation, which is called the *energy equation,* governs the flow of water through the system:

$$\frac{P_1}{\gamma} + \frac{V_1^2}{2g} + Z_1 - h_f + E_p = \frac{P_3}{\gamma} + \frac{V_3^2}{2g} + Z_3 \qquad (3\text{-}21)$$

in which p_i = pressure at point i, γ = specific weight of the fluid, V_i = velocity of the fluid at point i, h_f = friction losses, and E_p = energy head supplied by the pump. Each term in Eq. 3-21 has dimensions of length (i.e., feet or meters). For the system of Fig. 3-3, several assumptions can be made to simplify the energy equation of Eq. 3-21. First, since both reservoirs are open to the atmosphere, $P_1 = P_3$, so the pressure head terms can be dropped. If both reservoirs have large surface areas, the surface levels will fall and rise very slowly, so V_1 and V_3 are essentially zero. If we let $h = Z_3 - Z_1$, then Eq. 3-21 reduces to

$$E_p = h + h_f \qquad (3\text{-}22)$$

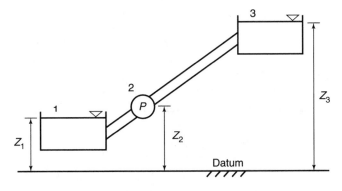

Figure 3-3   Schematic of a Pipe System

Equation 3-22 can be interpreted as follows: the pump must supply sufficient energy E_p to overcome the energy lost due to frictional forces h_f and the energy needed to increase the potential energy by h. We will assume that most of the physical characteristics of the system are known; this includes h, the length of the pipe L, the roughness of the pipe e_f, the viscosity of the fluid ν, and the specific weight of the fluid γ. Additionally, the flow rate Q through the pipe and the horsepower of the pump h_p are known. The horsepower input h_p is related to the energy head of the pump and the flow rate by

$$h_p = \frac{\gamma Q E_p}{550e} \tag{3-23}$$

in which e is the efficiency of the pump and motor. Friction losses in the system are related to the physical characteristics of the system and the velocity of the fluid in the pipe as

$$h_f = f\frac{L}{D}\left(\frac{V^2}{2g}\right) \tag{3-24}$$

in which D is the diameter of the pipe, f is the friction coefficient, V is the velocity of the flow in the pipe, and g is the gravitational acceleration. Based on the continuity equation $Q = AV$, where A is the cross-sectional area of the circular pipe, Eq. 3-24 can be rewritten as

$$h_f = f\frac{L}{D}\left(\frac{Q^2}{2gA^2}\right) = \frac{8fLQ^2}{\pi^2 gD^5} \tag{3-25}$$

Substituting Eqs. 3-25 and 3-23 into Eq. 3-22 yields

$$\frac{550eh_p}{\gamma Q} = h + \frac{8fLQ^2}{\pi^2 gD^5} \tag{3-26}$$

The goal of the problem is to find the diameter of the pipe that is necessary to pass flow rate Q for a specified pump. Equation 3-26 could be solved directly for D if the friction factor f were not a function of D. The relationship between f and D is nonlinearly related to the pipe roughness e_f and the Reynolds number, R as

$$R = \frac{VD}{\nu} \tag{3-27}$$

where ν = fluid viscosity. To simplify the solution, R is assumed to be large so that f is independent of R, but is a function of the ratio e_f/D (see Table 3-4). Thus the diameter D must be known to find f, but f must be known to find D. Thus an iterative solution is required. An initial estimate of D is assumed and used to find f from Table 3-4. Then the following equation, which is a rearrangement of Eq. 3-26, is used to compute D:

$$D = \left[\frac{8fLQ^2}{\pi^2 g\left(\dfrac{550h_p e}{\gamma Q} - h\right)}\right]^{0.2} = \left[\frac{8fL\gamma Q^3}{\pi^2 g(550h_p e - \gamma Qh)}\right]^{0.2} \tag{3-28}$$

TABLE 3-4   Relationship between Relative Roughness (e_f/D) and the Friction Factor (f) of a Pipe

e_f/D	f	e_f/D	f	e_f/D	f
0.05	0.072	0.01	0.038	0.0004	0.016
0.04	0.065	0.006	0.032	0.0002	0.0137
0.03	0.057	0.002	0.0235	0.0001	0.012
0.02	0.0485	0.001	0.0196	0.00005	0.0106
0.015	0.044	0.0008	0.0185	0.00001	0.00815

To illustrate the design process, consider a flow Q of 12 cfs of water through a 3000-ft length (L) of cast iron pipe (with e_f = 0.00085 ft). A 100-hp pump (h_p) with an efficiency e of 65% is used to pump the water from one reservoir to another that has a surface elevation 22 feet higher (h). Using g = 32.2 ft/sec² and g = 62.4 lb/ft³, Eq. 3-28 becomes

$$D = 3.3508\, f^{0.2} \qquad (3\text{-}29)$$

To begin the iterative process, a diameter of 1 ft will be assumed; thus e_f/D = 0.00085. From Table 3-4, the friction factor is 0.019; thus, from Eq. 3-29, the diameter is 1.52 ft. This gives a relative roughness e_f/D of 0.00056, which yields f = 0.017. With Eq. 3-29, we get a diameter of 1.48 ft. Another iteration yields e_f/D = 0.00057, f = 0.017, and D = 1.5 ft. Thus the engineer should specify an 18-in. pipe in the design plans.

3.6.2 Bending Moment for a Beam

A simply supported beam AB has an overall span of 9 m. The beam is loaded with a variable load intensity over a portion of the span of 6 m in length, as shown in Fig. 3-4a. The entire beam is considered to be a free body, and the following conditions of statics are used:

$$\sum M_A = 0 \qquad (3\text{-}30a)$$

$$\sum F_Y = 0 \qquad (3\text{-}30b)$$

The first condition states that the summation of moments about point A is zero, and the second condition states that the summation of forces along the y axis is zero. These two conditions can be used to obtain the reactions at the supports at A and B. The reaction R_B can be determined using Eq. 3-30a as

$$-R_B(9) + 0.5(6)(20)\left(\frac{2}{3}\right)(6) = 0$$

Therefore, R_B = 26.6667⁺ kN. The reaction R_A can be determined using Eq. 3-30b as

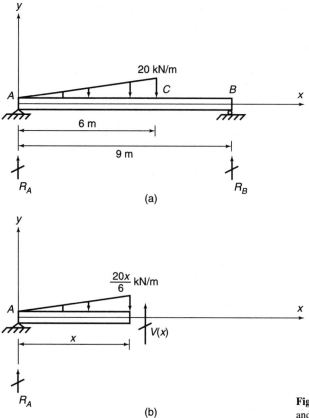

Figure 3-4 Simply Supported Beam: (a) Beam and Loading; (b) Free-body Diagram

$$R_A + R_B - (0.5)(6)(20) = 0$$

Therefore, $R_A = 33.3333^+$ kN. In structural engineering, the location of the maximum bending moment is commonly of interest. The maximum-moment location corresponds to the point of zero shear force. Numerical methods can be used to determine this point along the span of the beam. Although an analytical expression can be evaluated in this simple case, a numerical method is used to illustrate its use. For a more complex loading condition, the numerical method can still be used, whereas an analytical solution might not be possible.

The numerical solution starts by assuming a value for the location of the maximum bending moment, that is, zero shear force (V) location, say x_0. Using the condition of statics of Eq. 3-30b for the free-body diagram shown in Fig. 3-4b, the shear force $V(x_0)$ is

$$V(x_0) = R_A - \frac{1}{2}x_0\frac{x_0}{6}(20) \tag{3-31}$$

The shear force $V(x_0)$ based on this initial estimate of x_0 might not be zero. Then a new value for the location (distance), x_{i+1}, can be computed based on x_i (with a starting value of x_0) by adjusting the original distance using the following adjustment (Δx):

$$\Delta x = \frac{R_A - \frac{1}{2}x_i\frac{x_i}{6}(20)}{R_A} \tag{3-32}$$

Therefore, the new distance is

$$x_{i+1} = x_i + \Delta x = x_i + \frac{R_A - \frac{1}{2}x_i\frac{x_i}{6}(20)}{R_A} \tag{3-33}$$

The resulting new value from Eq. 3-33 (i.e., x_{i+1}), should be used again in Eq. 3-33 as an x_i value to obtain a better estimate of the distance. This process should be continued until some tolerance level (error level) is achieved. Starting with, for example, $x_0 = 4.5$ m, the resulting values for x_{i+1} are shown in Table 3-5.

TABLE 3-5   Iterations with an Initial Value of 4.5 m

Iteration i	x_i	Error in Percent
0	4.5	—
1	4.4875	0.34355
2	4.480617	0.189646
3	4.476821	0.104753
4	4.474725	0.057882
5	4.473567	0.031989
6	4.472927	0.017681
7	4.472573	0.009773
8	4.472378	0.005402
9	4.472269	0.002986
10	4.47221	0.001651
11	4.472177	0.000912
12	4.472159	0.000504
13	4.472148	0.000279
14	4.472143	0.000154
15	4.47214	8.52E-05

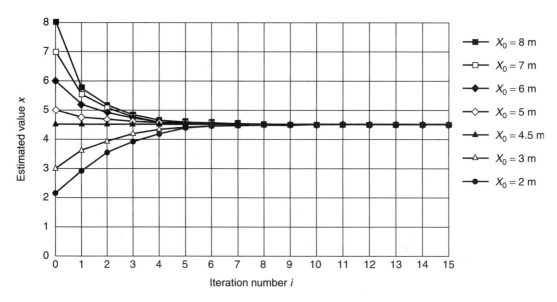

Figure 3-5   Effect of Initial Value x_0 on Estimated Value and Convergence

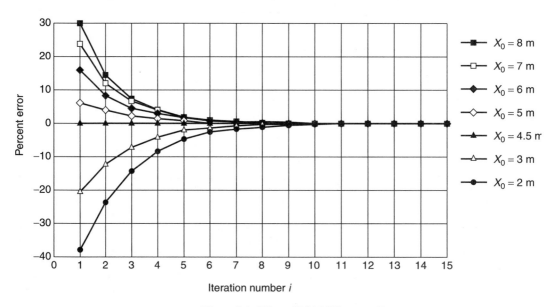

Figure 3-6   Effect of Initial Value x_0 on Error

The exact solution for the maximum-moment location can be computed analytically based on the condition that $V(x_0)$ from Eq. 3-31 equals zero at the location of maximum moment. Therefore, the exact or true distance (x_t) can be determined by solving for the distance from $V(x_0) = 0$, which is

$$x_t = \sqrt{2R_A\frac{6}{20}} = \sqrt{20} = 4.4721359 \; m \tag{3-34}$$

where $R_A = 33.3333^+ \; kN$. The percent of error is also shown in Table 3-5. The effect of the initial value x_0 on the convergence of the estimation process is shown in Fig. 3-5. The effect of the initial value on the percent of error is also shown in Fig. 3-6. In this case, it is evident from these two figures that the process converges to the true value (x_t) regardless of the initial value (x_0). However, the efficiency of the process, which is defined as the rate of convergence, increases as the initial value approaches the true value.

3.7 PROBLEMS

3-1. Formulate a numerical solution to find the cube root of a number x. Using an initial value of $x_0 = 3$, estimate the cube root of 31 with an accuracy of 0.000005.

3-2. Based on Eq. 3-2 and the equations for errors in Section 3.4, perform the following:

    a. Propose a general numerical formula for $\sqrt[n]{57}$ finding the nth root of a number. Test this formula by finding the value $\sqrt[4]{57}$ using an initial estimate of 1.0.

    b. Construct a computer subprogram that solves for nth root of the number x, with the following arguments in the definition statement: (1) the number x, (2) the value of n, (3) the initial estimate x_0, and (4) a tolerance describing the allowable change of Eq. 3-14. Include a WRITE (or PRINT) statement that prints out the value of x_i and Δe_i for each trial. Include a default option to stop iteration if the number of trials exceeds 50.

    c. Test the program of part b by finding $\sqrt[7]{817}$, with an initial estimate of 4 and a tolerance of 0.05.

3-3. A carnival employee has a booth where she estimates the weights of attendees. They get a prize if she is incorrect by more than 10 pounds. The following are 10 of the employee's guesses:

Guess	127	183	98	154	113	222	67	142	122	165
True value	134	196	101	147	121	219	75	154	127	176

    Provide a definition that the carnival owner could use to assess whether or not the employee provides accurate estimates of the weight of the attendees. Does the employee provide biased estimates? Assess the employee's prediction accuracy.

3-4. A chemical engineer has a model that predicts the temperature change in an experiment. The engineer conducts five experiments, measures the actual change, and compares the model-predicted values to the actual. The following data are provided:

Actual	47.2	53.3	64.4	58.7	43.8
Predicted	45.8	51.2	61.3	58.3	42.8

    Assess the accuracy and bias of the prediction model.

3-5. An environmental engineer creates a mixture that has a concentration of exactly 50 mg/L of a pollutant over a period of several months. Eight samples are sent to each of three laboratories for assessment, with the following results:

Lab 1	48.6	52.1	55.6	46.2	53.1	46.8	49.7	50.6
Lab 2	51.3	50.9	52.1	51.6	50.6	51.0	51.3	50.2
Lab 3	43.7	49.2	44.6	45.8	50.2	48.5	42.3	45.0

Compare the performance of the laboratories in terms of bias, precision, and accuracy.

3-6. If g is given to three significant digits and the height to two digits, find the accuracy of the terminal velocity V, where V is given by $V = \sqrt{2gh}$. If $g = 32.2$ ft/sec^2 and $h = 9.7$ ft, indicate the value of V and its expected accuracy.

3-7. The volume (V) of a block is given by $V = Lwh$. If L, w, and h are given to four significant digits, what is the accuracy of V? Find V and its accuracy if $L = 3.217$ m, $w = 0.7924$ m, and $h = 1.302$ m.

3-8. The surface area A of a right cylinder is $(2\pi r^2 + 2\pi rh)$, where r and h are the radius and height of the cylinder, respectively. If $r = 3.172$ in. and $h = 7.41$ in., find A and its accuracy.

3-9. An engineer is studying the effect of using fertilizer on crop yield. An experiment on eight fields, with the amount of fertilizer (X) varied, results in the following crop yields (Y) of the fields:

Field	1	2	3	4	5	6	7	8
X	0	0.5	1.4	2.9	4.1	6.2	7.8	9.3
Y	6.1	7.9	12.3	12.5	16.2	17.5	16.8	17.7

The engineer plots Y versus X, approximates a straight line through the points, and computes the following equation from the line: $\hat{Y} = 6 + 2X$. A colleague uses a statistical method for fitting a linear equation that results in the following model: $\hat{Y} = 8.7 + 1.16X$. Make two plots of the data, graph one equation on each plot, and discuss the representativeness of the two equations in terms of bias, precision, and accuracy. Use these assessments to discuss modeling error.

3-10. Compute the errors and relative errors for the two models and data of Problem 3-9. Also compute the average error, which is a mathematical estimate of the bias, and the standard deviation of the errors (use Eq. 2-62d), which is a measure of the accuracy. Compare the bias and accuracy of the two models.

3-11. For Example 3-4, find a valid solution to Eq. 3-17 using the numerical procedure of Eq. 3-18 and an initial estimate of $x_0 = -4$. Assume that convergence occurs when the change in the solution from one trial to the next is less than 0.05. What is the true error?

3-12. Construct a computer *FUNCTION* subprogram to solve Problem 3-1, and find solutions using the following initial estimates: (a) -6, (b) 0.1, (c) 2.7, (d) 2.8, and (e) 6.5. Assume a tolerance of 0.05.

4

Roots of Equations

4.1. INTRODUCTION

The solution of many scientific and engineering problems requires finding the roots of equations that can be nonlinear. Any nonlinear equation can, after suitable algebraic manipulation, be expressed as a function of the following form:

$$f(x) = 0 \qquad (4\text{-}1)$$

The values of x that satisfy the condition of Eq. 4-1 are termed the *roots* of the function; these roots are the solution to the equation $f(x) = 0$. Figure 4-1 shows graphically the roots for selected functions within an interval of interest $[A, B]$. Figure 4-1a shows a nonlinear function without roots since the function does not intersect the x axis. Figures 4-1b, c, and d show functions with one, two, and three roots, respectively.

For very simple functions, the roots can often be found analytically. For example, for the general second-order polynomial $ax^2 + bx + c = 0$, the roots x_1 and x_2 can be found analytically using the following equation:

$$x_1 = \frac{-b + \sqrt{b^2 - 4ac}}{2a} \qquad (4\text{-}2a)$$

$$x_2 = \frac{-b - \sqrt{b^2 - 4ac}}{2a} \qquad (4\text{-}2b)$$

The roots for a third-order polynomial can also be found analytically. However, no general solution exists for other higher-order polynomials.

It is more difficult to obtain analytical solutions for the roots of nonpolynomial, nonlinear functions. For example, the function $(x/2)^2 - \sin(x) = 0$ cannot be solved analytically. To obtain the roots of this function, some type of iterative numerical method must be employed. This, in general, requires a large number of calculations, particularly if the roots are to be determined to a high degree of precision. Thus the problem is well suited to numerical analysis.

x value

(a)

Root

x value

(b)

Roots

x value

(c)

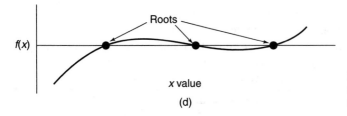

Roots

x value

(d)

Figure 4-1   Roots for Selected Functions:
(a) No Roots, (b) One Root, (c) Two Roots,
and (d) Three Roots

4.2 EIGENVALUE ANALYSIS

A large number of engineering problems require the determination of a set of values
called *eigenvalues* or *characteristic values*. These are values, usually denoted as λ, for
which the following matrix system has a nonzero (i.e., nontrivial) solution vector $\mathbf{X}$:

$$[\mathbf{A} - \lambda \mathbf{I}]\, \mathbf{X} = 0 \tag{4-3}$$

The matrices $\mathbf{A}$ and $\mathbf{I}$ are of the order $n \times n$, with $\mathbf{I}$ being an identity diagonal matrix. The
parameter λ is called the eigenvalue. The electrical engineer, for example, uses eigen-
value analysis in the solution of two-terminal-pair networks and in the optimization of
adjustments of a control system. The chemical engineer uses eigenvalue analysis in the

design of reactor systems. The engineering geologist must find eigenvalues in the analysis of sediment deposits in river beds. And the aeronautical engineer applies eigenvalue analyses in analyzing the flutter of an airplane wing. In each of these examples, the problem solution sets up the matrix **A** and seeks to find the solution of the following determinant equation:

$$|\mathbf{A} - \lambda\mathbf{I}| = 0 \tag{4-4}$$

where the vertical bars in the expression $|\mathbf{A} - \lambda\mathbf{I}|$ indicate the determinant of the matrix $[\mathbf{A} - \lambda\mathbf{I}]$. The determinant of Eq. 4-4 can be expanded into a polynomial called the *characteristic equation*, with the roots of the characteristic equation being the eigenvalues.

Consider the case where **X** is a third-order vector and **A** is a third-order matrix. Equation 4-4 then becomes

$$|\mathbf{A} - \lambda\mathbf{I}| = \begin{vmatrix} a_{11} - \lambda & a_{12} & a_{13} \\ a_{21} & a_{22} - \lambda & a_{23} \\ a_{31} & a_{32} & a_{33} - \lambda \end{vmatrix} = 0 \tag{4-5}$$

Expanding this determinant would provide a third-order polynomial (characteristic equation) having the form

$$(a_{11} - \lambda)[(a_{22} - \lambda)(a_{33} - \lambda) - a_{23}a_{32}] - a_{12}[a_{21}(a_{33} - \lambda) - a_{23}a_{31}] + a_{13}[a_{21}a_{32} - (a_{22} - \lambda)a_{31}] = 0 \tag{4-6a}$$

or

$$\lambda^3 + b_2\lambda^2 + b_1\lambda + b_0 = 0 \tag{4-6b}$$

where b_0, b_1, and b_2 are a function of the elements of **A**, which are denoted as a_{ij}. The solution (i.e., roots) of the characteristic equation would provide three eigenvalues.

Example 4-1 Soil Properties

If the solution of eigenvalue problems were the only reason, it would still be worthwhile knowing how to solve numerically nonlinear equations; however, other engineering problems also require such a solution. Consider the case of an engineer interested in identifying independent factors that affect sediment yield from agricultural lands. The eigenvalues of a matrix of correlation coefficients can be used to evaluate the degree of dependency between measured variables. Consider the simple case where the engineer measures a soil particle size index (x_1, the slope (x_2), and the mean annual sediment loss (y) from 50 agricultural fields, and the 50 sets of data yield the following correlation matrix:

	x_1	x_2	y	
x_1	1.00	−0.42	−0.61	
x_2	−0.42	1.00	0.37	(4-7)
y	−0.61	0.37	1.00	

Denoting the correlation matrix as **A**, the determinant $|\mathbf{A} - \lambda\,\mathbf{I}|$ becomes

$$\begin{vmatrix} 1 - \lambda & -0.42 & -0.61 \\ -0.42 & 1 - \lambda & 0.37 \\ -0.61 & 0.37 & 1 - \lambda \end{vmatrix} \tag{4-8}$$

Expanding the determinant yields the polynomial

$$(1 - \lambda)[(1 - \lambda)^2 - 0.37^2] + 0.42[-0.42(1 - \lambda) - 0.37(-0.61)] - 0.61[-0.42(0.37) + 0.61(1 - \lambda)]$$

$$(4\text{-}9a)$$

or

$$\lambda^3 - 3\lambda^2 + 2.3146\lambda - 0.504188 \qquad (4\text{-}9b)$$

The eigenvalues are found by setting the polynomial to zero and solving for the three roots. The roots (or λ's) provide useful information about the dependency of the soil particle size index, the slope, and the sediment yield.

The correlation matrix of Eq. 4-7 is a (3×3) matrix, which produced a third-order polynomial. Correlation matrices for engineering data are often (10×10) and larger, which produce high-order polynomials. Unlike second-order polynomials that can be solved using the quadratic formula of Eq. 4-2, such high-order polynomials cannot be solved analytically. It can be concluded from this discussion that an alternative to analytical solutions is required to find the roots of high-order polynomials that represent important engineering systems.

4.3 DIRECT-SEARCH METHOD

One obvious method for finding the roots of a general nonlinear function is a trial-and-error approach in which the function $f(x)$ is evaluated at many points over some range of x until $f(x)$ equals zero, that is, Eq. 4-1 is satisfied. Of course, we would have to be very lucky to guess the exact root of the function using this method. Even if the trial-and-error search is confined to a relatively small range for x, there are an infinite number of trial values for x to be considered. Fortunately, most problems do not require an *exact* value for the root; instead, an *estimate* of the root (to within some specified degree of *precision*) is often sufficient.

The direct-search method is perhaps the clearest, but it also can be the least efficient, technique for estimating the roots of functions. The steps in this method can be outlined as follows:

1. Specify a range or *interval* for x within which the root is assumed to occur. Some knowledge of the behavior of the function is required to specify this interval. Smaller initial intervals will require fewer calculations to obtain the root to the desired accuracy.

2. Divide the interval into smaller, uniformly spaced subintervals. The size of these subintervals will be dictated by the required precision for the estimate of the root. For example, if the root must be estimated to within ± 0.001, then the subintervals for the root will be 0.001 units long.

3. Search through all subintervals until the subinterval containing the root is located; this occurs when the equality of Eq. 4-1 exists within the interval.

The last step is the key to the method. A simple test for determining if the root occurs within a given subinterval is illustrated in Fig. 4-2. The function $f(x)$ is evaluated at the beginning and end points, A and B, respectively, of the subinterval. If a root does not lie within the subinterval, then the values of $f(A)$ and $f(B)$ will have the same sign, that is, either both negative or both positive (Figs. 4-2a and b). However, if a root does lie within the subinterval, then the values of $f(A)$ and $f(B)$ will have different signs (Figs. 4-2c and d). Alternatively stated, if the product of $f(A)$ and $f(B)$ is positive, then the function does not cross the x axis and a root does not lie within the subinterval; if the product is negative, then the function crosses the x axis and a root lies within the subinterval.

A FORTRAN program listing for the direct-search method for finding roots of functions is given in Fig. 4-3. This program will find all roots of $f(x)$ within the interval

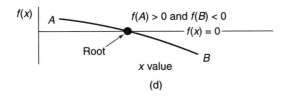

Figure 4-2   Direct-search Method: (a) no roots in the interval [A, B]; (b) no roots in the interval [**A, B**]; (c) one root in the interval [A, B]; (d) one root in the interval [A, B]

```
              PROGRAM ROOTS
c             Computes roots of a function using direct-search method
c             Read input data
              PRINT *, 'ENTER STARTING POINT FOR SEARCH INTERVAL:'
              READ *, A
              PRINT *, 'ENTER ENDING POINT FOR SEARCH INTERVAL:'
              READ *, B
              PRINT *, 'ENTER REQUIRED PRECISION:'
              READ *, PREC
c             Compute number and size of subintervals
              NINTVL = INT((B-A)/PREC) + 1
              DELTAX = (B-A)/REAL(NINTVL)
c             Initialize subinterval end points
              XB = A
              XE = XB + DELTAX
c             Loop over all subintervals
              DO 10 N = 1,NINTVL
                 FXB = F(XB)
                 FXE = F(XE)
                 IF (FXB*FXE.LE.0.0) THEN
                    PRINT 100, XB, XE
100                 FORMAT('A ROOT LIES BETWEEN X=', E12.4,'AND', E12.4)
                 END IF
                 XB = XE
                 XE = XB + DELTAX
10            CONTINUE
              END
              FUNCTION F(X)
c             Evaluates function at specified point
                 F = (X/2.0)**2 - SIN(X)
              RETURN
              END
```

Figure 4-3   Program Listing for Finding the Roots of a Function
Using the Direct-search Method

$A \leq x \leq B$ as long as the input precision value (that is, tolerance) TOL is sufficiently small that multiple roots will not occur within a single subinterval. Note that $f(x)$ is evaluated in a function subprogram; thus the program can be used to find the roots of any function simply be substituting the appropriate function subprogram for $f(x)$. In Fig. 4-3, the function $f(x) = (x/2)^2 - \sin(x)$ is used in the FUNCTION F(X) subprogram for the purpose of illustration.

The direct-search method will find the roots of any function as long as all the roots are real and within the specified interval. However, it should be obvious that this method is very cumbersome. For a high degree of precision, the subinterval size must be very small and a large number of calculations must be performed. To minimize the number of calculations, the subinterval size must be increased; but this obviously reduces the precision of the estimate roots. In addition, some roots might be missed entirely if the subinterval size is so large that more than one root occurs in a single subinterval. Nevertheless, the direct-search method is the most straightforward technique for determining all roots within a given interval.

The direct-search method assumes that there is one and only one root within each subinterval. If there is an even number of roots within the search interval, then $f(A)$ and $f(B)$ will have the same sign, and the search process will miss the roots within the interval. For example, given the function

$$f(x) = (x - 1.1)(x - 1.2)(x + 0.8) = x^3 - 1.52x - 0.52x + 1.056 \qquad (4\text{-}10)$$

the two positive roots will not be found if the search interval is 0.25, with $A = 1.0$ and $B = 1.25$. In this case, $f(A) = 0.036$ and $f(B) = 0.0154$, and so the search will proceed to the next search interval and miss the roots $X = 1.1$ and $X = 1.2$. This is especially critical where the function has two equal roots (called *multiple roots*), because the direct-search method will not identify any roots regardless of the size of the search subinterval. The multiple-root case is illustrated graphically in Fig. 4-4 as the tangent point to the x axis. This root cannot be determined by the direct-search method. An example function that has multiple roots (double roots) is

$$f(x) = (x - 1)^2 = x^2 - 2x + 1 \qquad (4\text{-}11)$$

Figure 4-4   Multiple Roots

Another case that might result in the failure of the direct-search method in finding the roots is shown in Fig. 4-5. In this case, the function has a discontinuity at a point within the interval [A, B] of interest.

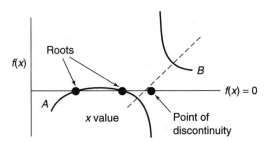

Figure 4-5   Point of Discontinuity

Example 4-2 Eigenvalue Problem

Consider the polynomial for Eq. 4-9a, which was the expansion of the determinant given in Section 4.1 as

$$\lambda^3 - 3\lambda^2 + 2.3146\lambda - 0.504188 = 0 \tag{4-9b}$$

The sum of the eigenvalues for a matrix with values of 1.0 on the principal diagonal equals the rank of the matrix, which equals the order of the characteristic polynomial; thus the three roots for this polynomial must be between 0.0 and 3.0, since all roots must be located in the range $0 \leq \lambda \leq 3$. We will assume for the purposes of illustration that an accuracy of 0.1 is sufficient. Thus the direct-search procedure can begin with a value of $\lambda = 0$ and continue in increments of 0.1 until estimates of all three roots are found. Table 4-1 gives the values of λ and $f(\lambda)$ using the direct-search method.

Table 4-1   Results of Direct-search Method

λ	$f(\lambda)$	λ	$f(\lambda)$	λ	$f(\lambda)$
0.0	−0.504	0.1	−0.302	0.2	−0.153
0.3	−0.053	0.4	0.006	0.5	0.199
0.6	0.021	0.7	−0.011	0.8	−0.061
0.9	−0.122	1.0	−0.190	1.1	−0.257
1.2	−0.319	1.3	−0.368	1.4	−0.400
1.5	−0.407	1.6	−0.385	1.7	−0.326
1.8	−0.226	1.9	−0.077	2.0	0.125

The smallest root occurs in the interval $0.3 < \lambda < 0.4$ since $f(0.3)$ is negative and $f(0.4)$ is positive. The second root occurs in the interval $0.6 < \lambda < 0.7$ since $f(0.6)$ is positive and $f(0.7)$ is negative. The largest root has a value between 1.9 and 2.0 since $f(1.9) < 0$ and $f(2.0) > 0$. We could assume roots of 0.4, 0.7, and 1.9, since these represent the values of λ where $f(\lambda)$ is closer to zero. Linear interpolation could be used to get better estimates, but we assumed that an accuracy of 0.1 was sufficient.

4.4 BISECTION METHOD

The bisection method is an extension of the direct-search method for cases when it is known that only one root occurs within a given interval of x. For the same level of precision, the bisection method will, in general, require fewer calculations than the direct-search method. Using Fig. 4-6 as a reference, the steps in the bisection method can be outlined as follows:

1. For the interval of x from the starting point x_s to the end point x_e, locate the midpoint x_m at the center of the interval. These points correspond to the starting and ending points of the half-intervals.

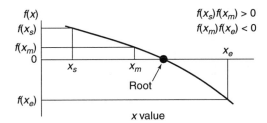

Figure 4-6   Schematic of Bisection Method

2. At the starting point x_s, midpoint x_m, and the ending point x_e of the interval, evaluate the function resulting into $f(x_s)$, $f(x_m)$, and $f(x_e)$, respectively.

3. Compute the products of the functions evaluated at the ends of the two half-intervals, that is, $f(x_s)f(x_m)$ and $f(x_m)f(x_e)$. The root lies in the interval for which the product is negative, and the midpoint x_m is used as the estimate of the root for this iteration.

4. Check for convergence as follows:

    a. If the convergence criterion (that is, tolerance) is satisfied, then use x_m as the final estimate of the root.

    b. If the tolerance has not been met, specify the ends of the half-interval in which the root is located as the starting and ending points for a new interval and return to step 1.

The bisection method will always converge on the root, provided that only one root lies within the starting interval for x.

4.4.1 Error Analysis and Convergence Criterion

Since numerical methods for finding the roots of functions are iterative, it is important to include a convergence criterion into the process of finding a root; otherwise, a computerized solution could continue to iterate indefinitely. Since the value of a root has significance to the engineering problem under consideration, the allowable error would depend on the tolerance permitted by the problem. Thus the absolute value between an estimate and the true value would be the ideal criterion to indicate convergence. But since the true value is not known, we need a surrogate value to represent the true value. In practice, the change in the estimate of the root from one trial to the next is used as the convergence criterion; this can be expressed as either the absolute value or as a percentage of the estimate. If the change between estimates on successive iterations is used as a convergence criterion, it is important to recognize that this criterion may not reflect the true accuracy of the estimate, that is, the absolute difference between the estimated value and the true value of the root.

To ensure closure of the iteration loop, a convergence criterion is needed to terminate the iterative procedure of a numerical method for finding the roots of a function. The convergence criterion used in step 4 of the bisection method can be expressed in terms of either the absolute value of the difference (ϵ_d) or the percent relative error (ϵ_r) for each iteration. The convergence criterion when the error is expressed as an absolute value is

$$\epsilon_d = |x_{m,i+1} - x_{m,i}| \qquad (4\text{-}12a)$$

For the case when the convergence criterion is expressed as a relative percent error, the criterion is

$$\epsilon_r = \left| \frac{x_{m,i+1} - x_{m,i}}{x_{m,i+1}} \right| 100 \qquad (4\text{-}12b)$$

where $x_{m,i}$ = the midpoint in the previous root-search iteration (*i*th iteration), and $x_{m,i+1}$ = the midpoint in a new root-search iteration ($i + 1$ iteration). The tolerance used with the convergence criterion of Eq. 4-12a can be specified depending on the accuracy that is needed for the specific problem. For some engineering problems, a tolerance of 5% according to Eq. 4-12b is adequate; other problems may require a tolerance of 0.01%.

The true accuracy of the solution at any iteration can be computed if the true solution (root x_t) is known. In such a case, the true error (ϵ_t) in the *i*th iteration is

$$\epsilon_t = \left| \frac{x_t - x_{m,i}}{x_t} \right| 100 \qquad (4\text{-}12c)$$

Equation 4-12c requires the knowledge of the true solution x_t. In practical use of numerical methods, the true solution is not known to us. Therefore, an approximate solution of the error can be evaluated using the relative error x_r as given in Eq. 4-12b.

Example 4-3 Roots of a Polynomial Using the Bisection Method

The following polynomial is used to illustrate the use of the bisection method for finding its roots:

$$f(x) = x^3 - x^2 - 10x - 8 = 0 \qquad (4\text{-}13)$$

We are interested in finding the roots within the interval $3.75 \le x \le 5.00$ to a relative accuracy as an absolute value between successive iterations of 0.01. Using the four steps of the bisection methods, the values of x_s, x_m and x_e were computed for each iteration *i*. The results are summarized in Table 4-2. Note that the midpoint is used as the estimate of the root for each iteration. The absolute value of the error is given in the last column of Table 4-2. The error decreases with each iteration, and on the seventh iteration the error equals the tolerance as the convergence is assumed. The final estimate of the root is 3.994.

It is evident that the root is near 4. Substituting $x = 4$ into Eq. 4-13 shows that the true value of the root is 4. Therefore, the true accuracy of the estimated root is 0.006 or, in relative terms according to Eq. 4-12c, 0.15%.

Example 4-4 Location of Maximum Bending Moment

In Section 3.6.2, the location of the maximum bending moment for a simply supported beam was determined based on the condition that the shear force at this location is zero. The exact location was also expressed based on the conditions of statics as the solution of the following equation:

$$f(x) = 20 - x^2 = 0 \qquad (4\text{-}14)$$

Table 4-2   Example Polynomial Solved Using the Bisection Method

Iteration i	x_s	x_m	x_e	$f(x_s)$	$f(x_m)$	$f(x_e)$	Subinterval Containing Root $f(x_s)f(x_m)$	$f(x_m)f(x_e)$	Error ϵ_d
1	3.750	4.375	5.000	−6.830	12.850	42.000	−	+	—
2	3.750	4.062	4.375	−6.830	1.903	12.850	−	+	0.313
3	3.750	3.906	4.062	−6.830	−2.724	1.903	+	−	0.156
4	3.906	3.984	4.062	−2.724	−0.477	1.903	+	−	0.078
5	3.984	4.023	4.062	−0.477	0.696	1.903	−	+	0.039
6	3.984	4.004	4.023	−0.477	0.120	0.696	−	+	0.019
7	3.984	3.994	4.004	−0.477	−0.180	0.120	+	−	0.010

The location of the maximum bending moment must lie within the span of the uniform load (i.e., $0 \le x \le 6\ m$). Therefore, the starting interval for the bisection method is [0, 6]. The computations of the bisection method are shown in Table 4-3. The estimated root (x_m) and

Table 4-3   Location of Maximum Bending Moment Using Bisection Method

Iteration i	x_s	x_m	x_e	$f(x_s)$	$f(x_m)$	$f(x_e)$	Absolute Error (%)	Relative Error (%)
1	0	3	6	20	11	−16	32.917960	—
2	3	4.5	6	11	−0.25	−16	0.623060	33.33333
3	3	3.75	4.5	11	5.9375	−0.25	16.147450	20.00000
4	3.75	4.125	4.5	5.9375	2.984375	−0.25	7.762196	9.09091
5	4.125	4.3125	4.5	2.984375	1.402344	−0.25	3.569568	4.34783
6	4.3125	4.40625	4.5	1.402344	0.584961	−0.25	1.473255	2.12766
7	4.40625	4.453125	4.5	0.584961	0.169678	−0.25	0.425098	1.05263
8	4.453125	4.476563	4.5	0.169678	−0.03961	−0.25	0.098980	0.52356
9	4.453125	4.464844	4.476563	0.169678	0.06517	−0.03961	0.163059	0.26247
10	4.464844	4.470703	4.476563	0.06517	0.012814	−0.03961	0.032039	0.13106
11	4.470703	4.473633	4.476563	0.012814	−0.01339	−0.03961	0.033470	0.06549
12	4.470703	4.472168	4.473633	0.012814	−0.00029	−0.01339	0.000720	0.03275
13	4.470703	4.471436	4.472168	0.012814	0.006264	−0.00029	0.015662	0.01638

errors in percent are shown in the table, as well as in Figs. 4-7 and 4-8. The two types of error are shown in Fig. 4-8, the absolute error and relative error. The relative error in an iteration was computed using Eq. 4-12b. The actual error for an iteration was computed according to Eq. 4-12c using the true root (x_t = 4.4721359).

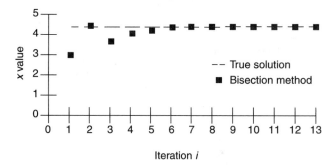

Figure 4-7   Location of Maximum Bending Moment

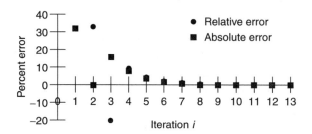

Figure 4-8   Errors in the Location of Maximum Bending Moment

4.5 NEWTON–RAPHSON ITERATION

Although the bisection method will always converge on the root, the rate of convergence is very slow. A faster method for converging on a single root of a function is the Newton–Raphson iteration method.

While the concept of a *Taylor series expansion* is formally introduced in Chapter 1, at this point it is sufficient to state that the equation that is used to develop the Newton–Raphson iteration method is the linear portion of a Taylor series:

$$f(x_1) = f(x_0) + \frac{df}{dx}\Delta x \tag{4-15}$$

where $f(x_1)$ is the value of the function at a specific value (x_1) of the independent variable x, $f(x_0)$ is the value at x_0, Δx equals the difference ($x_1 - x_0$), and df/dx is the derivative of the relationship. Since the root of the function relating $f(x)$ and x is the value of x where $f(x_1)$ equals zero, Eq. 4-15 can be expressed as

$$0 = f(x_0) + \frac{df}{dx}(x_1 - x_0) \tag{4-16}$$

where x_1 will now be the root of the function, since $f(x_1)$ was assumed to be zero. Solving Eq. 4-16 for the unknown x_1 yields

$$x_1 = x_0 - \frac{f(x_0)}{df/dx} \tag{4-17}$$

If the derivative df/dx is evaluated at x_0 and we denote this derivative as $f'(x_0)$, then Eq. 4-17 becomes

$$x_1 = x_0 - \frac{f(x_0)}{f'(x_0)} \tag{4-18a}$$

Thus Eq. 4-18a provides a revised approximation of the root (x_1) using some initial estimate of the root (x_0) and the values of both the function $f(x_0)$ and the derivative of the function $f'(x_0)$, both evaluated at x_0. Equation 4-18a can be iteratively evaluated as

$$x_{i+1} = x_i - \frac{f(x_i)}{f'(x_i)} \tag{4-18b}$$

If the function $f(x)$ is linear, then Eq. 4-18b will provide an exact solution on the first trial; however, if $f(x)$ is nonlinear, then the linear Taylor series of Eq. 4-15 is only valid over small ranges of Δx. The importance of the nonlinear terms that were truncated for Eq. 4-15 will determine the accuracy of the solution of Eqs. 4-18. In practice, Eqs. 4-18 are applied iteratively, with the first trial based on an estimate of x_0 and each subsequent trial based on the most recent estimate of the root. The solution procedure is illustrated in Fig. 4-9, where the function and independent variable are denoted as $f(x)$ and x, respectively. The subscript on x refers to the trial (or iteration) number, with x_0 being the initial estimate. It can be seen from Fig. 4-9 that Eqs. 4-18 can be obtained by expressing the slope $f'(x_i)$ over the interval $[x_i, x_{i+1}]$ as

$$f'(x_i) = \frac{f(x_i) - 0}{x_{i+1} - x_i} = \frac{f(x_i)}{x_{i+1} - x_i} \tag{4-19}$$

Therefore, Eqs. 4-18 can be obtained by rearranging the terms of Eq. 4-19.

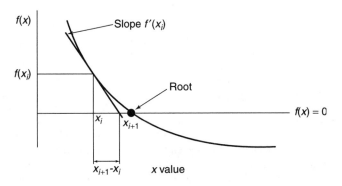

Figure 4-9   Newton–Raphson Method

Example 4-5 Third-order Polynomial

As an example of Newton–Raphson iteration, consider the following third-order polynomial:

$$f(x) = x^3 - x^2 - 10x - 8 \qquad (4\text{-}20a)$$

The derivative of this function is

$$f'(x) = 3x^2 - 2x - 10 \qquad (4\text{-}20b)$$

If an initial estimate of the root of 6.0 is assumed, then the first trial estimate is

$$x_1 = 6 - \frac{(6)^3 - (6)^2 - 10(6) - 8}{3(6)^2 - 2(6) - 10} = 6 - 1.3023 = 4.6977 \qquad (4\text{-}21a)$$

Using this value as the revised estimate of x_1, the second estimate would be obtained by

$$x_2 = 4.67977 - \frac{(4.6977)^3 - (4.6977)^2 - 10(4.6977) - 8}{3(4.6977)^2 - 2(4.6977) - 10} \qquad (4\text{-}21b)$$

$$= 4.6977 - 0.5688 = 4.1289$$

A third trial would yield an estimate of the root of 4.0057, and the fourth trial yields a value of 4.0000. For precision to five significant digits, the value does not change with additional iterations.

Example 4-6 Error Analysis for the Newton–Raphson Method

For the example of Eq. 4-20a, the difference in absolute value between both the estimated value of the x_i and the true root ($x_t = 4.000$), and the estimated value for a trial (x_i) and the previous trial (x_{i-1}) are shown in Table 4-4.

Table 4-4 Error Analysis for Example 4-5
Using Newton–Raphson Method

Iteration i	x_i	$x_i - x_{i-1}$	$x_i - x_t$
0	6.0000	—	2.0
1	4.6977	1.3023	0.6977
2	4.1289	0.5688	0.1289
3	4.0057	0.1232	0.0057
4	4.0000	0.0057	0.0000

For this case, the convergence criterion $x_i - x_{i-1}$ yields a conservative estimate of the true error. Thus it would be a useful convergence criterion for this case. For other problems, such as when the function $f(x)$ does not show large changes even for large changes in the estimate of the root, another criterion may be more appropriate to test for convergence. Regardless of the convergence criterion, the solution should be checked by substituting the calculated root into the function to ensure that the result is rational.

4.5.1 Nonconvergence

Newton–Raphson iteration usually converges to a root faster than does the bisection method. However, this increase in speed does not come without liabilities; under certain circumstances, Newton–Raphson iteration may fail to converge. Thus, when programming the method, the program should be written with an iteration counter and the option for discontinuing the search procedure when some preselected number of iterations has been exceeded. This procedure ensures that excessive iteration does not happen.

While nonconvergence is not common, it can occur if the initial estimate is selected such that the derivative of the function equals zero. In such cases, $f'(x_i)$ of Eq. 4-18b would be zero and $f(x_i)/f'(x_i)$ would equal infinity. Thus convergence would not be achieved and some error message indicating division by zero would be given. For example, a division by zero will occur if an initial estimate of $(2 \pm 124)/6$ were selected for the function for Eq. 4-20b. In this case, an iteration counter would not be necessary because program execution would be terminated by a division-by-zero error after the first trial. An example case of $f'(x_i) = 0$ is shown in Fig. 4-10a.

A second instance of nonconvergence would occur if $f(x_i)/f'(x_i)$ equals $-f(x_{i+1})/f'(x_{i+1})$. In this case, x_i would equal x_{i+2} (where i is the iteration number) and x_{i+1} would equal x_{i+3}. Thus the solution would involve iterating between values of x_i and x_{i+1}. Graphically this situation is shown in Figs. 4-10b and c. The problem of excessive iterations can be prevented by including an iteration counter with a specified maximum number of iterations in the program as a criterion.

A large number of iterations will be required if the value of $f'(x_i)$ is much larger than $f(x_i)$. In such cases, $f(x_i)/f'(x_i)$ is small, which leads to a small adjustment at each iteration. This situation can occur, for example, when the root of a polynomial is near zero. In such cases, the numerator in the second term of Eqs. 4-18 will be much smaller than the denominator, and thus the adjustment $f(x_i)/f'(x_i)$ will be small. For such cases, it is important to obtain a good initial estimate of the root or allow for a large number of iterations before discontinuing the search. An example case of this excessive iteration is shown in Fig. 4-10d.

Example 4-7 Eigenvalue Problem

The eigenvalue problem presented with Eq. 4-9b can be solved using the Newton–Raphson method. The eigenvalues are the roots of the polynomial. It is known from matrix theory that the sum of the eigenvalues equals the sum of the elements of the principal diagonal. For the correlation matrix, the sum would be 3; thus the sum of the eigenvalues will equal 3.0. Since all of the eigenvalues must be positive, it is reasonable to use a value of 2 as an initial estimate of λ. The derivative polynomial is

$$f'(\lambda) = 3\lambda^2 - 6\lambda + 2.3146 \tag{4-22}$$

Using Eqs. 4-18 with $\lambda_0 = 2$, the revised estimate of λ is given by

$$\lambda_1 = \lambda_0 - \frac{f(\lambda_0)}{f'(\lambda_0)} = 2 - \frac{(2)^3 - 3(2)^2 + 2.3146(2) - 0.504188}{3(2)^2 - 6(2) + 2.3146} \tag{4-23}$$

$$= 2 - \frac{0.125}{2.3146} = 1.94599$$

x value
(a)

x value
(b)

x value
(c)

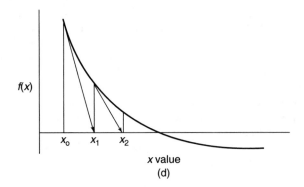

x value
(d)

Figure 4-10   Nonconvergence of the
Newton–Raphson Method: (a) $f'(x_i)$ is
approaching zero, (b) excessive iteration,
(c) excessive iteration, and (d) excessive
iteration

The result of the first iteration can be expressed to four significant digits as 1.946. A second
iteration yields a value of 1.942, and continued iteration does not result in any change. If $\lambda =$
1.942 is substituted into the third-order polynomial, it is evident that $f(\lambda) = 0$. This is shown
in Fig. 4-11.

Figure 4-11   Eigenvalue Solution

Example 4-8 Location of Maximum Bending Moment

Using the function of Example 4-4, $f(x) = 20 - x^2$, the Newton–Raphson method is used to solve for the unknown roots. This example illustrates the effect of the initial value on the solution for this method. Using $x_0 = 3$, the method converges as shown in Fig. 4-12 to the value 4.472136. Using $x_0 = 2$, the method also converges to the same answer. The figure shows other cases using other initial values. In Fig. 4-12, the results are connected by line segments to facilitate tracking the progress of the method for different initial values.

Figure 4-12   Location of Maximum Bending Moment

4.6 SECANT METHOD

The secant method is similar to the Newton–Raphson method, with the difference that the derivative $f'(x)$ is numerically evaluated, rather than computed analytically. The algorithm for the secant method can be developed using the geometric similarity of two triangles as shown in Fig. 4-13. A new estimate of x_{i+1} of the root can be obtained using values of the function $f(x_i)$ and $f(x_{i-1})$ at two other estimates x_i and x_{i-1} of the root, as shown in Fig. 4-13. A linear line segment between $[x_{i-1}, f(x_{i-1})]$ and $[x_i, f(x_i)]$ inter-sects the x axis at x_{i+1}. The following equality results from the triangle for the base $(x_{i+1} - x_{i-1})$ and the triangle for the base $(x_{i+1} - x_i)$:

$$\frac{f(x_{i-1})}{x_{i+1} - x_{i-1}} = \frac{f(x_i)}{x_{i+1} - x_i} \tag{4-24}$$

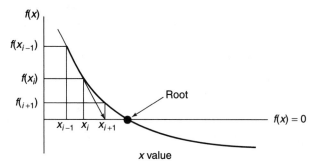

$f(x)$

$f(x_{i-1})$

$f(x_i)$

$f_{(i+1)}$

Root

$f(x) = 0$

x_{i-1} x_i x_{i+1}

x value

Figure 4-13   Secant Method

Solving for x_{i+1} yields

$$x_{i+1} = \frac{x_{i-1}f(x_i) - x_i f(x_{i-1})}{f(x_i) - f(x_{i-1})} \tag{4-25a}$$

or

$$x_{i+1} = x_i - \frac{f(x_i)[x_{i-1} - x_i]}{f(x_{i-1}) - f(x_i)} \tag{4-25b}$$

The quantity in the denominator of the second term on the right side of Eq. 4-25b divided by $[x_{i-1} - x_i]$ is a numerical approximation of the derivative $f'(x)$. Thus Eq. 4-25b is an approximation of Eq. 4-15. The secant method is especially useful when it is difficult to compute an analytical expression for the derivative. It should be noted that the secant method requires two initial estimates of a root.

Example 4-9 Secant Method

The secant method can be used to determine the root of the following equation:

$$f(x) = \exp(-x) - x^2 \tag{4-26}$$

Using initial estimates of $x_0 = 0$ and $x_1 = 1$, the next estimate of the root (x_2) can be deter-mined using Eq. 4-25b as

$$x_2 = 1 - \frac{f(1)[0-1]}{f(0) - f(1)} = 1 - \frac{[exp(-1) - 1^2][0-1]}{[exp(0) - 0^2] - [exp(-1) - 1^2]}$$

$$= 0.6127$$

Table 4-5 shows the results for several additional iterations. The method converges to a root of 0.703467.

Table 4-5   Secant Method for Example 4-9

Iteration i	$f(x_{i-1})$	x_i	Comments
0	1	0	Initial condition
1	−0.632120	1	Initial condition
2	0.166485	0.612700	
3	0.018994	0.693440	
4	−0.000710	0.703838	
5	2.81E-06	0.703466	
6	4.11E-10	0.703467	
7	−2.20E-16	0.703467	
8	0	0.703467	
9	0	0.703467	

4.7 POLYNOMIAL REDUCTION

After one root of a polynomial has been found, the process can be repeated using a new initial estimate. However, if proper consideration is not given to selecting the initial estimate of the second root, then application of, for example, Newton–Raphson iteration may result in the same root being found. Thus there is a need to eliminate the possibility of finding the same root on the second application.

Polynomial reduction can be used to eliminate a root from a polynomial, which then eliminates the possibility of converging to a root that was found on previous applications. Polynomial reduction is essentially a process of dividing the polynomial by the root, which reduces the original polynomial to a new polynomial that does not have the original root as a root (unless, of course, the polynomial has a pair of equal roots, that is, multiple roots).

Formally, polynomial reduction states that, if the polynomial $f(x)$ equals zero and root x_1 is a root of $f(x)$, then there is a reduced polynomial $f^*(x)$ such that $(x - x_1)f^*(x) = 0$, where $f^*(x) = f(x)/(x - x_1)$. Thus, if $f(x)$ is a polynomial of order n, then the reduced polynomial is a polynomial of order $n - 1$.

Example 4-10 Polynomial Reduction

For the example used to illustrate the Newton–Raphson method, one root of the polynomial of Eq. 4-20a was found to be $x = 4$. This polynomial can be reduced as

$$
\begin{array}{r}
x^2 + 3x\ +\ 2 \\
x - 4 \overline{\smash{\big)}\ x^3 -\ \ x^2 - 10x - 8} \\
\underline{x^3 - 4x^2} \\
3x^2 - 10x \\
\underline{3x^2 - 12x} \\
2x - 8 \\
\underline{2x - 8} \\
0 = \text{error}
\end{array}
\tag{4-27}
$$

Thus the reduced polynomial $x^2 + 3x + 2$ can be used to find additional roots of Eq. 4-20a. The reduced polynomial will equal zero for roots of the parent polynomial of Eq. 4-20a. The Newton–Raphson method can then be used to find a root of the reduced equation, and the polynomial can be reduced using polynomial reduction until all of the roots are found. It is important to note that any error in the value of x_1 will lead to error in the coefficients of the reduced equation and, thus, all roots found subsequently.

Example 4-11 Eigenvalue Problem

Polynomial reduction can be used to reduce the eigenvalue polynomial of Eq. 4-9b:

$$
\begin{array}{r}
\lambda^2 - 1.058\lambda\ + 0.2600 \\
\lambda - 1.942 \overline{\smash{\big)}\ \lambda^3 -\ \ \ \ \ \ 3\lambda^2 + 2.3146\lambda - 0.504188} \\
\underline{\lambda^3 - 1.942\lambda^2} \\
-1.058\lambda^2 + 2.3146\lambda \\
\underline{-1.058\lambda^2 + 2.0546\lambda} \\
0.2600\lambda - 0.504188 \\
\underline{0.2600\lambda - 0.50492} \\
0.000732 = \text{error}
\end{array}
\tag{4-28}
$$

Since the error is very small, $\lambda^2 - 1.058\lambda + 0.2600$ is the reduced polynomial. The roots of the reduced polynomial $\lambda^2 - 1.058\lambda + 0.2600$ are also the roots of the original third-order polynomial.

4.8 SYNTHETIC DIVISION

Polynomial reduction assumes that an estimate of the root is reasonably exact and that the aim is to reduce the polynomial. However, the concept underlying polynomial reduction can be used to find the value of a root. This method is called *synthetic division*.

Given the following nth-order polynomial $f_n(x)$, the polynomial can be reduced by dividing it by an initial estimate of the root x_0:

$$
f_n(x) = b_n x^n + b_{n-1} x^{n-1} + \cdots + b_1 x^1 + b_0
\tag{4-29}
$$

This produces a polynomial of order $n - 1$, which will be denoted as $h_{n-1}(x)$:

$$
\frac{f_n(x)}{x - x_0} = h_{n-1}(x) + \frac{R_0}{x - x_0}
\tag{4-30}
$$

where R_0 is the remainder. Equation 4-30 would be rewritten as

$$f_n(x) = (x - x_0)h_{n-1}(x) + R_0 \qquad (4\text{-}31)$$

The reduced polynomial $h_{n-1}(x)$ is given by:

$$h_{n-1}(x) = c_{n-1}x^{n-1} + c_{n-2}x^{n-2} + \cdots + c_1x + c_0 \qquad (4\text{-}32)$$

If $h_{n-1}(x)$ is also reduced using the estimate of the root, we get the following $(n-2)$th-order polynomial, which will be denoted as $g_{n-2}(x)$:

$$\frac{h_{n-1}(x)}{x - x_0} = g_{n-2}(x) + \frac{R_1}{x - x_0} \qquad (4\text{-}33)$$

where R_1 is the remainder. The reduced polynomial $g_{n-2}(x)$ can be written as

$$g_{n-2}(x) = d_{n-2}x^{n-2} + d_{n-3}x^{n-3} + \cdots + d_1x + d_0 \qquad (4\text{-}34)$$

Combining Eqs. 4-31 and 4-33 yields

$$f_n(x) = (x - x_0)^2 g_{n-2}(x) + (x - x_0)R_1 + R_0 \qquad (4\text{-}35)$$

The derivative of Eq. 4-35 with respect to x [i.e., $f'_n(x)$] is

$$f'_n(x) = 2(x - x_0)g_{n-2}(x) + (x - x_0)^2 g'_{n-2}(x) + R_1 \qquad (4\text{-}36)$$

where $g'_{n-2}(x) = $ the derivative of $g_{n-2}(x)$ with respect to x. At $x = x_0$, Eqs. 4-35 and 4-36 reduce to $f_n(x) = R_0$ and $f'_n(x) = R_1$. Thus, based on the Newton–Raphson method, we get

$$x_1 = x_0 - \frac{f_n(x)}{f'_n} = x_0 - \frac{R_0(x_i)}{R_1(x_i)} \qquad (4\text{-}37)$$

where $R_0(x_i)$ is the remainder term based on an initial polynomial reduction with a root of x_i, and $R_1(x_i)$ is the remainder term based on a second polynomial reduction using again x_i. Equation 4-37 indicates that, for an initial estimate of the root x_0, an improved estimate can be obtained with Eq. 4-37 after computing R_0 and R_1 by polynomial reduction. Therefore, the iterative root estimation equation for this method is

$$x_{i+1} = x_i - \frac{R_0(x_i)}{R_1(x_i)} \qquad (4\text{-}38)$$

4.8.1 Programming Considerations

The solution procedure for synthetic division is easily programmed using the general form of Eqs. 4-29 to 4-38. By dividing $f_n(x)$, Eq. 4-29, by $x - x_0$, we get

$$x - x_0 \overline{\left)\, b_n x^n \begin{array}{c} c_{n-1}x^{n-1} + c_{n-2}x^{n-2} + c_{n-3}x^{n-3} + \cdots + c_0 \\ + b_{n-1}x^{n-1} + b_{n-2}x^{n-2} + \cdots + b_0 \end{array}\right.} \qquad (4\text{-}39)$$

It can be shown that the coefficients of the reduced polynomial c_i are related to the b_i values by

$$c_n = b_n \qquad\qquad (4\text{-}40\text{a})$$

$$c_j = b_j + x_i c_{i+1} \qquad \text{for } j = (n-1), (n-2), \ldots, 1 \qquad (4\text{-}40\text{b})$$

where x_i is an estimate of the root in the ith iteration. Similarly, the coefficients of Eq. 4-34 are related to those of Eq. 4-32 by

$$d_n = c_n \qquad\qquad (4\text{-}41\text{a})$$

$$d_j = c_j + x_i d_{i+1} \qquad \text{for } j = (n-1), (n-2), \ldots, 1 \qquad (4\text{-}41\text{b})$$

The remainders R_0 and R_1 in the ith iteration are given by

$$R_0 = b_0 + x_i c_1 \qquad\qquad (4\text{-}42)$$

$$R_1 = c_1 + x_i d_2 \qquad\qquad (4\text{-}43)$$

Thus the procedure is as follows:

1. Input n, x_0, b_j for $j = 0, 1, \ldots, n$.
2. Compute the n values of c_j using Eq. 4-40.
3. Compute R_0.
4. Compute the $(n-1)$ values of d_j.
5. Compute R_1.
6. Use Eq. 4-38 to compute a revised estimate of x_{i+1} of the root x_i.
7. Check for convergence as follows:
   a. if $|x_{i+1} - x_i| \le$ tolerance, discontinue the iteration and use x_{i+1} as the best estimate of the root; or
   b. if $|x_{i+1} - x_i| >$ tolerance, set $x_i = x_{i+1}$ and return to step 2 and continue the iteration process.

Example 4-12 Synthetic Division

Using the example of Eq. 4-20a and an initial estimate x_0 of 6, the reductions indicated by Eqs. 4-30 and 4-33 are as follows:

$$
\begin{array}{r}
x^2 + 5x + 20 \\
x - 6 \,\overline{\smash{\big)}\, x^3 - x^2 - 10x - 8} \\
\underline{x^3 - 6x^2} \\
5x^2 - 10x \\
\underline{5x^2 - 30x} \\
20x - 8 \\
\underline{20x - 120} \\
112 = R_0
\end{array}
\qquad (4\text{-}44\text{a})
$$

$$\begin{array}{r}
x + 11 \\
x - 6 \overline{\smash{\big)}\, x^2 + 5x + 20} \\
\underline{x^2 - 16x} \\
11x + 20 \\
\underline{11x - 66} \\
86 = R_1
\end{array}$$

(4-44b)

Thus the revised estimate is

$$x_1 = 6 - \frac{112}{86} = 4.6977$$

(4-45)

This is the same value obtained on the first iteration of the Newton–Raphson method. While both methods give the same answer, Newton–Raphson iteration requires evaluation of the first derivative of the polynomial, whereas the synthetic division method requires two polynomial reductions. A second iteration could be performed using 4.6977 as the estimate of the root. The results for the first root are shown in Table 4-6. After six iterations, the root to seven significant digits is 4.

Table 4-6   Synthetic Division for Example 4-12

Parameter	Iteration 0	Iteration 1	Iteration 2	Iteration 3	Iteration 4	Iteration 5
x_i	6	4.69767	4.12889	4.00569	4.00001	4
R_0	112	26.62407	4.05173	0.17098	0.00036	1.54E–09
R_1	86	46.80909	32.88552	30.12522	30.00026	30
b_3	1	1	1	1	1	1
b_2	−1	−1	−1	−1	−1	−1
b_1	−10	−10	−10	−10	−10	−10
b_0	−8	−8	−8	−8	−8	−8
c_3	1	1	1	1	1	1
c_2	5	3.69767	3.12889	3.00569	3.00001	3
c_1	20	7.37047	2.91888	2.03984	2.00008	2
d_3	1	1	1	1	1	1
d_2	11	8.39535	7.25779	7.01138	7.00002	7
d_1	86	46.80909	32.88552	30.1252	30.0003	30

Example 4-13 Eigenvalues

The polynomial developed by expanding the determinant $|\mathbf{A} - \lambda\mathbf{I}|$ can be solved using synthetic division. The following is an example of a polynomial expansion of a determinant:

$$\lambda^3 - 3\lambda^2 + 2.3146\lambda - 0.504188 \qquad (4\text{-}46)$$

The values for b_j are $b_3 = 1$, $b_2 = -3$, $b_1 = 2.3146$, and $b_0 = -0.504188$. Using an initial estimate of 2 for the root, the values of c_j and d_j are computed as shown in Table 4-7. After two iterations the root converged to $\lambda = 1.942$, which is the value that was found using the Newton–Raphson method. The set of c_j values for the last iteration are $c_3 = 1$, $c_2 = -1.058$, and $c_1 = 0.2597$, which can be used to form the second-order polynomial:

$$\lambda^2 - 1.058\lambda + 0.2597 = 0 \qquad (4\text{-}47)$$

which is the same equation produced by polynomial reduction. This second-order polynomial can then be evaluated for a second root. Using these coefficients as the new b_j values and an initial estimate of 1.0, synthetic division yields the value of a second root in four iterations, $\lambda = 0.670$ (see Table 4-8). The c_j values are $c_2 = 1$ and $c_1 = -0.3860$, which give the polynomial

$$\lambda - 0.3860 = 0 \qquad (4\text{-}48)$$

This yields the last root of the original third-order polynomial, $\lambda = 0.3860$. The sum of the three roots equals 3, as expected.

Table 4-7   Example 4-13 Synthetic Division for a Third-order Polynomial

Parameter	Iteration 0	Iteration 1	Iteration 2	Iteration 3	Iteration 4	Iteration 5	Iteration 6
λ_i	2	1.94599	1.941691	1.941665	1.941665	1.941665	1.941665
R_0	0.12501	0.00859	5.24E–05	1.99E–09	5.55E–16	−3.30E–16	
R_1	2.31460	1.99929	1.974948	1.974798	1.974798	1.974798	
b_3	1	1	1	1	1	1	
b_2	−3	−3	−3	−3	−3	−3	
b_1	2.31460	2.31460	2.31460	2.31460	2.31460	2.31460	
b_0	−0.50419	−0.50419	−0.50419	−0.50419	−0.50419	−0.50419	
c_3	1	1	1	1	1	1	
c_2	−1	−1.05401	−1.05831	−1.05834	−1.05834	−1.05834	
c_1	0.31460	0.26351	0.25969	0.25967	0.25967	0.25967	
d_3	1	1	1	1	1	1	
d_2	1	0.89198	0.88338	0.88333	0.88333	0.88333	
d_1	2.31460	1.99929	1.97495	1.97480	1.97480	1.97480	

Table 4-8 Example 4-13 Synthetic Division for a Second-order Polynomial

Parameter	Iteration 0	Iteration 1	Iteration 2	Iteration 3	Iteration 4	Iteration 5	Iteration 6
λi	2	1.271502	0.914042	0.748042	0.685094	0.672387	0.671824
R_0	2.142998	0.530710	0.127778	0.027556	0.003963	0.000161	
R_1	2.941665	1.484668	0.769748	0.437750	0.311852	0.286439	
b_2	1	1	1	1	1	1	
b_1	-1.058340	-1.058340	-1.058340	-1.058340	-1.058340	-1.058340	
b_0	0.259668	0.259668	0.259668	0.259668	0.259668	0.259668	
c_2	1	1	1	1	1	1	
c_1	0.941665	0.213167	-0.144290	-0.310290	-0.373240	-0.385950	
d_2	1	1	1	1	1	1	
d_1	2.941665	1.484668	0.769748	0.437750	0.311852	0.286439	

4.9 MULTIPLE ROOTS

Figure 4-4 shows a case of multiple roots, where the function $f(x)$ is tangent to the x axis. This case corresponds to having two roots of the same magnitude and sign. In general, there can be triple, quadruple, . . . , or multiple roots for a function $f(x)$. It can be shown that even multiple roots result in a tangent $f(x)$ to the x axis, whereas odd multiple roots result in a function $f(x)$ that crosses the x axis with an inflection point at the root, that is, a point where the function changes curvature. For example, the following function has triple roots:

$$f(x) = (x - 1)(x - 1)(x - 1)(x - 5) \tag{4-49a}$$

This function can be expressed as

$$f(x) = x^4 - 8x^3 + 18x^2 - 16x + 5 \tag{4-49b}$$

The function in Eqs. 4-49 has triple roots ($x = 1$) and one root ($x = 5$) as shown in Fig. 4-14. The figure shows the inflection point at $x = 1$ that results in $f(x) = 0$.

Multiple roots pose difficulties for the previous methods. The bisection method has difficulties with multiple roots because the function does not change sign at even multiple roots. The Newton–Raphson and secant methods have difficulties because the derivative $f'(x)$ at a multiple root is zero. Since $f(x)$ reaches zero at a faster rate than $f'(x)$ as x approaches the multiple root, it is possible to check for the condition $f(x) = 0$ and terminate the computations before reaching $f'(x) = 0$.

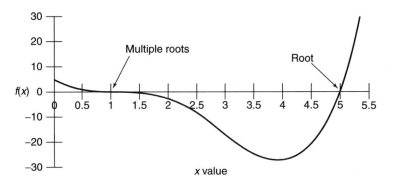

Figure 4-14   Multiple Roots at $x = 1$, and One Root at $x = 5$

4.10 SYSTEMS OF NONLINEAR EQUATIONS

The methods introduced so far for finding the roots of a function deal with single-variable equations of the type $f(x) = 0$. Some engineering problems have two or more variables for which roots are needed. For the two-variable problem, this would have the form $f_i(x, y) = 0$, where the subscript i denotes the equation number, and both x and y are independent variables. The number of equations should be equal to the number of variables. The following is an example of a two-variable system of equations:

$$x^3 - 3x^2 + xy = 0 \tag{4-50a}$$

$$4x^2 - 4xy^2 + 3y^2 = 0 \tag{4-50b}$$

The methods used for single-variable functions cannot be applied directly to find the values of x and y for multiple-variable functions. Instead, the *Jacobi method* is adapted to solve such equations. In this case, each of the i equations is solved for one of the i variables. Using initial estimates, the transformed equations are solved iteratively until the solution converges. The solution procedure is best illustrated by an example. Equations 4-50 are used for this purpose. Although Eq. 4-50a can be reduced to $x^2 - 3x + y = 0$ by dividing it by x, the original equation can be used without any effect on the final answer. These equations are solved for x and y, respectively:

$$x = (3x^2 - xy)^{1/3} \tag{4-51a}$$

$$y = \left(\frac{4x^2 + 3y^2}{4x}\right)^{0.5} \tag{4-51b}$$

Using initial estimates that $x = y = 3$, Eq. 4-51a yields

$$x = [3(3)^2 - (3)(3)]^{1/3} = 2.621$$

Using $x = 2.621$ and $y = 3$ with Eq. 4-51b gives

$$y = \left[\frac{4(2.621)^2 + 3(3)^2}{4(2.621)} \right]^{0.5} = 2.2796$$

For the second iteration, $x = 2.621$ and $y = 2.2796$ with Eq. 4-51a yields

$$x = [3(2.621)^2 - (2.621)(2.2796)]^{1/3} = 2.4458$$

Using $x = 2.4458$ and $y = 2.2796$ with Eq. 4-51b yields

$$y = \left[\frac{4(2.4458)^2 + 3(2.2796)^2}{4(2.4458)} \right]^{0.5} = 2.0098$$

If the iterations were continued, the solution would converge to the following roots: $x = 2.16$ and $y = 1.82$. The iterations are shown in Table 4-9 and Fig. 4-15.

Table 4-9   Results for Example 4-13

Iteration i	x	y
1	2.620741	2.279550
2	2.445807	2.009789
3	2.353163	1.908024
4	2.297176	1.867022
5	2.259935	1.848446
6	2.233679	1.838727
7	2.214562	1.832913
8	2.200396	1.829070
9	2.189798	1.826369
10	2.181825	1.824402
11	2.175806	1.822942
12	2.171253	1.821849
13	2.167804	1.821025
14	2.165189	1.820402
15	2.163204	1.819931
16	2.161698	1.819573
17	2.160554	1.819302
18	2.159684	1.819096
19	2.159024	1.818939
20	2.158522	1.818820

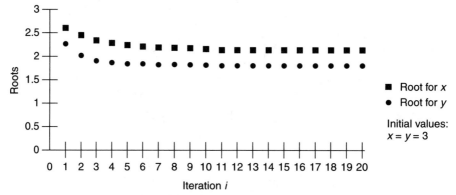

Figure 4-15   Results for Example 4-13

Example 4-14 Roots for a System of Equations

The solution procedure for a system of equations is illustrated using a pair of linear equations. While an analytical solution can easily be performed for these equations, the numerical solution is used for the purpose of demonstration. In Example 8-9, the following equations were developed to demonstrate the Galerkin method:

$$\frac{1}{4}b_1 + \frac{7}{15}b_2 = \frac{1}{3} \tag{4-52a}$$

$$\frac{2}{15}b_1 + \frac{1}{3}b_2 = \frac{1}{4} \tag{4-52b}$$

Solving for b_1 and b_2 yields

$$b_1 = \frac{4}{3} - \frac{28}{15}b_2$$

$$b_2 = \frac{3}{4} - \frac{6}{15}b_1$$

Using initial estimates of $b_1 = b_2 = 1$, the first iteration is

$$b_1 = \frac{4}{3} - \frac{28}{15}(1) = -0.53333$$

$$b_2 = \frac{3}{4} - \frac{6}{15}(-0.5333) = 0.96333$$

The second iteration is

$$b_1 = \frac{4}{3} - \frac{28}{15}(0.96333) = -0.46489$$

$$b_2 = \frac{3}{4} - \frac{6}{15}(-0.46489) = 0.93596$$

Table 4-10 and Figure 4-16 show summaries of the iterations. Comparing the values computed for the twentieth iteration with the analytically derived solution of Eq. 8-52 (i.e., $b_1 = -0.26316$ and $b_2 = 0.85526$) indicates that the method does not converge rapidly; however, it does converge. The problem of convergence for the Jacobi method is discussed in Section 5.5.

Table 4-10   Results for Example 4-14

Iteration i	b_1	b_2
1	−0.53333	0.963333
2	−0.46489	0.935956
3	−0.41378	0.915513
4	−0.37563	0.900250
5	−0.34713	0.888853
6	−0.32586	0.880344
7	−0.30998	0.873990
8	−0.29811	0.869246
9	−0.28926	0.865704
10	−0.28265	0.863059
11	−0.27771	0.861084
12	−0.27402	0.859609
13	−0.27127	0.858508
14	−0.26922	0.857686
15	−0.26768	0.857072
16	−0.26654	0.856614
17	−0.26568	0.856272
18	−0.26504	0.856016
19	−0.26456	0.855825
20	−0.26421	0.855683

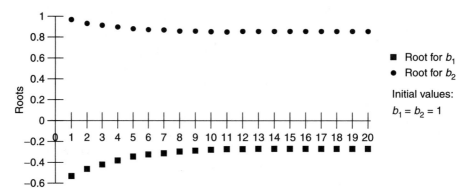

Figure 4-16 Results for Example 4-14

4.11 APPLICATIONS

4.11.1 Pipe Flow Evaluation

In Section 3.6.1, a method for designing a pipe system was introduced. The same concepts can be used to evaluate the flow rate for a given system. Equation 3-26 can be rearranged to the following third-order polynomial:

$$\frac{8fL}{\pi^2 gD^5}Q^3 + hQ - \frac{550eh_p}{\gamma} = 0 \tag{4-53}$$

where f = friction coefficient of pipe, L = length of pipe, g = gravitational acceleration, D = diameter of pipe, Q = VA in which V = flow velocity and A = cross-sectional area, h = potential head, e = efficiency of pump, h_p = horsepower of pump, and γ = specific weight of fluid. Two of the three roots of Eq. 4-53 are imaginary; the third root of the polynomial is the flow rate for the characteristics of the reservoir system, fluid, and pump.

Consider the case of an 800-ft pipe section (with pipe roughness coefficient e_f = 0.0004) connecting two reservoirs containing oil (specific gravity = 0.82). The 6-in. pipe system includes a 6-hp pump (with e = 0.6). The difference in the surface elevations of the two reservoirs is h = 5 ft. For the 6-in. pipe, e_f/D = 0.0008 ft, which gives a friction coefficient f of 0.0185 (see Table 3-4). Thus Eq. 4-53 is

$$\frac{8(0.0185)(800)}{\pi^2(32.2)(0.5)^5}Q^3 + 5Q - \frac{550(0.6)(6)}{0.82(62.4)} = 0 \tag{4-54a}$$

$$11.922Q^3 + 5Q - 38.696 = 0 \tag{4-54b}$$

It is evident that $Q = 1$ would produce a negative value for the left side of Eq. 4-54, but $Q = 2$ would produce a positive value. Thus a Q of 1.25 cfs will be used as the initial estimate. The Newton–Raphson method can be used. The derivative of Eq. 4-54 is

$$f'(Q) = 35.766Q^2 + 5 \tag{4-55}$$

Thus Newton's correction formula is

$$Q_{i+1} = Q_i - \frac{11.922Q_i^3 + 5Q_i - 38.696}{35.766Q_i^2 + 5} \tag{4-56}$$

The corrected estimate of Q is

$$1.25 - \frac{11.922(1.25)^3 + 5(1.25) - 38.696}{35.766(1.25)^2 + 5} = 1.40 \text{ cfs} \tag{4-57}$$

A second iteration yields

$$1.40 - \frac{11.922(1.40)^3 + 5(1.40) - 38.696}{35.766(1.40)^2 + 5} = 1.386 \text{ cfs} \tag{4-58}$$

A third iteration yields 1.386 cfs, which is of sufficient accuracy for all practical purposes.

4.11.2 Gas Law

The ideal gas law $PV = nRT$ is a basic concept covered in introductory chemistry classes. In practice, more accurate P–V–T relationships of gases are available. The following Beattie–Bridgeman equation is one example:

$$P = \frac{RT}{V} + \frac{a_1}{V^2} + \frac{a_2}{V^3} + \frac{a_3}{V^4} \tag{4-59}$$

This can be rearranged to the form of a fourth-order polynomial:

$$PV^4 - RTV^3 - a_1V^2 - a_2V - a_3 = 0 \tag{4-60}$$

For values of R, a_1, a_2, and a_3, the volume V for any temperature T and pressure P can be determined from the real roots of the polynomial.

For a particular gas, $a_1 = -1.06$, $a_2 = 0.057$, and $a_3 = -0.00011$. If the pressure P equals 25 atm, then the temperature T equals 293K, and $R = 0.082$ liter-atm/K-g mole. Then Eq. 4-60 becomes

$$f(V) = 25V^4 - 24.03V^3 + 1.06V^2 - 0.057V + 0.00011 = 0 \tag{4-61}$$

An initial approximation of V can be made with the following ideal gas law:

$$V_0 = \frac{RT}{P} = \frac{0.082(293)}{25} = 0.961 \tag{4-62}$$

Substituting the initial estimate V_0 into Eq. 4-61 yields 0.92, not the zero value that would be expected if V_0 were the correct volume. The derivative of Eq. 4-61 is

$$f'(V) = 100V^3 - 72.09V^2 + 2.12V - 0.057 \tag{4-63}$$

Newton's formula is

$$V_{i+1} = V_i - \frac{f(V)}{f'(V)} \tag{4-64}$$

Using the initial estimate of $V_0 = 0.961$ and Eqs. 4-61 and 4-63 with Eq. 4-64 yields the results in Table 4-11. Thus three rounds are necessary for the solution to converge to a volume of 0.918.

Table 4-11   Results of Gas Law Application

Iteration	V_i	$f(V)$	$f'(V)$
0	0.961	0.9200	18.75
1	0.912	−0.1050	17.76
2	0.918	0.0039	18.49
3	0.918		

4.12 PROBLEMS

The polynomial

$$x^3 + ax^2 + bx + c = 0$$

can be rearranged and solved for x as

$$x^2 + ax + b + \frac{c}{x} = 0$$

$$x^2 = -ax - b - \frac{c}{x}$$

$$x = \sqrt{-ax - b - \frac{c}{x}}$$

A root can be found using an initial estimate of x and iterating until the expression on the right of the equal sign equals the estimate of the root. Solve Problems 4-1 through 4-3 for all roots using this iteration technique.

4-1. $x^3 + 0.2x^2 - 0.67x - 0.364 = 0$

4-2. $x^3 - 17.2x^2 - 81.97x - 83.79 = 0$

4-3. $x^3 + 7.2x^2 + 13.55x + 3.444 = 0$

4-4. Consider the four correlation matrices

$$\mathbf{R}_1 = \begin{bmatrix} 1 & 0 \\ 0 & 1 \end{bmatrix}, \qquad \mathbf{R}_2 = \begin{bmatrix} 1 & 0.2 \\ 0.2 & 1 \end{bmatrix}$$

$$\mathbf{R}_3 = \begin{bmatrix} 1 & 0.7 \\ 0.7 & 1 \end{bmatrix}, \qquad \mathbf{R}_4 = \begin{bmatrix} 1 & 1 \\ 1 & 1 \end{bmatrix}$$

The off-diagonal elements are the intercorrelations between the two predictor variables. Evaluate the eigenvalues of the four matrices and develop general guidelines for interpreting the eigenvalues of a correlation matrix.

4-5. The following function contains three real roots between 0 and 2.5:

$$f(x) = x^3 - 3.5x^2 + 3.28x - 0.924 = 0$$

a. Using an increment of 0.25, determine the approximate location of the roots using the direct-search method. Discuss the problem identified from this analysis.

b. Using the increment $\Delta x = 0.25$, plot $f(x)$ versus x and discuss how the graph can be used to suggest the most feasible intervals to search for the roots.

4-6. The following function contains one real root in the interval from 0 to 1:

$$f(x) = x^4 - 1.74x^3 + 0.9419x^2 - 0.4252x + 0.0027 = 0$$

a. Using an increment of 0.10, determine the approximate location of the roots using the direct-search method. Discuss the problem identified from this analysis.

b. Using the interval $\Delta x = 0.10$, plot $f(x)$ versus x and discuss how the graph can be used to suggest the most feasible intervals to search for the roots.

4-7. Estimate the root of the polynomial in Problem 4-5 that lies in the interval 2 to 2.5 using the bisection method to a precision of (a) 0.02 and (b) 1%.

4-8. Estimate the root of the polynomial in Problem 4-6 that lies in the interval 0 to 0.02 using the bisection method to a precision of (a) 0.002 and (b) 0.1%.

The functions given in Problems 4-9 through 4-18 possess real roots. Using a precision of 2% of a root, find the estimates of all roots using the Newton–Raphson method. After finding a root, reduce the polynomial using polynomial reduction before finding the next root with the Newton–Raphson method.

4-9. $x^3 - 0.39x + 0.07 = 0$

4-10. $x^3 - 1.7x^2 + 0.84x - 0.108 = 0$

4-11. $x^3 - 2x^2 - 5x + 6 = 0$

4-12. $x^3 + x^2 - 17x + 15 = 0$

4-13. $x^3 - 6.1x^2 + 11.26x - 6.336 = 0$

4-14. $x^3 + 3x^2 - 4 = 0$

4-15. $x^4 - 15x^2 - 6x + 24 = 0$

4-16. $x^4 - 10x^2 + 9 = 0$

4-17. $x^4 - 2x^3 - 15x^2 - 4x + 20 = 0$

4-18. $x^4 - 5.6x^3 - 39.97x^2 - 12.33x + 24.288 = 0$

For the functions given in Problems 4-19 through 4-22, determine the interval for each root in which any initial estimate in the interval will converge to the root.

4-19. $x^3 - 2x^2 - 5x + 6 = 0$

4-20. $x^3 - 3.5x^2 + 3.28x - 0.924 = 0$

4-21. $x^4 - 10x^2 + 9 = 0$

4-22. $x^4 - 15x^2 - 6x + 24 = 0$

The functions given in Problems 4-23 through 4-26 possess real roots. Using a precision of 1% of a root, find the estimates of all roots using the Secant method. After finding a root, reduce the polynomial using polynomial reduction before finding the next root with the secant method.

4-23. Use the function of Problem 4-10.

4-24. Use the function of Problem 4-12.

4-25. Use the function of Problem 4-13.

4-26. Use the function of Problem 4-18.

The functions given in Problems 4-27 through 4-34 possess real roots. Using a precision of 2% of a root, find estimates of all roots by synthetic division.

4-27. $x^3 - 0.6x^2 + 0.11x - 0.006 = 0$

4-28. $x^3 + 0.9x^2 - 10.56x - 16.94 = 0$

4-29. $x^3 + 0.6x^2 - 4.77x + 2.43 = 0$

4-30. $x^3 - 13.1x^2 + 48.48x - 46.62 = 0$

4-31. $x^4 - 8.37x^2 - 10.336x - 1.734 = 0$

4-32. $x^4 - 7.8x^3 + 22.08x - 26.738x + 11.6127 = 0$

4-33. $x^4 + 4.5x^3 + 2.64x^2 - 4.528x + 1.0752 = 0$

4-34. $x^4 + 7.4x^3 + 17.99x^2 + 15.31x + 2.4 = 0$

Using any of the methods introduced in this chapter, determine the eigenvalues of the matrices of Problems 4-35 through 4-38.

4-35. $\begin{bmatrix} 1.00 & 0.73 & -0.24 \\ 0.73 & 1.00 & 0.74 \\ -0.24 & 0.74 & 1.00 \end{bmatrix}$

4-36. $\begin{bmatrix} 1.00 & -0.28 & 0.84 \\ -0.28 & 1.00 & -0.62 \\ 0.84 & -0.62 & 1.00 \end{bmatrix}$

4-37. $\begin{bmatrix} 1.00 & 0.54 & -0.67 \\ 0.38 & 1.00 & 0.55 \\ 0.92 & -0.24 & 1.00 \end{bmatrix}$

4-38. $\begin{bmatrix} 2.0 & 1.2 & 0.8 \\ 0.6 & 3.0 & 1.5 \\ 1.1 & 1.4 & 1.0 \end{bmatrix}$

4-39. The cube root of a positive number c is equivalent to the solution of the equation $x^3 - c = 0$. Develop an algorithm based on Newton–Raphson's method for finding the cube root. Test the algorithm using $c = 15$ with an initial estimate of the cube root of 5.

4-40. Write a computer program for the synthetic division method as outlined in Section 4.8. Test the program using Examples 4-12 and 4-13.

4-41. The function given in Eq. 4-49 possess multiple roots. Using a precision of 2% of a root, find the estimates of all roots using the Newton–Raphson method. The function is given by

$$f(x) = x^4 - 8x^3 + 18x^2 - 16x + 5$$

4-42. Obtain the solution, if any, for the following system of nonlinear equations:

$$x^4 - 8yx^3 + 18yx^2 - 16x = 0$$
$$x^3 + 18xy^2 - 16y = 0$$

5

Simultaneous Linear Equations

5.1 INTRODUCTION

Many engineering and scientific problems can be formulated in terms of systems of simultaneous linear equations. When these systems consist of only a few equations, a solution can be found analytically using the standard methods from algebra, such as substitution. However, complex problems may involve a large number of equations that cannot realistically be solved using analytical methods. In these cases, we will need to find the solution numerically using computers.

Before introducing numerical methods for solving large systems of simultaneous linear equations, we will do two things. First, we will show how systems of simultaneous linear equations arise from physical engineering problems. Second, we will review the substitution method for obtaining an analytical solution in order to illustrate the need for numerical methods. The two example cases introduced in this section are used as motivation for the numerical methods that follow in subsequent sections.

Example 5-1 Material Purchasing for Manufacturing

Let us assume that a manufacturer is marketing a product made of an alloy material meeting a certain specified composition. The three critical ingredients of the alloy are manganese, silicon, and copper. The specifications require 15 pounds of manganese, 22 pounds of silicon, and 39 pounds of copper for each ton of alloy to be produced. This mix of ingredients requires the manufacturer to obtain inputs from three different mining suppliers. Ore from the different suppliers has different concentrations of the alloy ingredients, as detailed in Table 5-1. Given this information, the manufacturer must determine how much ore to purchase from each supplier so that there is no waste of the alloy ingredients. A solution to this problem can be found by defining the following variables:

$$X_j = \text{amount of ore purchased from supplier } j \tag{5-1a}$$

$$C_i = \text{amount of ingredient } i \text{ required per ton of alloy} \tag{5-1b}$$

$$a_{ij} = \text{amount of ingredient } i \text{ contained in each ton of ore shipped from supplier } j \tag{5-1c}$$

TABLE 5-1 Concentration of Alloy Ingredients for Three Suppliers

	Supplier 1 (lb/ton of ore)	Supplier 2 (lb/ton of ore)	Supplier 3 (lb/ton of ore)
Manganese	1	3	2
Silicon	2	4	3
Copper	3	4	7

Using these notations, we can formulate a general equation that defines (1) the relationships among the compositions of the ore shipped by the different suppliers, (2) the amount of ore needed from each supplier, and (3) the required composition of the final alloy as

$$\sum_{j=1}^{n} a_{ij}X_j = C_i \qquad \text{for } i = 1, 2, \ldots, m \tag{5-2}$$

in which m is the number of ingredients and n is the number of suppliers. For the case under consideration, both m and n equal 3. More specifically, the problem can be represented by the following system of linear simultaneous equations:

$$X_1 + 3X_2 + 2X_3 = 15 \tag{5-3a}$$

$$2X_1 + 4X_2 + 3X_3 = 22 \tag{5-3b}$$

$$3X_1 + 4X_2 + 7X_3 = 39 \tag{5-3c}$$

The solution to these equations determines the amount of ore to purchase from each supplier by the alloy manufacturer.

Example 5-2 Electrical Circuit Analysis

Current flows in circuits are governed by Kirchhoff's laws. Kirchhoff's first law states that the algebraic sum of the currents flowing into a junction of a circuit must equal zero. Kirchhoff's second law states that the algebraic sum of the electromotive forces around a closed circuit must equal the sum of the voltage drops around the circuit, where a voltage drop equals the product of the current and the resistance. In solving circuit problems for which Kirchhoff's laws apply, the circuit network is separated into individual sections called loops. For the example network shown in Fig. 5-1, there are two separate networks, loop $acdb$ and loop $aefb$. Applying Kirchhoff's first law at junction c yields the linear equation

$$I_1 + I_2 - I_3 = 0 \tag{5-4}$$

Applying Kirchhoff's second law to network loop $acdb$ yields the following linear equation:

$$V_1 = R_1I_1 + R_3I_3 \tag{5-5}$$

Applying Kirchhoff's second law to network loop $aefb$ yields the following linear equation:

$$V_1 - V_2 = R_1I_1 - R_2I_2 \tag{5-6}$$

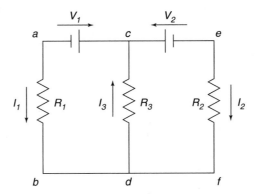

Figure 5-1   Example Electrical Circuit

Using the voltages and resistances shown in Fig. 5-1 (with assumed $R_1 = 2$, $R_2 = 4$, $R_3 = 5$, $V_1 = 6$, and $V_2 = 2$), the above circuit equations governing the current flows in the circuit can be written as the following system of three linear simultaneous equations:

$$I_1 + I_2 - I_3 = 0 \tag{5-7a}$$

$$2I_1 \quad\quad + 5I_3 = 6 \tag{5-7b}$$

$$2I_1 - 4I_2 \quad\quad = 4 \tag{5-7c}$$

The solution to these three equations produces the current flows in the network.

5.1.1 General Form for a System of Equations

The previous examples are illustrations of the following general form for a set of simultaneous linear equations:

$$a_{11}X_1 + a_{12}X_2 + \cdots + a_{1n}X_n = C_1$$
$$a_{21}X_1 + a_{22}X_2 + \cdots + a_{2n}X_n = C_2$$
$$\vdots \tag{5-8}$$
$$a_{n1}X_1 + a_{n2}X_2 + \cdots + a_{nn}X_n = C_n$$

in which the a_{ij} terms are the known coefficients of the equations, the X_j terms are the unknown variables, and the C_i terms are the known constants. Since values for both the a_{ij} and C_i terms will be known for any problem, the system of equations in Eqs. 5-8 represents n linear equations with n unknowns. It is important to emphasize that the solution methods provided in this chapter assume that the number of unknowns equals the number of equations. Also, the equations must be linearly independent; that is, any one equation cannot be produced by a linear combination of any of the other equations.

5.1.2 Solution of Two Equations

If n equals 2, then Eqs. 5-8 reduce to

$$a_{11}X_1 + a_{12}X_2 = C_1 \tag{5-9a}$$

$$a_{21}X_1 + a_{22}X_2 = C_2 \tag{5-9b}$$

For simple systems such as the one in Eqs. 5-9, a solution can be obtained by substitution. Solving Eq. 5-9a for X_1 yields

$$X_1 = \frac{C_1 - a_{12}X_2}{a_{11}} \tag{5-10a}$$

which can be substituted into Eq. 5-9b:

$$a_{21}\frac{C_1 - a_{12}X_2}{a_{11}} + a_{22}X_2 = C_2 \tag{5-10b}$$

Equation 5-10b is a single equation with one unknown, X_2. Thus Eq. 5-10b can be solved for X_2 as

$$X_2 = \frac{a_{11}C_2 - a_{21}C_1}{a_{11}a_{22} - a_{21}a_{12}} \tag{5-11a}$$

Equation 5-11a can be substituted back into Eq. 5-10a, producing the following equation for X_1:

$$X_1 = \frac{a_{22}C_1 - a_{12}C_2}{a_{11}a_{22} - a_{21}a_{12}} \tag{5-11b}$$

Thus Eqs. 5-11 provide a means of determining the values of the unknowns of Eqs. 5-9 in terms of the known values of the a_{ij} coefficients and the constants C_i.

While the solution procedure based on substitution appears simple to apply, one should try to imagine the effort that would be required to solve a set of 10 simultaneous equations using the substitution procedure that led to Eqs. 5-11. Many complex engineering problems involve hundreds or even thousands of simultaneous equations. The need for alternative solution procedures is obvious.

Given the availability of computers, numerical solution methods are widely used. There are three general types of numerical procedures: elimination methods, iteration methods, and the method of determinants. Before discussing the advantages of each, the individual computational methods will be introduced.

5.1.3 Classification of Systems of Equations Based on Graphical Interpretation

Systems of equations can be classified based on their solutions to the following three types:

1. Systems that have solutions
2. Systems without solutions
3. Systems with an infinite number of solutions

These three types can be illustrated graphically using systems of two equations. For example, the following four systems are shown in Fig. 5-2:

$$\text{System 1:} \quad 2X_1 + 3X_2 = 6 \tag{5-12a}$$

$$2X_1 + 9X_2 = 12 \tag{5-12b}$$

$$\text{System 2:} \quad 3X_1 + 9X_2 = 5 \tag{5-12c}$$

$$X_1 + 3X_2 = 6 \tag{5-12d}$$

$$\text{System 3:} \quad 2X_1 + 3X_2 = 4 \tag{5-12e}$$

$$4X_1 + 6X_2 = 8 \tag{5-12f}$$

$$\text{System 4:} \quad 2X_1 + 2.2X_2 = 5.7 \tag{5-12g}$$

$$2X_1 + 2X_2 = 5.5 \tag{5-12h}$$

The first system has a solution according to Eqs. 5-11a and 5-11b of

$$X_1 = 1.5 \quad \text{and} \quad X_2 = 1 \tag{5-13}$$

This solution is uniquely defined as the intersection point of Eqs. 5-12a and b as shown in Fig. 5-2a. System 2 does not have a solution, since the use of Eqs. 5-11 results in a divi-

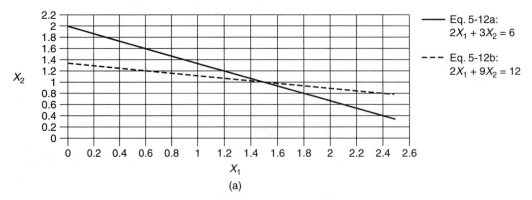

Figure 5-2   Three Systems of Equations: (a) system 1 with a solution;

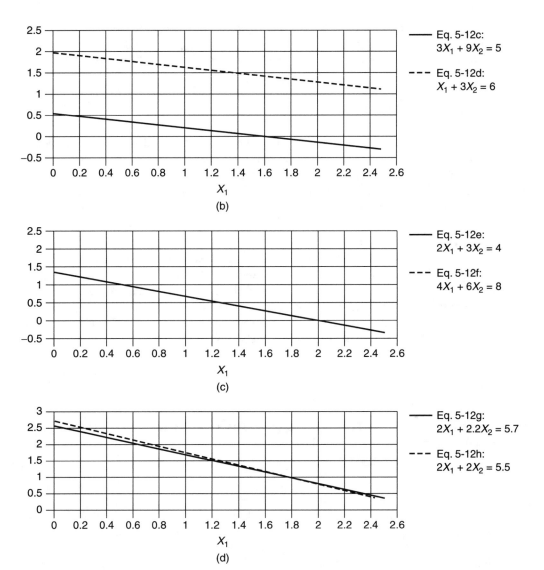

Figure 5-2   (Continued) (b) system 2 without a solution; (c) system 3 with an infinite number
of solutions; and (d) system 4 with ill-conditioned parameters

sion by zero. According to Fig. 5-2b, the equations of system 2 have the same slope but
different intercepts; that is, they are parallel. Therefore, they do not intersect. System 3
results also in a division by zero according to Eqs. 5-11. The equations in this case are
shown in Fig. 5-2c as parallel lines with the same slope and intercept, that is, the same
line. The intersection of these two lines is defined by the entire line; therefore, there are
an infinite number of solutions in this case. The two equations in system 3 are not unique,
since one equation can be produced by the other by multiplying it by a constant. System 4

(Eqs. 5-12g and h) has ill-conditioned parameters that result in two lines with about the same slope as shown in Fig. 5-2d. Therefore, the solution defined as the intersection point of the two lines can result in computational difficulties according to numerical methods.

5.2 GAUSSIAN ELIMINATION

Gaussian elimination is similar to the substitution procedure used in deriving expressions for X_1 and X_2 of Eqs. 5-11a and b, respectively, from Eqs. 5-9. Generally, the elimination procedure involves performing a set of operations on a system of equations until each equation is reduced to an identity for a single unknown. Details of the procedure are provided following a discussion of the *permissible* operations for a set of equations.

5.2.1 Permissible Operations

Mathematical operations on equations in a set of simultaneous equations that do not alter the solution are termed *permissible* operations. Three permissible operations are of interest here; they are described in the context of simple, two-equation examples.

First, the solution to a set of simultaneous equations does not change if the order (that is, sequence) in which the equations are written is changed. For example, the solution to the equations

$$2X_1 + 3X_2 = 1 \tag{5-14a}$$

$$-4X_1 + X_2 = 5 \tag{5-14b}$$

can be determined from Eqs. 5-11; the solution is $X_1 = -1$ and $X_2 = 1$. This solution is identical to the solution for the following rearranged equations:

$$-4X_1 + X_2 = 5 \tag{5-15a}$$

$$2X_1 + 3X_2 = 1 \tag{5-15b}$$

This can be verified by substituting $X_1 = -1$ and $X_2 = 1$ into Eqs. 5-15. Although the order of the equations has no influence on the solution, it is important to note that the order of the equations can in some instances influence both the likelihood of finding a solution or the accuracy of a numerical solution. In cases where the coefficients are of different orders of magnitude, round-off and truncation errors can lead to inaccurate estimates of the unknowns. This aspect of the solution is discussed and illustrated later.

Second, any one of the simultaneous equations can be multiplied or divided by a nonzero constant without changing the solution. For example, if Eq. 5-14a is multiplied by 2, we would have the following set of simultaneous equations:

$$4X_1 + 6X_2 = 2 \tag{5-16a}$$

$$-4X_1 + X_2 = 5 \tag{5-16b}$$

The solutions for Eqs. 5-14 and Eqs. 5-16 are identical; this can be verified by substituting the solution for Eqs. 5-14 into Eqs. 5-16.

Third, it is permissible to add two equations together and use the resulting equation to replace either of the two original equations. For example, the addition of Eq. 5-14a to Eq. 5-14b results in

$$-2X_1 + 4X_2 = 6 \tag{5-17}$$

The solution of the two simultaneous equations as given by Eqs. 5-14a and 5-17 is the same as for Eqs. 5-14; also, the solution for Eqs. 5-14b and 5-17 is the same as for Eqs. 5-14.

The permissible operations can be applied in any combination to achieve some computational interests. For example, using the second and third operations, we can multiply an equation by a constant and add it to another equation, with the resulting equation replacing one of the original equations. This type of operation forms the basis for the Gaussian elimination method.

5.2.2 Matrix Representation of the System of Equations

While a solution of two or three simultaneous equations does not present a problem of notation, the solution of a set of n simultaneous equations can be simplified by presenting the equations in terms of matrices. The fundamentals of matrix algebra are provided in Chapter 2. Using the notation of Eq. 5-8, we can express the set of simultaneous equations as the following matrix relation:

$$\begin{bmatrix} a_{11} & a_{12} & \cdots & a_{1n} \\ a_{21} & a_{22} & \cdots & a_{2n} \\ \vdots & \vdots & \vdots & \vdots \\ a_{n1} & a_{n2} & \cdots & a_{nn} \end{bmatrix} \begin{bmatrix} X_1 \\ X_2 \\ \vdots \\ X_n \end{bmatrix} = \begin{bmatrix} C_1 \\ C_2 \\ \vdots \\ C_n \end{bmatrix} \tag{5-18}$$

Equation 5-18 can be expressed in a more compact and convenient form by dropping the unknown X_i terms and incorporating the constant C_i terms as an additional column in the coefficient matrix:

$$\begin{bmatrix} a_{11} & a_{12} & \cdots & a_{1n} & C_1 \\ a_{21} & a_{22} & \cdots & a_{2n} & C_2 \\ \vdots & \vdots & \vdots & \vdots & \vdots \\ a_{n1} & a_{n2} & \cdots & a_{nn} & C_n \end{bmatrix} \tag{5-19}$$

Using this notation, the two simultaneous equations of Eqs. 5-14 would appear as

$$\begin{bmatrix} 2 & 3 & 1 \\ -4 & 1 & 5 \end{bmatrix} \tag{5-20}$$

In this notation, the ith row corresponds to the coefficients and constants for the ith equation. The jth column contains the coefficients corresponding to X_j, except for the last column, which contains the vector of the C_i constants. This simplified form is acceptable because row operations act only on the coefficients a_{ij} and constants C_i (but not the X_j values), so the augmented form of Eq. 5-19 gives a compact representation by row of only those elements that are actually going to be modified.

5.2.3 Gaussian Elimination Procedure

The Gaussian elimination procedure can be separated into two parts, which are often referred to as the *forward pass* and *back substitution*. The objective of the forward pass is to apply the three permissible operations to transform Eq. 5-19 to an upper-triangular matrix having the form

$$
\begin{bmatrix}
1 & d_{12} & d_{13} & \cdots & d_{1n} & e_1 \\
0 & 1 & d_{23} & \cdots & d_{2n} & e_2 \\
0 & 0 & 1 & \cdots & d_{3n} & e_3 \\
\vdots & \vdots & \vdots & \vdots\vdots\vdots & \vdots & \vdots \\
0 & 0 & 0 & \cdots & 1 & e_n
\end{bmatrix}
\tag{5-21}
$$

In Eq. 5-21, all off-diagonal elements of the lower half of the coefficient matrix equal zero and all the diagonal terms equal 1.

After the forward pass is completed, the unknowns in the equations are found in the back substitution procedure. The back substitution procedure can be better illustrated if the matrix of Eq. 5-21 is rewritten in the equivalent form in terms of individual equations:

$$X_1 + d_{12}X_2 + d_{13}X_3 + \cdots + d_{1,n-2}X_{n-2} + \quad d_{1,n-1}X_{n-1} + \quad d_{1,n}X_n = e_1 \tag{5-22a}$$

$$X_2 + d_{23}X_3 + \cdots + d_{2,n-2}X_{n-2} + \quad d_{2,n-1}X_{n-1} + \quad d_{2,n}X_n = e_2 \tag{5-22b}$$

$$X_3 + \cdots + d_{3,n-2}X_{n-2} + \quad d_{3,n-1}X_{n-1} + \quad d_{3,n}X_n = e_3 \tag{5-22c}$$

$$X_{n-2} + d_{n-2,n-1}X_{n-1} + d_{n-2,n}X_n = e_{n-2} \tag{5-22d}$$

$$X_{n-1} + d_{n-1,n}X_n = e_{n-1} \tag{5-22e}$$

$$X_n = e_n \tag{5-22f}$$

From Eqs. 5-22, which represent the system of equations after the forward pass, we can easily obtain the solution for the X_i terms. The last equation (5-22f) involves only a single unknown; thus the value of X_n is given by

$$X_n = e_n \tag{5-23}$$

The next to last equation (5-22e) involves only the unknowns X_{n-1} and X_n. However, X_n has already been found with Eq. 5-22f. Thus, the unknown X_{n-1} can be found by rearranging the terms in Eq. 5-22e and substituting this known value for X_n as follows:

$$X_{n-1} = e_{n-1} - d_{n-1,n}X_n \tag{5-24}$$

Similarly, Eq. 5-22d can be rearranged for X_{n-2} in terms of the now known values of X_{n-1} and X_n. The back substitution step thus involves working upward through the set of reduced equations produced by the forward pass and solving for each unknown in turn.

Example 5-3 Gaussian Elimination Procedure

The Gaussian elimination procedure is best illustrated with an example. Consider the following set of four simultaneous equations:

$$2X_1 + 3X_2 - 2X_3 - X_4 = -2 \tag{5-25a}$$

$$2X_1 + 5X_2 - 3X_3 + X_4 = 7 \tag{5-25b}$$

$$-2X_1 + X_2 + 3X_3 - 2X_4 = 1 \tag{5-25c}$$

$$-5X_1 + 2X_2 - X_3 + 3X_4 = 8 \tag{5-25d}$$

These equations can be represented in the matrix form of Eq. 5-19 as

$$\begin{bmatrix} 2 & 3 & -2 & -1 & -2 \\ 2 & 5 & -3 & 1 & 7 \\ -2 & 1 & 3 & -2 & 1 \\ -5 & 2 & -1 & 3 & 8 \end{bmatrix} \tag{5-26}$$

As the first step in the forward pass, we will convert the element a_{11} of the first transformed matrix to 1 and eliminate, that is, set to zero, all the other elements in the first column. Thus our objective of the first step of the forward pass is to derive a matrix with the following form:

$$\begin{bmatrix} 1 & d'_{12} & d'_{13} & d'_{14} & e'_1 \\ 0 & d'_{22} & d'_{23} & d'_{24} & e'_2 \\ 0 & d'_{32} & d'_{33} & d'_{34} & e'_3 \\ 0 & d'_{42} & d'_{43} & d'_{44} & e'_4 \end{bmatrix} \tag{5-27}$$

Equation 5-27 can be derived by applying the permissible operations to our system of equations. Note that the values of the terms in Eqs. 5-26 will change after these operations; hence we will use different symbols, d'_{ij} and e'_i, for the modified equation coefficients and constants. In Eq. 5-27, the prime notation means that the resulting values after one operation are not the final values as given by Eq. 5-21. To obtain Eqs. 5-27 from Eqs. 5-26, we first normalize row 1 by dividing the entire row by a_{11}, which equals 2 in our example. Element a_{11} is often termed the *pivot* element for row 1. We then loop over each row below the pivot element and eliminate the elements a_{i1} by multiplying each element of row 1 by $-a_{i1}$ and adding it to the corresponding element a_{ij} in row i. This is shown as follows:

		Original Matrix				Operation		Resultant Matrix			

$$\begin{array}{ccc}
\textit{Original Matrix} & \textit{Operation} & \textit{Resultant Matrix} \\[4pt]
\begin{bmatrix} 2 & 3 & -2 & -1 & -2 \\ 2 & 5 & -3 & 1 & 7 \\ -2 & 1 & 3 & -2 & 1 \\ -5 & 2 & -1 & 3 & 8 \end{bmatrix}
&
\begin{matrix} R'_1 = \frac{R_1}{2} \\ R'_2 = R_2 - 2R'_1 \\ R'_3 = R_3 + 2R'_1 \\ R'_4 = R_4 + 5R'_1 \end{matrix}
&
\begin{bmatrix} 1 & \frac{3}{2} & -1 & -\frac{1}{2} & -1 \\ 0 & 2 & -1 & 2 & 9 \\ 0 & 4 & 1 & -3 & -1 \\ 0 & \frac{19}{2} & -6 & \frac{1}{2} & 3 \end{bmatrix}
\end{array} \tag{5-28}$$

In the above notation, the *Operation* column describes the row operations performed on each row R_i, where R_i is the vector of row values and R'_i is the resulting value. The process is then repeated starting with row 2, with coefficient d'_{22} becoming the pivot element. The objective of this second step of the forward pass is to reduce the equations further to the following form:

$$
\begin{bmatrix}
1 & f_{12} & f_{13} & f_{14} & g_1 \\
0 & 1 & f_{23} & f_{24} & g_2 \\
0 & 0 & f_{33} & f_{34} & g_3 \\
0 & 0 & f_{43} & f_{44} & g_4
\end{bmatrix}
\tag{5-29}
$$

The same procedure is used as in the first step in Eqs. 5-28. Specifically, the pivot row, row 2, is normalized by dividing the new pivot, d_{22} (which equals 2 in the matrix on the right side of Eqs. 5-28), and the d_{i2} coefficients are set to zero by performing the following permissible operations on rows 3 and 4:

Starting Matrix   *Operation*   *Resultant Matrix*

$$
\begin{bmatrix}
1 & \frac{3}{2} & -1 & -\frac{1}{2} & -1 \\
0 & 2 & -1 & 2 & 9 \\
0 & 4 & 1 & -3 & -1 \\
0 & \frac{19}{2} & -6 & \frac{1}{2} & 3
\end{bmatrix}
\quad
\begin{array}{l}
R'_1 = R_1 \\
R'_2 = \dfrac{R_2}{2} \\
R'_3 = R_3 - 4R'_2 \\
R'_4 = R_4 - \dfrac{19}{2}R'_2
\end{array}
\quad
\begin{bmatrix}
1 & \frac{3}{2} & -1 & -\frac{1}{2} & -1 \\
0 & 1 & -\frac{1}{2} & 1 & \frac{9}{2} \\
0 & 0 & 3 & -7 & -19 \\
0 & 0 & -\frac{5}{4} & -9 & -\frac{159}{4}
\end{bmatrix}
\tag{5-30}
$$

Note that the starting matrix (left side of Eqs. 5-30) for step 2 of the forward pass is the same as the ending matrix (right side of Eqs. 5-28) from step 1. The resultant matrix in Eqs. 5-30 has the same form as the matrix of Eqs. 5-29, which was the objective of the operations in this step.

In steps 3 and 4 of the forward pass, the third and fourth rows are reduced using the same procedure. The third row is normalized by dividing the row by the pivot element, f_{33}, in the previous Eq. 5-29. The revised third row is then used to eliminate f_{43} from row 4. Row 4 is then normalized by dividing the row by its pivot element. The following matrices illustrate these operations:

Step 3:

Starting Matrix   *Operation*   *Resultant Matrix*

$$
\begin{bmatrix}
1 & \frac{3}{2} & -1 & -\frac{1}{2} & -1 \\
0 & 1 & -\frac{1}{2} & 1 & \frac{9}{2} \\
0 & 0 & 3 & -7 & -19 \\
0 & 0 & -\frac{5}{4} & -9 & -\frac{159}{4}
\end{bmatrix}
\begin{array}{l}
R'_1 = R_1 \\
R'_2 = R_2 \\
R'_3 = \dfrac{R_3}{3} \\
R'_4 = R_4 + \dfrac{5}{4}R'_3
\end{array}
\begin{bmatrix}
1 & \frac{3}{2} & -1 & -\frac{1}{2} & -1 \\
0 & 1 & -\frac{1}{2} & 1 & \frac{9}{2} \\
0 & 0 & 1 & -\frac{7}{3} & -\frac{19}{3} \\
0 & 0 & 0 & -\frac{143}{12} & -\frac{572}{12}
\end{bmatrix}
\tag{5-31}
$$

Step 4:

Starting Matrix   *Operation* *Resultant Matrix*

$$
\begin{bmatrix}
1 & \frac{3}{2} & -1 & -\frac{1}{2} & -1 \\
0 & 1 & -\frac{1}{2} & 1 & \frac{9}{2} \\
0 & 0 & 1 & -\frac{7}{3} & -\frac{19}{3} \\
0 & 0 & 0 & -\frac{143}{12} & -\frac{572}{12}
\end{bmatrix}
\begin{array}{l}
R'_1 = R_1 \\
R'_2 = R_2 \\
R'_3 = R_3 \\
R'_4 = \dfrac{R_4}{-143/12}
\end{array}
\begin{bmatrix}
1 & \frac{3}{2} & -1 & -\frac{1}{2} & -1 \\
0 & 1 & -\frac{1}{2} & 1 & \frac{9}{2} \\
0 & 0 & 1 & -\frac{7}{3} & -\frac{19}{3} \\
0 & 0 & 0 & 1 & \frac{572}{143}
\end{bmatrix}
\tag{5-32}
$$

The resultant matrix in Eq. 5-32 represents a system of four equations with four unknowns:

$$X_1 + \frac{3}{2}X_2 - X_3 - \frac{1}{2}X_4 = -1 \tag{5-33a}$$

$$X_2 - \frac{1}{2}X_3 + X_4 = \frac{9}{2} \tag{5-33b}$$

$$X_3 - \frac{7}{3}X_4 = -\frac{19}{3} \tag{5-33c}$$

$$X_4 = \frac{572}{143} = 4 \tag{5-33d}$$

Equations 5-33 represent the results from the forward pass in the Gaussian elimination method. The back substitution procedure can now be applied to these equations to solve for the unknown X_i values.

In step 1 of the backward substitution, the objective is to eliminate the known value for X_4 from the other three equations. In matrix form, this step can be expressed as

$$
\begin{bmatrix}
1 & \frac{3}{2} & -1 & -\frac{1}{2} & -1 \\
0 & 1 & -\frac{1}{2} & 1 & \frac{9}{2} \\
0 & 0 & 1 & -\frac{7}{3} & -\frac{19}{3} \\
0 & 0 & 0 & 1 & 4
\end{bmatrix}
\quad
\begin{matrix}
R_1' = R_1 + \frac{R_4'}{2} \\
R_2' = R_2 - R_4' \\
R_3' = R_3 + \frac{7R_4'}{3} \\
R_4' = R_4
\end{matrix}
\quad
\begin{bmatrix}
1 & \frac{3}{2} & -1 & 0 & 1 \\
0 & 1 & -\frac{1}{2} & 0 & \frac{1}{2} \\
0 & 0 & 1 & 0 & 3 \\
0 & 0 & 0 & 1 & 4
\end{bmatrix}
\tag{5-34}
$$

Row 3 of the revised matrix represents the equation $X_3 = 3$. Given the value for X_3, it can be removed from the first and second rows of the matrix; this process represents step 2 of the backward substitution analysis, which has the objective of obtaining zeros in the first and second rows in column 3:

$$
\begin{bmatrix}
1 & \frac{3}{2} & -1 & 0 & 1 \\
0 & 1 & -\frac{1}{2} & 0 & \frac{1}{2} \\
0 & 0 & 1 & 0 & 3 \\
0 & 0 & 0 & 1 & 4
\end{bmatrix}
\quad
\begin{matrix}
R_1' = R_1 + R_3' \\
R_2' = R_2 + \frac{R_3'}{2} \\
R_3' = R_3 \\
R_4' = R_4
\end{matrix}
\quad
\begin{bmatrix}
1 & \frac{3}{2} & 0 & 0 & 4 \\
0 & 1 & 0 & 0 & 2 \\
0 & 0 & 1 & 0 & 3 \\
0 & 0 & 0 & 1 & 4
\end{bmatrix}
\tag{5-35}
$$

Step 2 of the backward substitution yields $X_2 = 2$. The solution is completed by eliminating X_2 from the first equation as step 3:

$$
\begin{bmatrix}
1 & \frac{3}{2} & 0 & 0 & 4 \\
0 & 1 & 0 & 0 & 2 \\
0 & 0 & 1 & 0 & 3 \\
0 & 0 & 0 & 1 & 4
\end{bmatrix}
\quad
\begin{matrix}
R_1' = R_1 - \frac{3}{2}R_2' \\
R_2' = R_2 \\
R_3' = R_3 \\
R_4' = R_4
\end{matrix}
\quad
\begin{bmatrix}
1 & 0 & 0 & 0 & 1 \\
0 & 1 & 0 & 0 & 2 \\
0 & 0 & 1 & 0 & 3 \\
0 & 0 & 0 & 1 & 4
\end{bmatrix}
\tag{5-36}
$$

This completes the back substitution phase of the Gaussian elimination solution procedure; the solution to the original simultaneous equations of Eq. 5-25 is $X_1 = 1$, $X_2 = 2$, $X_3 = 3$, and $X_4 = 4$.

Example 5-4 Gaussian Elimination for Material Purchasing

A problem of material purchasing of ores was discussed in the introduction to this chapter. The statement of the problem was transformed into a solution of three linear equations with three unknowns (Eqs. 5-3). The equations can be expressed in matrix form, with the forward elimination steps as follows:

Forward Elimination: Step 1

$$
\begin{bmatrix} 1 & 3 & 2 & 15 \\ 2 & 4 & 3 & 22 \\ 3 & 4 & 7 & 39 \end{bmatrix}
\quad
\begin{matrix} R_1' = R_1 \\ R_2' = R_2 - 2R_1' \\ R_3' = R_3 - 3R_1' \end{matrix}
\quad
\begin{bmatrix} 1 & 3 & 2 & 15 \\ 0 & -2 & -1 & -8 \\ 0 & -5 & 1 & -6 \end{bmatrix}
\tag{5-37a}
$$

Forward Elimination: Step 2

$$
\begin{bmatrix} 1 & 3 & 2 & 15 \\ 0 & -2 & -1 & -8 \\ 0 & -5 & 1 & -6 \end{bmatrix}
\quad
\begin{matrix} R_1' = R_1 \\ R_2' = \frac{R_2}{-2} \\ R_3' = R_3 + 5R_2' \end{matrix}
\quad
\begin{bmatrix} 1 & 3 & 2 & 15 \\ 0 & 1 & \frac{1}{2} & 4 \\ 0 & 0 & \frac{7}{2} & 14 \end{bmatrix}
\tag{5-37b}
$$

Forward Elimination: Step 3

$$
\begin{bmatrix} 1 & 3 & 2 & 15 \\ 0 & 1 & \frac{1}{2} & 4 \\ 0 & 0 & \frac{7}{2} & 14 \end{bmatrix}
\quad
\begin{matrix} R_1' = R_1 \\ R_2' = R_2 \\ R_3' = \frac{2R_3}{7} \end{matrix}
\quad
\begin{bmatrix} 1 & 3 & 2 & 15 \\ 0 & 1 & \frac{1}{2} & 4 \\ 0 & 0 & 1 & 4 \end{bmatrix}
\tag{5-37c}
$$

The last matrix represents the following set of equations:

$$
X_1 + 3X_2 + 2X_3 = 15
\tag{5-38a}
$$

$$
X_2 + \frac{1}{2}X_3 = 4
\tag{5-38b}
$$

$$
X_3 = 4
\tag{5-38c}
$$

The following backward substitution process can be used to obtain values for the unknowns:

Backward Substitution: Step 1

$$
\begin{bmatrix} 1 & 3 & 2 & 15 \\ 0 & 1 & \frac{1}{2} & 4 \\ 0 & 0 & 1 & 4 \end{bmatrix}
\quad
\begin{matrix} R_1' = R_1 - 2R_3' \\ R_2' = R_2 - \frac{1}{2}R_3' \\ R_3' = R_3 \end{matrix}
\quad
\begin{bmatrix} 1 & 3 & 0 & 7 \\ 0 & 1 & 0 & 2 \\ 0 & 0 & 1 & 4 \end{bmatrix}
\tag{5-39a}
$$

Backward Substitution: Step 2

$$
\begin{bmatrix} 1 & 3 & 0 & 7 \\ 0 & 1 & 0 & 2 \\ 0 & 0 & 1 & 4 \end{bmatrix}
\quad
\begin{matrix} R_1' = R_1 - 3R_2' \\ R_2' = R_2 \\ R_3' = R_3 \end{matrix}
\quad
\begin{bmatrix} 1 & 0 & 0 & 1 \\ 0 & 1 & 0 & 2 \\ 0 & 0 & 1 & 4 \end{bmatrix}
\tag{5-39b}
$$

Thus the manufacturer should purchase ore from the three suppliers in the ratio of 1, 2, and 4.

Example 5-5 Gaussian Elimination for an Electrical Circuit

In the introduction to this chapter, the formulation of a set of simultaneous equations was illustrated using an electrical circuit with two loops (Eqs. 5–7). The resulting equations can be placed in matrix form and solved using the following Gaussian elimination:

Forward Pass: Step 1

$$\begin{bmatrix} 1 & 1 & -1 & 0 \\ 2 & 0 & 5 & 6 \\ 2 & -4 & 0 & 4 \end{bmatrix} \quad \begin{matrix} R_1' = R_1 \\ R_2' = R_2 - 2R_1' \\ R_3' = R_3 - 2R_1' \end{matrix} \quad \begin{bmatrix} 1 & 1 & -1 & 0 \\ 0 & -2 & 7 & 6 \\ 0 & -6 & 2 & 4 \end{bmatrix} \quad \text{(5-40a)}$$

Forward Pass: Step 2

$$\begin{bmatrix} 1 & 1 & -1 & 0 \\ 0 & -2 & 7 & 6 \\ 0 & -6 & 2 & 4 \end{bmatrix} \quad \begin{matrix} R_1' = R_1 \\ R_2' = \frac{R_2}{-2} \\ R_3' = R_3 + 6R_2' \end{matrix} \quad \begin{bmatrix} 1 & 1 & -1 & 0 \\ 0 & 1 & -\frac{7}{2} & -3 \\ 0 & 0 & -19 & -14 \end{bmatrix} \quad \text{(5-40b)}$$

Forward Pass: Step 3

$$\begin{bmatrix} 1 & 1 & -1 & 0 \\ 0 & 1 & -\frac{7}{2} & -3 \\ 0 & 0 & -19 & -14 \end{bmatrix} \quad \begin{matrix} R_1' = R_1 \\ R_2' = R_2 \\ R_3' = \frac{R_3}{-19} \end{matrix} \quad \begin{bmatrix} 1 & 1 & -1 & 0 \\ 0 & 1 & -\frac{7}{2} & -3 \\ 0 & 0 & 1 & \frac{14}{19} \end{bmatrix} \quad \text{(5-40c)}$$

Backward Substitution: Step 1

$$\begin{bmatrix} 1 & 1 & -1 & 0 \\ 0 & 1 & -\frac{7}{2} & -3 \\ 0 & 0 & 1 & \frac{14}{19} \end{bmatrix} \quad \begin{matrix} R_1' = R_1 + R_3' \\ R_2' = R_2 + \frac{7}{2}R_3' \\ R_3' = R_3 \end{matrix} \quad \begin{bmatrix} 1 & 1 & 0 & \frac{14}{19} \\ 0 & 1 & 0 & -\frac{8}{19} \\ 0 & 0 & 1 & \frac{14}{19} \end{bmatrix} \quad \text{(5-40d)}$$

Backward Substitution: Step 2

$$\begin{bmatrix} 1 & 1 & 0 & \frac{14}{19} \\ 0 & 1 & 0 & -\frac{8}{19} \\ 0 & 0 & 1 & \frac{14}{19} \end{bmatrix} \quad \begin{matrix} R_1' = R_1 - R_2' \\ R_2' = R_2 \\ R_3' = R_3 \end{matrix} \quad \begin{bmatrix} 1 & 0 & 0 & \frac{22}{19} \\ 0 & 1 & 0 & -\frac{8}{19} \\ 0 & 0 & 1 & \frac{14}{19} \end{bmatrix} \quad \text{(5-40e)}$$

The solution can be verified by substituting the solution $\left(\text{i.e., } I_1 = \frac{22}{19}, I_2 = -\frac{8}{19}, \text{ and } I_3 = \frac{14}{19}\right)$ into the original set of three simultaneous equations. The negative value of I_2 indicates that the assumed direction of flow was incorrect and that the current flows in a clockwise direction (that is, from c to e to f to d).

5.3 GAUSS–JORDAN ELIMINATION

The Gaussian elimination procedure requires a forward pass to transform the coefficient matrix into an *upper-triangle* form. That is, the coefficient matrix was transformed to a matrix having all terms below the principal diagonal equal to zero. After the upper-triangular coefficient matrix of Eq. 5-21 was obtained, the back substitution pass solved for the unknowns. An alternative process of elimination in which *all* coefficients in a column except for the pivot element are eliminated can also be used to obtain a solution. This alternative procedure is referred to as Gauss–Jordan elimination.

In Gauss–Jordan elimination, the solution is obtained directly after the forward pass; there is no back substitution phase. While the Gauss–Jordan method will involve more computational effort for solving a single problem, it is more efficient when we have several right-side vectors with the coefficient C_i that we need to solve for the same coefficient matrix a_{ij}.

Whereas the rows above the pivot row are not modified during the forward substitution of Gauss elimination, they are transformed in the Gauss–Jordan elimination. The reduction of the rows above the pivot row in the Gauss–Jordan scheme follows the same procedure used to eliminate values from rows below the pivot row in the regular Gauss elimination. The following steps show the Gauss–Jordan procedure for the system of equations of Eqs. 5-25:

Step 1

Original Matrix

$$\begin{bmatrix} 2 & 3 & -2 & -1 & -2 \\ 2 & 5 & -3 & 1 & 7 \\ -2 & 1 & 3 & -2 & 1 \\ -5 & 2 & -1 & 3 & 8 \end{bmatrix}$$

Operation

$$R'_1 = \frac{R_1}{2}$$
$$R'_2 = R_2 - 2R'_1$$
$$R'_3 = R_3 + 2R'_1$$
$$R'_4 = R_4 + 5R'_1$$

Resultant Matrix

$$\begin{bmatrix} 1 & \frac{3}{2} & -1 & -\frac{1}{2} & -1 \\ 0 & 2 & -1 & 2 & 9 \\ 0 & 4 & 1 & -3 & -1 \\ 0 & \frac{19}{2} & -6 & \frac{1}{2} & 3 \end{bmatrix}$$

(5-41a)

Step 2

Starting Matrix

$$\begin{bmatrix} 1 & \frac{3}{2} & -1 & -\frac{1}{2} & -1 \\ 0 & 2 & -1 & 2 & 9 \\ 0 & 4 & 1 & -3 & -1 \\ 0 & \frac{19}{2} & -6 & \frac{1}{2} & 3 \end{bmatrix}$$

Operation

$$R'_1 = R_1 - \frac{3}{2}R'_2$$
$$R'_2 = \frac{R_2}{2}$$
$$R'_3 = R_3 - 4R'_2$$
$$R'_4 = R_4 - \frac{19}{2}R'_2$$

Resultant Matrix

$$\begin{bmatrix} 1 & 0 & -\frac{1}{4} & -2 & -\frac{31}{4} \\ 0 & 1 & -\frac{1}{2} & 1 & \frac{9}{2} \\ 0 & 0 & 3 & -7 & -19 \\ 0 & 0 & -\frac{5}{4} & -9 & -\frac{159}{4} \end{bmatrix}$$

(5-41b)

Step 3

Starting Matrix

$$\begin{bmatrix} 1 & 0 & -\frac{1}{4} & -2 & -\frac{31}{4} \\ 0 & 1 & -\frac{1}{2} & 1 & \frac{9}{2} \\ 0 & 0 & 3 & -7 & -19 \\ 0 & 0 & -\frac{5}{4} & -9 & -\frac{159}{4} \end{bmatrix}$$

Operation

$$R'_1 = R_1 + \frac{1}{4}R'_3$$
$$R'_2 = R_2 + \frac{1}{2}R'_3$$
$$R'_3 = \frac{R_3}{3}$$
$$R'_4 = R_4 + \frac{5}{4}R'_3$$

Resultant Matrix

$$\begin{bmatrix} 1 & 0 & 0 & -\frac{31}{12} & -\frac{112}{12} \\ 0 & 1 & 0 & -\frac{1}{6} & \frac{4}{3} \\ 0 & 0 & 1 & -\frac{7}{3} & -\frac{19}{3} \\ 0 & 0 & 0 & -\frac{143}{12} & -\frac{572}{12} \end{bmatrix}$$

(5-41c)

Step 4

Starting Matrix	Operation	Resultant Matrix

$$
\begin{bmatrix}
1 & 0 & 0 & -\frac{31}{12} & -\frac{112}{12} \\
0 & 1 & 0 & -\frac{1}{6} & \frac{4}{3} \\
0 & 0 & 1 & -\frac{7}{3} & -\frac{19}{3} \\
0 & 0 & 0 & -\frac{143}{12} & -\frac{572}{12}
\end{bmatrix}
\quad
\begin{aligned}
R_1' &= R_1 + \tfrac{31}{12}R_4' \\
R_2' &= R_2 + \tfrac{1}{6}R_4' \\
R_3' &= R_3 + \tfrac{7}{3}R_4' \\
R_4' &= \frac{R_4}{-143/12}
\end{aligned}
\quad
\begin{bmatrix}
1 & 0 & 0 & 0 & 1 \\
0 & 1 & 0 & 0 & 2 \\
0 & 0 & 1 & 0 & 3 \\
0 & 0 & 0 & 1 & 4
\end{bmatrix}
\qquad (5\text{-}41d)
$$

In each of the four matrix transformations, the pivot row is normalized first and then used to reduce the other rows. For example, rows 1, 2, 3, and 4 were first normalized in the matrices of Eqs. 5-41a, b, c, and d, respectively, and then used to reduce the other rows.

Example 5-6 Gauss–Jordan Elimination for Material Purchasing

Gauss–Jordan elimination can be used to find the solution to the three simultaneous equations for the material purchasing example formulated in the introduction to this chapter, that is, Eqs. 5-3. The equations are placed in matrix form and solved as follows:

Step 1

$$
\begin{bmatrix}
1 & 3 & 2 & 15 \\
2 & 4 & 3 & 22 \\
3 & 4 & 7 & 39
\end{bmatrix}
\quad
\begin{aligned}
R_1' &= R_1 \\
R_2' &= R_2 - 2R_1' \\
R_3' &= R_3 - 3R_1'
\end{aligned}
\quad
\begin{bmatrix}
1 & 3 & 2 & 15 \\
0 & -2 & -1 & -8 \\
0 & -5 & 1 & -6
\end{bmatrix}
\qquad (5\text{-}42a)
$$

Step 2

$$
\begin{bmatrix}
1 & 3 & 2 & 15 \\
0 & -2 & -1 & -8 \\
0 & -5 & 1 & -6
\end{bmatrix}
\quad
\begin{aligned}
R_1' &= R_1 - 3R_2' \\
R_2' &= \frac{R_2}{-2} \\
R_3' &= R_3 + 5R_2'
\end{aligned}
\quad
\begin{bmatrix}
1 & 0 & \frac{1}{2} & 3 \\
0 & 1 & \frac{1}{2} & 4 \\
0 & 0 & \frac{7}{2} & 14
\end{bmatrix}
\qquad (5\text{-}42b)
$$

Step 3

$$
\begin{bmatrix}
1 & 0 & \frac{1}{2} & 3 \\
0 & 1 & \frac{1}{2} & 4 \\
0 & 0 & \frac{7}{2} & 14
\end{bmatrix}
\quad
\begin{aligned}
R_1' &= R_1 - \tfrac{1}{2}R_3' \\
R_2' &= R_2 - \tfrac{1}{2}R_3' \\
R_3' &= \frac{R_3}{-7/2}
\end{aligned}
\quad
\begin{bmatrix}
1 & 0 & 0 & 1 \\
0 & 1 & 0 & 2 \\
0 & 0 & 1 & 4
\end{bmatrix}
\qquad (5\text{-}42c)
$$

The solution of $X_1 = 1$, $X_2 = 2$, and $X_3 = 4$ indicates that the ore should be purchased from suppliers 1, 2, and 3 in the ratios of 1 : 2 : 4. This material purchase policy will result in the optimum composition with no waste.

5.4 ADDITIONAL CONSIDERATIONS FOR ELIMINATION PROCEDURES

5.4.1 Accumulated Round-off Errors

An important consideration in solving systems of simultaneous equations using either Gaussian or Gauss–Jordan elimination is a loss of precision caused by accumulated round-off errors in the numerical calculations. Problems with round-off and truncation are most likely to occur when the coefficients in the equations differ by several orders of magnitude. In the example of Eq. 5-25, all the coefficients were of the same order of magnitude; thus the resulting solution was not affected by round-off and the solution was *exact*, at least within the finite precision of the numerical calculations.

We can minimize round-off problems by rearranging the equations such that the largest coefficient in each equation is placed on the principal diagonal of the coefficient matrix, using one of the permissible operations that allows equations to be rearranged without changing the solution. Additionally, the equations should be ordered such that the equation having the largest pivot is reduced first, followed by the equation having the next largest pivot, and so on. These manipulations of equations have been shown by experience to minimize the effect of accumulated round-off errors on the solution. These manipulations in pivoting are called partial pivoting.

5.4.2 Zero Pivot Element

In many problems, one or more of the coefficients will equal zero in the initial matrix. If any of the pivot elements equal zero, a *division by zero* error will occur when that pivot equation is normalized during the forward pass of the elimination procedure. However, in most cases the equations can be rearranged to eliminate the division by zero. The fact that a matrix begins with a zero element on the principal diagonal does not mean that a solution does not exist.

In some cases, the system of equations may include linearly dependent equations, that is, equations that are a linear transformation or combination of one or more other equations in the system. This condition produces what is termed a *singular* coefficient matrix. In this case, one or more of the pivot elements in the system will always equal zero regardless of how the equations are written or rearranged. For example, the following pair of equations is singular:

$$2X_1 + 3X_2 = 7 \tag{5-43a}$$

$$4X_1 + 6X_2 = 14 \tag{5-43b}$$

The matrix solution is

$$\begin{bmatrix} 2 & 3 & 7 \\ 4 & 6 & 14 \end{bmatrix} \quad \begin{matrix} R_1' = \frac{R_1}{2} \\ R_2' = R_2 - 4R_1' \end{matrix} \quad \begin{bmatrix} 1 & \frac{3}{2} & \frac{7}{2} \\ 0 & 0 & 0 \end{bmatrix} \tag{5-44}$$

Thus there is more than one solution to the equations. In fact, any pair of values of X_1 and X_2 that satisfies either equation is a valid solution. For example, each of the following three pairs of values satisfies Eqs. 5-43: $(2, 1)$, $\left(\frac{1}{2}, 2\right)$, and $\left(0, \frac{7}{3}\right)$.

5.4.3 Considerations in Programming

Computers are most efficient when repetition exists in the solution procedure. It should be evident from Eqs. 5-28, 5-30 through 5-32, and 5-34 through 5-36 that the procedure underlying Gaussian elimination (and also Gauss–Jordan elimination) involves much repetition. In the forward pass, the algorithm loops over all the rows, with each row successively being designated as the pivot row. Operations are not performed on rows above the pivot row. The elements of the pivot row are normalized by dividing the row by the pivot element a_{ii}. The rows below the pivot row are then reduced so that the column elements below the pivot are zero. The forward pass can thus be programmed using the following steps:

1. Loop over each row i, making each row i in turn the pivot row.
2. Normalize the elements of the pivot row (row i) by dividing each element in the row by a_{ii} as follows:

$$a_{ij} = \frac{a_{ij}}{a_{ii}} \qquad \text{for } j = (i + 1), (i + 2), \ldots, (n + 1) \tag{5-45a}$$

$$C_i = \frac{C_i}{a_{ii}} \tag{5-45b}$$

$$a_{ii} = 1 \tag{5-45c}$$

3. Loop over rows $(i + 1)$ to n below the pivot row and reduce the elements in each row as follows:

$$a_{kj} = a_{kj} - a_{ki}a_{ij} \qquad \text{for } j = i, \ldots, n \tag{5-45d}$$

$$C_k = C_k - a_{ki}C_i \qquad \text{for } k = (i + 1), \ldots, n \tag{5-45e}$$

The back substitution loops over the results from the forward pass, working from the last equation to the first as follows:

1. For the last row n:

$$X_n = C_n \tag{5-45f}$$

2. For rows $(n - 1)$ through 1,

$$X_i = C_i - \sum_{j=i+1}^{n} a_{ij}X_j \qquad \text{for } j = (i + 1), \ldots, n \text{ and } i = (n - 1), \ldots, 1 \tag{5-45g}$$

The outlined procedure in Eqs. 5-45 can be programmed. In developing a computer program for this method, the programmer needs to consider the recommendations in Sections 5.4.1 and 5.4.2.

5.5 LU DECOMPOSITION

5.5.1 General Case

As discussed in Section 5.2, the Gaussian elimination method consists of two stages: forward pass and back substitution. The forward pass results in the upper triangular matrix given by Eq. 5-21. This matrix can be denoted as **U** and is given by

$$\mathbf{U} = \begin{bmatrix} 1 & u_{12} & u_{13} & \cdots & u_{1n} \\ 0 & 1 & u_{23} & \cdots & u_{2n} \\ 0 & 0 & 1 & \cdots & u_{3n} \\ \vdots & \vdots & \vdots & \vdots & \vdots \\ 0 & 0 & 0 & \cdots & 1 \end{bmatrix} \tag{5-46}$$

The values of this matrix can be related to the values in Eq. 5-21 as

$$u_{ij} = d_{ij} \quad \text{for } i = 1, 2, \ldots, n \text{ and } j = 1, 2, \ldots, n \tag{5-47}$$

This upper-triangular matrix **U** can be related to the coefficient matrix **A** as

$$\mathbf{LU} = \mathbf{A} \tag{5-48}$$

where **L** is a lower-triangular matrix that can be viewed to represent the back substitution aspect of the Gaussian elimination method. The general expression for **L** is

$$\mathbf{L} = \begin{bmatrix} l_{11} & 0 & 0 & \cdots & 0 \\ l_{21} & l_{22} & 0 & \cdots & 0 \\ l_{31} & l_{32} & l_{33} & \cdots & 0 \\ \vdots & \vdots & \vdots & \vdots & \vdots \\ l_{n1} & l_{n2} & l_{n3} & \cdots & l_{nn} \end{bmatrix} \tag{5-49}$$

Therefore, Eq. 5-48 can be expressed as

$$\begin{bmatrix} l_{11} & 0 & 0 & \cdots & 0 \\ l_{21} & l_{22} & 0 & \cdots & 0 \\ l_{31} & l_{32} & l_{33} & \cdots & 0 \\ \vdots & \vdots & \vdots & \vdots & \vdots \\ l_{n1} & l_{n2} & l_{n3} & \cdots & l_{nn} \end{bmatrix} \begin{bmatrix} 1 & u_{12} & u_{13} & \cdots & u_{1n} \\ 0 & 1 & u_{23} & \cdots & u_{2n} \\ 0 & 0 & 1 & \cdots & u_{3n} \\ \vdots & \vdots & \vdots & \vdots & \vdots \\ 0 & 0 & 0 & \cdots & 1 \end{bmatrix} = \begin{bmatrix} a_{11} & a_{12} & a_{13} & \cdots & a_{1n} \\ a_{21} & a_{22} & a_{23} & \cdots & a_{2n} \\ a_{31} & a_{32} & a_{33} & \cdots & a_{3n} \\ \vdots & \vdots & \vdots & \vdots & \vdots \\ a_{n1} & a_{n2} & a_{n3} & \cdots & a_{nn} \end{bmatrix} \tag{5-50}$$

Equation 5-50 states that the coefficient matrix **A** can be decomposed to two triangular matrices **L** and **U**. These matrices can be determined, without use of the Gaussian elimination method, by performing the matrix multiplication **LU** in Eq. 5-50 and equating the resulting terms with the corresponding coefficients in the matrix **A**. For example, the multiplication of the first row of **L** with the first column of **U** results in the following value that is equal to a_{11}:

$$l_{11}(1) = a_{11} \tag{5-51}$$

Performing other multiplication operations of rows and columns produces the following for the elements' values of **L** and **U**:

$$l_{i1} = a_{i1} \qquad \text{for } i = 1, 2, \ldots, n \tag{5-52a}$$

$$u_{1j} = \frac{a_{1j}}{l_{11}} \qquad \text{for } j = 2, 3, \ldots, n \tag{5-52b}$$

$$l_{ij} = a_{ij} - \sum_{k=1}^{j-1} l_{ik}u_{kj} \qquad \text{for } j = 2, 3, \ldots, n-1 \text{ and } i = j, j+1, \ldots, n \qquad (5\text{-}52\text{c})$$

$$u_{ji} = \frac{a_{ji} - \sum_{k=1}^{j-1} l_{jk}u_{ki}}{l_{jj}} \qquad \text{for } j = 2, 3, \ldots, n-1 \text{ and } i = j+1, j+2, \ldots, n \qquad (5\text{-}52\text{d})$$

and

$$l_{nn} = a_{nn} - \sum_{k=1}^{n-1} l_{nk}u_{kn} \qquad (5\text{-}52\text{e})$$

Once the matrix $\mathbf{A}$ is decomposed into $\mathbf{L}$ and $\mathbf{U}$, these matrices can be used to solve a system of equations with constants C_i, $i = 1, 2, \ldots, n$. The solution can be obtained in two steps, forward pass and back substitution. The forward substitution produces the e_i values, $i = 1, 2, \ldots, n$, of Eq. 5-21 as follows:

$$e_1 = \frac{C_1}{l_{11}} \qquad (5\text{-}53\text{a})$$

$$e_i = \frac{C_i - \sum_{j=1}^{i-1} l_{ij}e_j}{l_{ii}} \qquad \text{for } i = 2, 3, \ldots, n \qquad (5\text{-}53\text{b})$$

The back substitution results in the X_i values, $i = 1, 2, \ldots, n$, as follows:

$$X_n = e_n \qquad (5\text{-}54\text{a})$$

$$X_i = e_i - \sum_{j=i+1}^{n} u_{ij}X_j \qquad \text{for } i = n-1, n-2, \ldots, 1 \qquad (5\text{-}54\text{b})$$

The considerations discussed in Section 5.4 for partial pivoting are applicable for **LU** decomposition and should be used. Since partial pivoting results in rearranging the rows of the $\mathbf{A}$ matrix, the constants in the $\mathbf{C}$ vector need to be arranged to maintain their correspondence with the rows of $\mathbf{A}$. The rearranged $\mathbf{C}$ vector should be used in Eqs. 5-53 and 5-54 to obtain the solution of a system of linear equations.

Example 5-7 LU Decomposition for Material Purchasing

The **LU** decomposition method can be applied to the three simultaneous equations for the material purchasing example formulated according to Eqs. 5-3. The coefficient matrix is given by

$$\mathbf{A} = \begin{bmatrix} 1 & 3 & 2 \\ 2 & 4 & 3 \\ 3 & 4 & 7 \end{bmatrix} \qquad (5\text{-}55)$$

Using Eqs. 5-52, the **L** and **U** matrices can be obtained. The first column of the **L** matrix can be computed using Eq. 5-52a as follows:

$$l_{11} = a_{11} = 1 \tag{5-56a}$$

$$l_{21} = a_{21} = 2 \tag{5-56b}$$

$$l_{31} = a_{31} = 3 \tag{5-56c}$$

Equation 5-52b can be used to compute the first row of the **U** matrix as

$$u_{12} = \frac{a_{12}}{l_{11}} = \frac{3}{1} = 3 \tag{5-57a}$$

$$u_{13} = \frac{a_{13}}{l_{11}} = \frac{2}{1} = 2 \tag{5-57b}$$

The second column of the **L** matrix can now be computed using Eq. 5-52c as follows:

$$l_{22} = a_{22} - \sum_{k=1}^{2-1} l_{2k}u_{k2} = 4 - 2(3) = -2 \tag{5-58a}$$

$$l_{32} = a_{32} - \sum_{k=1}^{2-1} l_{3k}u_{k2} = 4 - 3(3) = -5 \tag{5-58b}$$

Using Eq. 5-52d, the value of the last element of **U** can be computed as

$$u_{23} = \frac{a_{23} - \sum_{k=1}^{2-1} l_{2k}u_{k3}}{l_{22}} = \frac{3 - 2(2)}{-2} = 0.5 \tag{5-59}$$

The last element of the **L** matrix can be computed using Eq. 5-52e as

$$l_{33} = a_{33} - \sum_{k=1}^{3-1} l_{3k}u_{k3} = 7 - (3)(2) - (-5)(0.5) = 3.5 \tag{5-60}$$

Therefore, the **L** and **U** matrices are

$$\mathbf{L} = \begin{bmatrix} 1 & 0 & 0 \\ 2 & -2 & 0 \\ 3 & -5 & 3.5 \end{bmatrix} \tag{5-61a}$$

$$\mathbf{U} = \begin{bmatrix} 1 & 3 & 2 \\ 0 & 1 & 0.5 \\ 0 & 0 & 1 \end{bmatrix} \tag{5-61b}$$

The forward substitution can be performed using Eq. 5-53 as follows:

$$e_1 = \frac{C_1}{l_{11}} = \frac{15}{1} = 15 \tag{5-62a}$$

$$e_2 = \frac{C_2 - \sum\limits_{j=1}^{2-1} l_{2j} e_j}{l_{22}} = \frac{22 - 2(15)}{-2} = 4 \tag{5-62b}$$

$$e_3 = \frac{C_3 - \sum\limits_{j=1}^{3-1} l_{3j} e_j}{l_{33}} = \frac{39 - 3(15) - (-5)(4)}{3.5} = 4 \tag{5-62c}$$

The back substitution can be performed using Eq. 5-54 as follows:

$$X_3 = e_3 = 4 \tag{5-63a}$$

$$X_2 = e_2 - \sum\limits_{j=2+1}^{3} u_{2j} X_j = 4 - 0.5(4) = 2 \tag{5-63b}$$

$$X_1 = e_1 - \sum\limits_{j=1+1}^{3} u_{1j} X_j = 15 - 3(2) - (2)(4) = 1 \tag{5-63c}$$

The result obtained with the **LU** decomposition method is the same as what was previously obtained.

5.5.2 Banded Matrices

In Section 2.1.3, a banded matrix was defined as a square matrix with elements of zero except on the principal diagonal and the values in the positions adjacent to the principal diagonal. For example, a tridiagonal matrix is the special case of a banded matrix that has zeros except in the three diagonals as follows:

$$\begin{bmatrix} a_{11} & a_{12} & 0 & 0 & 0 \\ a_{21} & a_{22} & a_{23} & 0 & 0 \\ 0 & a_{32} & a_{33} & a_{34} & 0 \\ 0 & 0 & a_{43} & a_{44} & a_{45} \\ 0 & 0 & 0 & a_{54} & a_{55} \end{bmatrix} \tag{5-64}$$

This matrix is described to have a band width of 3, in reference to the three nonzero diagonals. Alternatively, the matrix is described to have a half-band width of 1, in reference to the number of nonzero diagonals on one side of the principal diagonal, that is, not including the diagonal where $i = j$ for a_{ij}. Therefore, the following relationship between the band width (b_{w1}) and half-band width (b_{w2}) can be obtained:

$$b_{w1} = 2b_{w2} + 1 \tag{5-65}$$

The elimination methods previously described can be used to solve simultaneous equations with banded coefficient matrices. The efficiency of these methods can be enhanced by recognizing the nonnecessity of pivoting for the zero elements that do not fall within the bands of the matrices. The computational algorithm can, therefore, be developed to reduce the number of pivotings by performing the pivotings for element a_{ij} within the bands. Also, banded matrices can be stored more efficiently by storing the banded elements only, thereby reducing storage requirements for a solution.

5.5.3 Symmetric Matrices

A symmetric square matrix is defined as a matrix where $a_{ij} = a_{ji}$. In engineering, it is common to deal with symmetric matrices. For example, the stiffness properties of a structure can be described using a symmetric stiffness matrix, and correlations among the structural variables can be described using a symmetric correlation matrix. A symmetric matrix $\mathbf{A}$ has the following property:

$$\mathbf{A} = \mathbf{A}^T \tag{5-66}$$

where $\mathbf{A}^T$ = the transpose of the matrix. For a symmetric matrix, the decomposition of $\mathbf{A}$ can be expressed as

$$\mathbf{A} = \mathbf{LL}^T \tag{5-67}$$

As a result, the $\mathbf{LU}$ decomposition can be computed more effectively using a recurrence method called the *Cholesky decomposition method*. According to this method, l_{11} is given by

$$l_{11} = \sqrt{a_{11}} \tag{5-68a}$$

The rest of the ith row of the $\mathbf{L}$ matrix is

$$l_{ii} = \sqrt{a_{ii} - \sum_{k=1}^{i-1} l_{ik}^2} \qquad \text{for } i = 2, 3, \ldots, n \tag{5-68b}$$

$$l_{ij} = \frac{a_{ij} - \sum_{k=1}^{j-1} l_{ik} l_{jk}}{l_{jj}} \qquad \text{for } j = 2, 3, \ldots, i-1, \text{ and } j < i \tag{5-69}$$

These equations can be used to replace Eqs. 5-52 in the $\mathbf{LU}$ decomposition method.

Example 5-8 Cholesky Decomposition Method

The computational aspects of the Cholesky decomposition method are illustrated in this example using the following symmetric matrix:

$$\mathbf{A} = \begin{bmatrix} 1 & 2 & 3 \\ 2 & 8 & 10 \\ 3 & 10 & 22 \end{bmatrix} \tag{5-70}$$

Using Eq. 5-69, the first row ($i = 1$) can be computed as follows:

$$l_{11} = \sqrt{a_{11}} = \sqrt{1} = 1 \tag{5-71a}$$

The second row ($i = 2$) can be computed using Eqs. 5-68b and 5-69 as

$$l_{21} = \frac{a_{21}}{l_{11}} = \frac{2}{1} = 2 \tag{5-71b}$$

$$l_{22} = \sqrt{a_{22} - \sum_{k=1}^{2-1} l_{2k}^2} = \sqrt{8 - 2^2} = 2 \tag{5-71c}$$

The third row ($i = 3$) is

$$l_{31} = \frac{a_{31}}{l_{11}} = \frac{3}{1} = 3 \tag{5-71d}$$

$$l_{32} = \frac{a_{32} - \sum_{k=1}^{2-1} l_{3k}l_{2k}}{l_{22}} = \frac{10 - (3)(2)}{2} = 2 \qquad \text{for } j = 1, 2, 3, \ldots, i - 1 \tag{5-71e}$$

$$l_{33} = \sqrt{a_{33} - \sum_{k=1}^{3-1} l_{3k}^2} = \sqrt{22 - 3^2 - (2)^2} = 3 \tag{5-71f}$$

Therefore, the **L** matrix is

$$\mathbf{L} = \begin{bmatrix} 1 & 0 & 0 \\ 2 & 2 & 0 \\ 3 & 2 & 3 \end{bmatrix} \tag{5-71g}$$

The validity of Eq. 5-67 can be verified as

$$\mathbf{A} = \mathbf{L}\mathbf{L}^T = \begin{bmatrix} 1 & 0 & 0 \\ 2 & 2 & 0 \\ 3 & 2 & 3 \end{bmatrix} \cdot \begin{bmatrix} 1 & 2 & 3 \\ 0 & 2 & 2 \\ 0 & 0 & 3 \end{bmatrix} = \begin{bmatrix} 1 & 2 & 3 \\ 2 & 8 & 10 \\ 3 & 10 & 22 \end{bmatrix} \tag{5-71h}$$

5.6 ITERATIVE EQUATION-SOLVING METHODS

Elimination methods like the Gaussian elimination procedure are often called *direct* equation-solving methods because the solution is found after a fixed, predictable number of operations. As an alternative, however, we can solve sets of simultaneous equations by using a *trial-and-error* procedure. In this procedure, we can assume a solution, that is, a set of estimates for the unknowns, and successively refine our estimate of the solution through some set of rules. This approach is the basis for *iterative* methods for solving simultaneous equations. Iterative methods, in general, produce an exact solution for the problem; the precision of the solution depends on our patience in performing the iterations, that is, on the number of trial-and-error or iteration cycles that we perform. Obviously, the number of operations required to obtain a given solution using iterative methods is not fixed; this, too, is a function of the number of iteration cycles that we choose to perform. However, a major advantage of iterative methods is that they can be used to solve nonlinear simultaneous equations, a task that is not possible using direct elimination methods. The use of these methods for solving nonlinear simultaneous equations is discussed in Section 4.10.

A number of iteration methods are available for solving simultaneous equations. Two of the most common methods, the Jacobi and Gauss–Seidel procedures, are presented herein.

5.6.1 Jacobi Iteration

The general procedure for Jacobi iteration is illustrated using the general notation for the simultaneous equations in Eqs. 5-8. As a first step, each equation is rearranged to produce an expression for a single unknown. For example, if we solve each row i for the unknown X_i, then Eqs. 5-8 transform to

$$X_1 = \frac{C_1 - a_{12}X_2 - a_{13}X_3 - \cdots - a_{1n}X_n}{a_{11}} \tag{5-72a}$$

$$X_2 = \frac{C_2 - a_{21}X_1 - a_{23}X_3 - \cdots - a_{2n}X_n}{a_{22}} \tag{5-72b}$$

$$\vdots$$

$$X_n = \frac{C_n - a_{n1}X_1 - a_{n2}X_2 - \cdots - a_{n-1,n}X_{n-1}}{a_{nn}} \tag{5-72c}$$

To start the iterative calculations, an initial solution estimate for the X_i unknowns is required. The initial estimates for all the X_i are substituted into the right sides of Eqs. 5-72 to obtain a new set of calculated (left side) values for the X_i's. These new values are substituted into the right side of Eqs. 5-72 and another new set of values is obtained for the X_i's. This *iteration* process continues until the calculated values for the X_i terms converge to the true solution. Convergence can be defined in many ways. For now, it is sufficient to say that convergence is obtained when the changes in all the X_i values between successive iteration cycles are acceptably small; that is, we can assume that the solution has converged when the values of the X_i for one iteration are not much different from the values of the previous iteration.

Example 5-9 Jacobi Iteration

Given the following set of equations, we wish to solve for values of the unknowns using Jacobi iteration:

$$3X_1 + X_2 - 2X_3 = 9 \tag{5-73a}$$

$$-X_1 + 4X_2 - 3X_3 = -8 \tag{5-73b}$$

$$X_1 - X_2 + 4X_3 = 1 \tag{5-73c}$$

As a first step, we will rearrange Eqs. 5-73 such that each equation is an expression for a single, different unknown:

$$X_1 = \frac{9 - X_2 + 2X_3}{3} \tag{5-74a}$$

$$X_2 = \frac{-8 + X_1 + 3X_3}{4} \qquad (5\text{-}74b)$$

$$X_3 = \frac{1 - X_1 + X_2}{4} \qquad (5\text{-}74c)$$

Next, we must assume an initial estimate for the solution; values of $X_1 = X_2 = X_3 = 1$ for this initial estimate are assumed. We can now substitute these values into the appropriate locations on the right side of Eqs. 5-74 to obtain a new set of values for the unknowns, as follows:

$$X_1 = \frac{9 - 1 + 2(1)}{3} = \frac{10}{3} \qquad (5\text{-}75a)$$

$$X_2 = \frac{-8 + 1 + 3(1)}{4} = -1 \qquad (5\text{-}75b)$$

$$X_3 = \frac{1 - 1 + 1}{4} = \frac{1}{4} \qquad (5\text{-}75c)$$

These new values for X_1, X_2, and X_3 are then used as the new solution estimate. These values are substituted into the right side of Eqs. 5-74 to obtain a second set of computed values for the unknowns:

$$X_1 = \frac{9 - (-1) + 2\left(\frac{1}{4}\right)}{3} = \frac{7}{2} \qquad (5\text{-}76a)$$

$$X_2 = \frac{-8 + \frac{10}{3} + 3\left(\frac{1}{4}\right)}{4} = -\frac{47}{48} \qquad (5\text{-}76b)$$

$$X_3 = \frac{1 - \frac{10}{3} + (-1)}{4} = -\frac{5}{6} \qquad (5\text{-}76c)$$

This process is repeated until the differences between the previous values and the new values are small. The values for the first 13 iteration cycles are shown in Table 5-2. The accuracy of the estimates of the unknowns depends on the number of iteration cycles performed. To illustrate the convergence of the iterative calculations, the change in absolute value of the calculated X_i's from each iteration cycle to the next is also shown in Table 5-2. If we had chosen a convergence criterion for the value of any unknown based on a maximum absolute change of less than 0.05, then the iteration would have been discontinued after the seventh cycle, when the value of $|\Delta X_1|$ (the largest $|\Delta X_i|$) equals 0.049. At this point, the true errors for X_1, X_2, and X_3 (that is, the differences between the seventh cycle estimates and the true solution of $X_1 = 3$, $X_2 = -2$, and $X_3 = -1$) are 0.028, 0.025, and 0.004, respectively. The true solution, or at least the estimated solution matching the true solution to three decimal places, is not obtained until the thirteenth iteration cycle.

TABLE 5-2   Example of Jacobi Iteration

Iteration	X_1	$|\Delta X_1|$	X_2	$|\Delta X_2|$	X_3	$|\Delta X_3|$
0	1	—	1	—	1	—
1	3.333	2.333	−1.000	2.000	0.250	0.750
2	3.500	0.167	−0.979	0.021	−0.833	1.083
3	2.771	0.729	−1.750	0.771	−0.870	0.036
4	3.003	0.233	−1.960	0.210	−0.880	0.010
5	3.066	0.063	−1.909	0.050	−0.991	0.111
6	2.976	0.090	−1.976	0.067	−0.994	0.003
7	2.996	0.020	−2.001	0.025	−0.988	0.006
8	3.008	0.012	−1.992	0.009	−0.999	0.011
9	2.998	0.011	−1.997	0.005	−1.000	0.001
10	2.999	0.001	−2.001	0.003	−0.999	0.001
11	3.001	0.002	−1.999	0.001	−1.000	0.001
12	3.000	0.001	−2.000	0.000	−1.000	0.001

5.6.2 Gauss–Seidel Iteration

A careful examination of the calculation steps used in our Jacobi iteration example suggests that we may not have been as efficient as possible in our iterative solution. For example, in Eq. 5-75b we used our initial estimate of $X_1 = 1$ on the right side when, in fact, we had already computed an updated estimate of $X_1 = \frac{10}{3}$ in Eq. 5-75a. Similarly, for Eq. 5-75c the original solution estimates for X_1 and X_2 were used, rather than the revised estimates obtained from Eqs. 5-75a and b. One could reasonably argue that, if supposedly better estimates are available, then they should be used as soon as possible. This argument is the basis of the Gauss–Seidel iteration method.

In the Jacobi iteration procedure, we always complete a full iteration cycle over all the equations before updating our solution estimates. In the Gauss–Seidel iteration procedure, however, we update each unknown as soon as a new estimate of that unknown is computed. The philosophy here is that the most recent estimate of the unknown is usually the best estimate and, therefore, should be used as soon as it is available. However, in practice, the most recent estimate of any one unknown is not always the best. For example, the calculations in Table 5-2 show that the initial estimate of X_1 from iteration 1 is better than the revised estimate from iteration 2. Nevertheless, we can show that, in general, the procedure converges much faster using the most recent estimate (i.e.,

Gauss–Seidel iteration), rather than the values from the previous Jacobi iteration. Thus the Gauss–Seidel iteration method is more efficient than the Jacobi method.

Example 5-10 Gauss–Seidel Iteration

The following system of equations as provided in Eqs. 5-73 was solved using the Gauss–Seidel iteration method:

$$3X_1 + X_2 - 2X_3 = 9$$
$$-X_1 + 4X_2 - 3X_3 = -8$$
$$X_1 - X_2 + 4X_3 = 1$$

Again, we will assume an initial solution estimate of $X_1 = X_2 = X_3 = 1$. The rearranged Eqs. 5-74 can be used to perform the first iteration cycle, expressed as

$$X_1 = \frac{9 - 1 + 2(1)}{3} = 3.333 \qquad (5\text{-}77a)$$

$$X_2 = \frac{-8 + 3.333 + 3(1)}{4} = -0.417 \qquad (5\text{-}77b)$$

$$X_3 = \frac{1 - 3.333 + (-0.417)}{4} = -0.688 \qquad (5\text{-}77c)$$

Note that in Eq. 5-77a the initial assumed estimates for X_2 and X_3 are used in the right side of the equation. However, in Eq. 5-77b the initial assumed estimate for X_1 is not used; instead, the just-calculated updated estimate of X_1 from Eq. 5-77a is substituted in the right side. The initial assumed estimate for X_3 is still used in Eq. 5-77b because no more recent estimate has yet been computed. Similarly, in Eq. 5-77c the updated values for X_1 and X_2 computed in Eqs. 5-77a and b are used in the right side of the equation.

The second iteration cycle is computed as

$$X_1 = \frac{9 - (-0.417) + 2(-0.688)}{3} = 2.680 \qquad (5\text{-}78a)$$

$$X_2 = \frac{-8 + 2.680 + 3(-0.688)}{4} = -1.845 \qquad (5\text{-}78b)$$

$$X_3 = \frac{1 - 2.680 + (-1.845)}{4} = -0.882 \qquad (5\text{-}78c)$$

Again, note that the most recently computed values for the X_i terms are always used in the right side of Eqs. 5-78. The calculations for the remaining iteration cycles are summarized in Table 5-3.

For a convergence criterion requiring a maximum absolute difference between successive estimates to be less than 0.05, the Gauss–Seidel procedure (Table 5-3) converges in five cycles, versus seven cycles for the Jacobi iteration method (Table 5-2). Whereas the Jacobi method requires 13 iteration cycles before all values of the unknowns are accurate to three decimal places, the Gauss–Seidel method requires only seven iteration cycles.

TABLE 5-3 Example of Gauss–Seidel Iteration

Iteration	X_1	$|\Delta X_1|$	X_2	$|\Delta X_2|$	X_3	$|\Delta X_3|$
0	1	—	1	—	1	—
1	3.333	2.333	−0.417	1.417	−0.688	1.688
2	2.680	0.348	−1.845	1.428	−0.882	0.194
3	3.027	0.346	−1.904	0.059	−0.983	0.101
4	2.979	0.048	−1.992	0.088	−0.993	0.010
5	3.002	0.023	−1.994	0.002	−0.999	0.006
6	2.999	0.003	−2.000	0.006	−1.000	0.001
7	3.000	0.001	−2.000	0.000	−1.000	0.000
8	3.000	0.000	−2.000	0.000	−1.000	0.000

5.6.3 Convergence Considerations of the Iterative Methods

Although both the Jacobi and Gauss–Seidel iterative methods converged for our example problem, in the general case either one or both of the methods may diverge; that is, the differences between successive estimates of the solution become *larger* with each iteration cycle. In fact, the example problem can be made divergent by interchanging the order of the starting Eqs. 5-73a and b. That is, Eq. 5-73a is used to solve for X_2 and Eq. 5-73b is used to solve for X_1. In this case, the following set of equations is obtained:

$$X_1 = 8 + 4X_2 - 3X_3 \tag{5-79a}$$

$$X_2 = 9 - 3X_1 + 2X_3 \tag{5-79b}$$

$$X_3 = \frac{1 - X_1 + X_2}{4} \tag{5-79c}$$

The computations for three iteration cycles are shown in Table 5-4 using the Gauss–Seidel method with the same initial estimates as used for the calculations in Tables 5-2 and 5-3. It is evident that the solution is diverging and that, even with more iteration cycles, the solution would not converge. Since the solution converged when the equations were written in the order of Eqs. 5-74 but diverged for the order in Eqs. 5-79, we must conclude that convergence can depend on the order in which the equations are processed.

It is important to note that the divergence of the iterative calculations in Table 5-4 does not imply that there is no solution to Eqs. 5-79. Recall that it is a permissible operation to interchange the order in which a set of equations is written; by the definition of a permissible operation, this means that the solution should be unaffected by the reordering. However, although the solution may still exist, the method for calculating this solu-

TABLE 5-4   Divergence of Gauss–Seidel Iteration

Iteration	X_1	$|\Delta X_1|$	X_2	$|\Delta X_2|$	X_3	$|\Delta X_3|$
0	1	—	1	—	1	—
1	9	8	-16	17	-6	7
2	-38	47	111	127	37.5	43.5
3	339.5	377.5	-934.5	1045.5	-318.25	355.75

tion may not work after the equations have been reordered. This is exactly what has happened in this example problem.

A means of ordering the equations so that the solution procedure converges can be demonstrated using two simultaneous equations ($n = 2$ in Eq. 5-8):

$$a_{11}X_1 + a_{12}X_2 = C_1 \tag{5-80a}$$

$$a_{21}X_1 + a_{22}X_2 = C_2 \tag{5-80b}$$

If we solve both equations individually for X_1 we get

$$X_1 = \frac{C_1}{a_{11}} - \frac{a_{12}}{a_{11}}X_2 \tag{5-81a}$$

$$X_1 = \frac{C_2}{a_{21}} - \frac{a_{22}}{a_{21}}X_2 \tag{5-81b}$$

If we refer to Eqs. 5-81a and b as f_1 and f_2, respectively, and plot Eqs. 5-81 on an (X_1, X_2) axis system as shown in Fig. 5-3, the solution occurs at the intersection of the two functions f_1 and f_2. It is easy to show by equating Eqs. 5-81a and b and solving for X_2 that the intersection of f_1 and f_2 is given by Eqs. 5-11a and b. But, since we do not know the values of X_1 and X_2, we must make an initial estimate. If we select a value of X_2, say $X_2^{(1)}$, and substitute it into f_1, we get a value for X_1, say $X_1^{(1)}$. For cycle 2, we substitute $X_1^{(1)}$ into f_2 and get a revised value of X_2, which is denoted as $X_2^{(2)}$. The value $X_2^{(2)}$ is closer to the intersection of f_1 and f_2 than $X_1^{(1)}$. If we continue this iterative process, we will converge on the intersection of f_1 and f_2. This convergence occurs whether $X_2^{(1)}$ is greater than or less than the true value of X_2.

However, if the slope of the line of f_1 is greater than the slope of the line of f_2, then we have the situation shown in Fig. 5-4. We begin with an estimate of X_2, say $X_2^{(1)}$, and use f_1 to obtain a value $X_1^{(1)}$. This value is then used with f_2 to get a value of X_2, say $X_2^{(2)}$. It is evident from Fig. 5-4 that the initial estimate of X_2 was a better estimate of X_2 than is $X_2^{(2)}$. The same result occurs when we use f_1 and $X_2^{(2)}$ to obtain $X_1^{(2)}$. This process, if continued, results in a diverging solution, with the differences between successive estimates of X_1 and X_2 changing sign and increasing in magnitude. Again, divergence occurs regardless of the initial value $X_2^{(1)}$, unless the value of $X_2^{(1)}$ exactly equals the true value of X_2.

In comparing the configurations in Figs. 5-3 and 5-4, it should be evident that for two simultaneous equations the iterative solution converges only when the initial value of

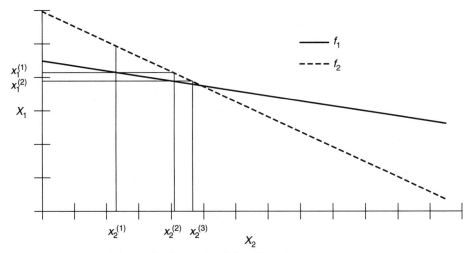

Figure 5-3   Converging Order of Equations

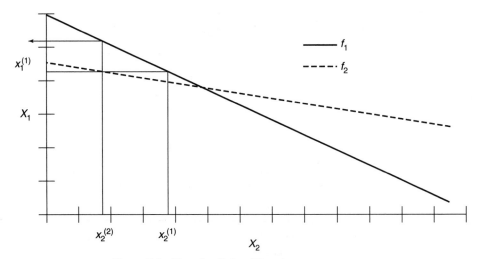

Figure 5-4   Diverging Order of Equations

$X_2^{(1)}$ is substituted into the equation that has the smaller slope, that is, when the slope of f_1 is less than the slope of f_2. Thus the equations should be arranged such that X_1 is expressed in terms of X_2 with the following conditions:

$$\left| \frac{a_{12}}{a_{11}} \right| < \left| \frac{a_{22}}{a_{21}} \right| \tag{5-82}$$

It should be evident from Eq. 5-82 that this inequality holds if we make both a_{11} and a_{22} as large as possible. The term on the left side of the inequality of Eq. 5-82 most probably is smaller if the equation having the largest value of a_{11} is selected as the first equation and the other equation as the second equation. A system of equations in which the coefficients with the largest absolute values appear on the diagonal is referred to as being diagonally dominant.

While it is not practical to try to show the convergence–divergence analysis for systems that include more than two equations, the underlying concepts are the same. We can increase the likelihood of convergence of our iterative solution scheme by selecting the equation with the largest coefficient for X_1 as the first equation, the equation with the largest coefficient for X_2 in the remaining equations as the second equation, and so on.

Example 5-11 Gauss–Seidel Solution for an Electrical Circuit

A system of three linear equations was developed to represent the distribution of current flow in the circuit of Fig. 5-1. These equations were rearranged so that the matrix is diagonally dominant:

$$I_1 + I_2 - I_3 = 0 \qquad (5\text{-}83\text{a})$$

$$2I_1 - 4I_2 \quad = 4 \qquad (5\text{-}83\text{b})$$

$$2I_1 + 5I_3 \quad = 6 \qquad (5\text{-}83\text{c})$$

Solving these equations for the variable on the diagonal yields

$$I_1 = I_3 - I_2 \qquad (5\text{-}84\text{a})$$

$$I_2 = \frac{2I_1 - 4}{4} \qquad (5\text{-}84\text{b})$$

$$I_3 = \frac{6 - 2I_1}{5} \qquad (5\text{-}84\text{c})$$

Initial estimates of $I_1 = 1$ ampere (A), $I_2 = 0$, and $I_3 = 1$ A were selected; these initial estimates satisfy Kirchhoff's first law, which is represented by the first of the three linear equations. The Gauss–Seidel method was applied, with a partial listing shown in Table 5-5. Because of the slopes of the planes representing the equation, convergence is a slow process. It takes 20 cycles before the largest error in absolute value is 0.05 or less. After 31 cycles, the process indicates the following solution: $I_1 = 1.15$ A, $I_2 = -0.42$ A, and $I_3 = 0.74$ A. This solution results in absolute value errors of 0.01, 0.02, and 0.00 for the three unknown currents.

5.6.4 Considerations in Programming

Both the Jacobi and the Gauss–Seidel iteration methods are based on Eqs. 5-72. Each of these has the same general form:

$$X_i = \frac{C_i - \sum_{\substack{j=1 \\ j \neq i}}^{n} a_{ij} X_j}{a_{ii}} \qquad \text{for } i = 1, \ldots, n \qquad (5\text{-}85)$$

The summation includes all the n terms except for the case where $i = j$. One iteration cycle requires n evaluations of Eq. 5-85, with i varying from 1 to n. Thus the general computational procedure consists of the following steps:

1. For iteration 1, assume initial estimates for all X_i.

TABLE 5-5   Solution for Electrical Circuit

Iteration	I_1	$|\Delta I_1|$	I_2	$|\Delta I_2|$	I_3	$|\Delta I_3|$
0	1	—	0	—	1	—
1	1	0	−0.5	0.50	0.8	0.20
2	1.3	0.30	−0.35	0.15	0.68	0.12
3	1.03	0.27	−0.49	0.14	0.47	0.11
4	1.27	0.24	−0.36	0.13	0.69	0.10
5	1.05	0.22	−0.47	0.11	0.78	0.09
6	1.25	0.20	−0.37	0.10	0.70	0.08
7	1.07	0.18	−0.46	0.09	0.77	0.07
10	1.22	0.13	−0.39	0.07	0.71	0.05
15	1.12	0.08	−0.44	0.04	0.75	0.03
16	1.19	0.07	−0.40	0.04	0.72	0.03
20	1.18	0.05	−0.41	0.02	0.73	0.02
21	1.14	0.04	−0.43	0.02	0.74	0.01
25	1.15	0.02	−0.43	0.02	0.74	0.01
26	1.17	0.02	−0.42	0.01	0.73	0.01
30	1.17	0.02	−0.42	0.01	0.73	0.01
31	1.15	0.02	−0.42	0.00	0.74	0.01

2. For each subsequent iteration cycle, perform the following steps:
   a. Loop over all values of X_i from $i = 1$ to n, solving Eq. 5-85 for each new value of X_i.
   b. Check for convergence, and if the solution has converged, exit from the loop; otherwise, perform another cycle starting with step 2a.

These steps can be used to solve a system of equations by either Jacobi or Gauss–Seidel iteration. The only difference between the two methods is the handling of the X_i. For the Jacobi method, the values computed from Eq. 5-85 should be stored in a vector labeled differently than the X_i, say Y_i. Thus Eq. 5-85 becomes

$$Y_i = \frac{C_i - \sum_{\substack{j=1 \\ j \neq i}}^{n} a_{ij} X_j}{a_{ii}} \qquad \text{for } i = 1, \ldots, n \tag{5-86}$$

After completing each iteration cycle, the values of Y_i must be transferred into the X_i vector for the start of the next iteration cycle.

A pseudocode outline for the Jacobi iteration procedure is given in Fig. 5-5. The corresponding outline for the Gauss–Seidel iteration procedure is given in Fig. 5-6.

```
        Assign Initial Solution Estimate to Xᵢ
   10 DO FOR I = 1 TO N                    [ALL N EQUATIONS]
          CALCULATE Yᵢ = NEW SOLUTION
          ESTIMATE BASED ON Xᵢ VALUES
      END DO
      DO FOR I = 1 TO N
          CALCULATE ERROR = |Yᵢ − Xᵢ|
          FIND MAXIMUM ERROR IN SYSTEM
      END DO
      IF (MAXIMUM ERROR IS GREATER THAN CONVERGENCE
          TOLERANCE) THEN
      DO FOR I = 1 to N
          Xᵢ = Yᵢ
      END DO
      GO TO 10
      ELSE
      PRINT RESULTS
      STOP
      END IF
      END
```

Figure 5-5   Pseudocode Outline for Jacobi Iteration Method

```
        Assign initial solution estimate to all Xᵢ
   10 ERROR = 0.0
      DO FOR I = 1 TO N                [ALL N EQUATIONS]
          XTEMP = Xᵢ
          CALCULATE NEW Xᵢ FROM REARRANGED EQUATIONS AND
          OLD Xᵢ VALUES
          DELTAX = Xᵢ − XTEMP
          IF (DELTAX > ERROR) ERROR = DELTAX
      END DO
      IF (ERROR > CONVERGENCE TOLERANCE) THEN
      GO TO 10
      ELSE
      PRINT RESULTS
      STOP
      ENDIF
      END
```

Figure 5-6   Pseudocode Outline for Gauss–Seidel Iteration Method

5.7 USE OF DETERMINANTS

A determinant is a unique scalar number that can be used to represent a square matrix. For the matrix $\mathbf{A}$ the determinant is denoted as $|\mathbf{A}|$. To solve a system of linear equations using determinants, it is necessary to view the equations (Eqs. 5-8) as a coefficient matrix $\mathbf{A}$ and the column vector of constants $\mathbf{C}$:

$$\mathbf{A} = \begin{bmatrix} a_{11} & a_{12} & \cdots & a_{1n} \\ a_{21} & a_{22} & \cdots & a_{2n} \\ \vdots & \vdots & \ddots & \vdots \\ a_{n1} & a_{n2} & \cdots & a_{nn} \end{bmatrix} \tag{5-87a}$$

and

$$\mathbf{C} = \begin{bmatrix} C_1 \\ C_2 \\ \vdots \\ C_n \end{bmatrix} \tag{5-87b}$$

The value of X_i is obtained using Cramer's rule as

$$X_i = \frac{|\mathbf{A}_i|}{|\mathbf{A}|} \tag{5-88}$$

where $|\mathbf{A}|$ is the determinant of matrix $\mathbf{A}$ and $|\mathbf{A}_i|$ is the determinant of a matrix formed by replacing column i of $\mathbf{A}$ with the column vector of Eq. 5-87b. For example, $|\mathbf{A}_2|$ would be given by

$$|\mathbf{A}_2| = \begin{bmatrix} a_{11} & C_1 & a_{13} & \cdots & a_{1n} \\ a_{21} & C_2 & a_{23} & \cdots & a_{2n} \\ \vdots & \vdots & \vdots & \ddots & \vdots \\ a_{n1} & C_n & a_{n3} & \cdots & a_{nn} \end{bmatrix} \tag{5-89}$$

Example 5-12 Three Linear Equations

The system of three linear equations given by Eq. 5-73 is easily solved using Eq. 5-88. The determinant of the coefficient matrix is given by

$$|\mathbf{A}| = \begin{vmatrix} -1 & 4 & -3 \\ 3 & 1 & -2 \\ 1 & -1 & 4 \end{vmatrix} = (-1)[1(4) - (-2)(-1)] - 4[(3(4) - (-2)(1)] - 3[3(-1) - (1)(1)] = -46 \tag{5-90}$$

The determinants for the matrices for the numerator of Eq. 5-88 are given by

$$|\mathbf{A}_1| = \begin{vmatrix} -8 & 4 & -3 \\ 9 & 1 & -2 \\ 1 & -1 & 4 \end{vmatrix} = -138 \tag{5-91}$$

$$|\mathbf{A_2}| = \begin{vmatrix} -1 & -8 & -3 \\ 3 & 9 & -2 \\ 1 & 1 & 4 \end{vmatrix} = 92 \tag{5-92}$$

$$|\mathbf{A_3}| = \begin{vmatrix} -1 & 4 & -8 \\ 3 & 1 & 9 \\ 1 & -1 & 1 \end{vmatrix} = 46 \tag{5-93}$$

Thus the solution is

$$X_1 = \frac{|\mathbf{A_1}|}{|\mathbf{A}|} = \frac{-138}{-46} = 3 \tag{5-94a}$$

$$X_2 = \frac{|\mathbf{A_2}|}{|\mathbf{A}|} = \frac{92}{-46} = -2 \tag{5-94b}$$

$$X_3 = \frac{|\mathbf{A_3}|}{|\mathbf{A}|} = \frac{46}{-46} = -1 \tag{5-94c}$$

Thus the solution of Eq. 5-94 is the same solution that was determined for Eqs. 5-73 using the Jacobi iteration method.

Example 5-13 Method of Determinants for Material Purchasing

The method of determinants could be used to obtain a solution to the manufacturing problem (Eqs. 5-3). Using the system of linear equations and Cramer's rule, we get the following solution:

$$|\mathbf{X}| = \begin{vmatrix} 1 & 3 & 3 \\ 2 & 4 & 3 \\ 3 & 4 & 7 \end{vmatrix} = -7 \tag{5-95a}$$

and

$$X_1 = \frac{\begin{vmatrix} 15 & 3 & 2 \\ 22 & 4 & 3 \\ 39 & 4 & 7 \end{vmatrix}}{-7} \tag{5-95b}$$

$$X_2 = \frac{\begin{vmatrix} 1 & 15 & 2 \\ 2 & 22 & 3 \\ 3 & 39 & 7 \end{vmatrix}}{-7} \tag{5-95c}$$

$$X_3 = \frac{\begin{vmatrix} 1 & 3 & 15 \\ 2 & 4 & 22 \\ 3 & 4 & 39 \end{vmatrix}}{-7} \tag{5-95d}$$

Therefore, the solution is

$$X_1 = \frac{-7}{-7} = 1 \tag{5-96a}$$

$$X_2 = \frac{-14}{-7} = 2 \tag{5-96b}$$

$$X_3 = \frac{-28}{-7} = 4 \tag{5-96c}$$

The solution is the same as that obtained using Gaussian elimination (Example 5-4).

Example 5-14 Method of Determinants for Electrical Circuit

The Gauss–Seidel solution for Example 5-11 required a large number of iterations to achieve convergence. The method of determinants would, thus, be easier to apply. The solution is as follows:

$$|\mathbf{I}| = \begin{vmatrix} 1 & 1 & -1 \\ 2 & 0 & 5 \\ 2 & -4 & 0 \end{vmatrix} = 38 \tag{5-97}$$

and

$$I_1 = \frac{\begin{vmatrix} 0 & 1 & -1 \\ 6 & 0 & 5 \\ 4 & -4 & 0 \end{vmatrix}}{38} \tag{5-98a}$$

$$I_2 = \frac{\begin{vmatrix} 1 & 0 & -1 \\ 2 & 6 & 5 \\ 2 & 4 & 0 \end{vmatrix}}{38} \tag{5-98b}$$

$$I_3 = \frac{\begin{vmatrix} 1 & 1 & 0 \\ 2 & 0 & 6 \\ 2 & -4 & 4 \end{vmatrix}}{38} \tag{5-98c}$$

Therefore, the solution is

$$I_1 = \frac{44}{38} = 1.158 \text{ A} \tag{5-99a}$$

$$I_2 = \frac{-16}{38} = -0.421 \text{ A} \tag{5-99b}$$

$$I_3 = \frac{28}{38} = 0.737 \text{ A} \tag{5-99c}$$

The solution is exact for the number of significant digits shown. Comparing this solution with the results of the Gauss–Seidel solution, the errors for the Gauss–Seidel solution were less than 0.01 A in absolute value.

5.7.1 Considerations in Programming

The solution of Eq. 5-88 for each X_i requires a subprogram that determines the determinant of a matrix. Given such a subprogram, the determinant $|\mathbf{A}|$ can be computed. Then it is only necessary to iterate over the X_i values from $i = 1$ to n. For each step, the vector of constants $\mathbf{C}$ is used to replace column i in matrix $\mathbf{A}$, which is the iteration loop counter, of the coefficient matrix; this replacement is accomplished using another iteration loop. The determinant $|\mathbf{A}_i|$ is then determined using the subprogram, and the value of X_i is computed.

5.8 MATRIX INVERSION

Matrix inversion is frequently necessary in the solution of engineering problems, as was discussed in Chapter 2. For example, in fitting a linear multiple regression equation (see Chapter 10), the vector of unknowns ($\mathbf{t}$) is related to the correlation matrix ($\mathbf{R}$) by

$$\mathbf{t} = \mathbf{R}_{11}^{-1}\mathbf{R}_{12} \tag{5-100}$$

where $\mathbf{R}_{11}$ is the matrix of correlation coefficients between the predictor (independent) variables, and $\mathbf{R}_{12}$ is the vector of correlation coefficients between the dependent variable and each predictor variable. To solve Eq. 5-100, it is necessary to find the inverse of $\mathbf{R}_{11}$, that is, $\mathbf{R}_{11}^{-1}$.

The inverse of a matrix $\mathbf{P}$ is defined by the following equation (see Section 2-2):

$$\mathbf{P}^{-1}\mathbf{P} = \mathbf{I} \tag{5-101}$$

in which $\mathbf{I}$ is the identity or unit matrix and both $\mathbf{P}$ and $\mathbf{I}$ are square matrices. The unit matrix is a square matrix with values of 1.0 on the principal diagonal and values of zero for all other elements. In Eq. 5-101, we consider $\mathbf{P}$ and $\mathbf{I}$ to be given, with the elements of $\mathbf{P}^{-1}$ the unknowns. The values of $\mathbf{P}^{-1}$ can be computed by expanding Eq. 5-101 into a set of n^2 simultaneous equations, where n is the order of the matrix. The solution of

Eq. 5-101 for the elements of $\mathbf{P}^{-1}$ can be demonstrated using the case where $n = 3$. Letting the elements of $\mathbf{P}^{-1}$ and $\mathbf{P}$ be denoted as q_{ij} and p_{ij}, respectively, Eq. 5-101 becomes

$$
\begin{bmatrix} q_{11} & q_{12} & q_{13} \\ q_{21} & q_{22} & q_{23} \\ q_{31} & q_{32} & q_{33} \end{bmatrix} \cdot \begin{bmatrix} p_{11} & p_{12} & p_{13} \\ p_{21} & p_{22} & p_{23} \\ p_{31} & p_{32} & p_{33} \end{bmatrix} = \begin{bmatrix} 1 & 0 & 0 \\ 0 & 1 & 0 \\ 0 & 0 & 1 \end{bmatrix} \tag{5-102}
$$

Expanding Eq. 5-102 yields the following set of $n^2 = 9$ simultaneous equations:

$$
\begin{aligned}
p_{11}q_{11} + p_{21}q_{12} + p_{31}q_{13} &= 1 \\
p_{12}q_{11} + p_{22}q_{12} + p_{32}q_{13} &= 0 \\
p_{13}q_{11} + p_{23}q_{12} + p_{33}q_{13} &= 0 \\
p_{11}q_{21} + p_{21}q_{22} + p_{31}q_{23} &= 0 \\
p_{12}q_{22} + p_{22}q_{22} + p_{32}q_{23} &= 1 \\
p_{13}q_{23} + p_{23}q_{22} + p_{33}q_{23} &= 0 \\
p_{11}q_{31} + p_{21}q_{32} + p_{31}q_{33} &= 0 \\
p_{12}q_{31} + p_{22}q_{32} + p_{32}q_{33} &= 0 \\
p_{13}q_{31} + p_{23}q_{32} + p_{33}q_{33} &= 1
\end{aligned} \tag{5-102}
$$

Equations 5-102 include nine simultaneous equations with nine q_{ij} unknowns (that is, q_{ij} for $i = 1$, 2 and 3 and $j = 1$, 2, and 3). The solution yields the values of the inverse $\mathbf{P}^{-1}$.

It should be evident from Eq. 5-102 that the solution could be found by considering the set of nine equations to consist of three sets of three equations. The unknowns for row i of $\mathbf{P}^{-1}$ are independent in Eqs. 5-102 of the unknowns for every other row. The solution could be made by solving n sets of n simultaneous equations as given by

$$
\mathbf{P}^T \mathbf{R}_i = I_i \qquad \text{for } i = 1, 2, \ldots, n \tag{5-103}
$$

where $\mathbf{P}^T$ is the transpose of the $(n \times n)$ $\mathbf{P}$ matrix of Eq. 5-102, $\mathbf{R}_i$ is the nth-order column vector that consists of the elements of the ith row of the matrix $\mathbf{P}^{-1}$, and $\mathbf{I}_i$ is the nth-order column vector of the identity matrix.

Example 5-15 Inverse of a Matrix

To illustrate the estimation of the inverse of a matrix, consider the following correlation matrix $\mathbf{P}$:

$$
\mathbf{P} = \begin{bmatrix} 1 & -0.42 \\ -0.42 & 1 \end{bmatrix} \tag{5-104}
$$

If we were interested in developing an equation to predict X_3 for values of X_1 and X_2, then we would be interested in the inverse of the correlation matrix for X_1 and X_2, which is

$$
\begin{bmatrix} q_{11} & q_{12} \\ q_{21} & q_{22} \end{bmatrix} \begin{bmatrix} 1 & -0.42 \\ -0.42 & 1 \end{bmatrix} = \begin{bmatrix} 1 & 0 \\ 0 & 1 \end{bmatrix} \tag{5-105}
$$

Based on Eq. 5-103, the values of q_{11} and q_{12} can be found by solving the following two simultaneous equations:

$$q_{11} - 0.42q_{12} = 1 \qquad (5\text{-}106a)$$

$$-0.42q_{11} + q_{12} = 0 \qquad (5\text{-}106b)$$

Equations 5-106 can be solved using any of the methods previously introduced. The method of determinants yields

$$q_{11} = \frac{\begin{vmatrix} 1 & -0.42 \\ 0 & 1 \end{vmatrix}}{\begin{vmatrix} 1 & -0.42 \\ -0.42 & 1 \end{vmatrix}} = 1.214 \qquad (5\text{-}107a)$$

and

$$q_{12} = \frac{\begin{vmatrix} 1 & 1 \\ -0.42 & 0 \end{vmatrix}}{\begin{vmatrix} 1 & -0.42 \\ -0.42 & 1 \end{vmatrix}} = 0.510 \qquad (5\text{-}107b)$$

For the elements of the second row of Eq. 5-105, we get the following two simultaneous equations

$$q_{21} - 0.42q_{22} = 0 \qquad (5\text{-}108a)$$

$$-0.42q_{21} + q_{22} = 1 \qquad (5\text{-}108b)$$

The method of determinants yields

$$q_{21} = \frac{\begin{vmatrix} 0 & -0.42 \\ 1 & 1 \end{vmatrix}}{\begin{vmatrix} 1 & -0.42 \\ -0.42 & 1 \end{vmatrix}} = 0.510 \qquad (5\text{-}109a)$$

and

$$q_{22} = \frac{\begin{vmatrix} 1 & 0 \\ -0.42 & 1 \end{vmatrix}}{\begin{vmatrix} 1 & -0.42 \\ -0.42 & 1 \end{vmatrix}} = 1.214 \qquad (5\text{-}109b)$$

Thus the inverse matrix is

$$\mathbf{P}^{-1} = \begin{bmatrix} 1.214 & 0.510 \\ 0.510 & 1.214 \end{bmatrix} \qquad (5\text{-}110)$$

5.9 APPLICATIONS

5.9.1 Flexibility and Stiffness Analyses of a Beam

The cantilever beam shown in Fig. 5-7 can be loaded with a concentrated force (P) or moment (M) at the free end. The loads result in deformations at the free end in the form of a vertical deflection (y) and rotation (θ). The deformations due to a unit concentrated force applied at the free end are

$$y_P = \frac{L^3}{3EI} \tag{5-111a}$$

$$\theta_P = \frac{L^2}{2EI} \tag{5-111b}$$

where E = modulus of elasticity, L = span length of the beam, and I = moment of inertia of the beam's cross-sectional area. The deformations due to a unit concentrated moment applied at the free end are

$$y_M = \frac{L^2}{2EI} \tag{5-112a}$$

$$\theta_M = \frac{L}{EI} \tag{5-112b}$$

The deformation expressions provided by Eqs. 5-111 and 5-112 are called the flexibility coefficients. They can be used to assemble the flexibility matrix ($\mathbf{F}$) as follows:

$$\mathbf{F} = \begin{bmatrix} \frac{L^3}{3EI} & \frac{L^2}{2EI} \\ \frac{L^2}{2EI} & \frac{L}{EI} \end{bmatrix} \tag{5-113}$$

The flexibility matrix has the symmetry property. The flexibility matrix can be used to compute the deformations due to any combination of P and M as

$$\begin{bmatrix} y \\ \theta \end{bmatrix} = \begin{bmatrix} \frac{L^3}{3EI} & \frac{L^2}{2EI} \\ \frac{L^2}{2EI} & \frac{L}{EI} \end{bmatrix} \begin{bmatrix} P \\ M \end{bmatrix} \tag{5-114}$$

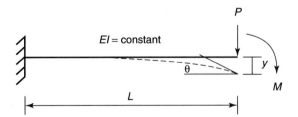

Figure 5-7 Cantilever Beam

The inverse of Eq. 5-114 is sometimes needed for the analysis of statically indeterminate structures; that is,

$$\begin{bmatrix} P \\ M \end{bmatrix} = \begin{bmatrix} \frac{L^3}{3EI} & \frac{L^2}{2EI} \\ \frac{L^2}{2EI} & \frac{L}{EI} \end{bmatrix}^{-1} \begin{bmatrix} y \\ \theta \end{bmatrix}$$

(5-115)

where the inverse of the flexibility matrix is called the stiffness matrix (**S**). The relationship between **F** and **S** according to Eq. 5-102 can be expressed as

$$\begin{bmatrix} S_{11} & S_{12} \\ S_{21} & S_{22} \end{bmatrix} \begin{bmatrix} \frac{L^3}{3EI} & \frac{L^2}{2EI} \\ \frac{L^2}{2EI} & \frac{L}{EI} \end{bmatrix} = \begin{bmatrix} 1 & 0 \\ 0 & 1 \end{bmatrix}$$

(5-116)

Therefore, the following four equations can be obtained:

$$\frac{L^3}{3EI}S_{11} + \frac{L^2}{2EI}S_{12} = 1$$

(5-117a)

$$\frac{L^2}{2EI}S_{11} + \frac{L}{EI}S_{12} = 0$$

(5-117b)

$$\cdot \frac{L^3}{3EI}S_{21} + \frac{L^2}{2EI}S_{22} = 0$$

(5-117c)

$$\frac{L^2}{2EI}S_{21} + \frac{L}{EI}S_{22} = 1$$

(5-117d)

Equations 5-117a and b can be simultaneously solved for S_{11} and S_{12} using Eqs. 5-11 to obtain

$$S_{11} = \frac{12EI}{L^3}$$

(5-118a)

$$S_{12} = -\frac{6EI}{L^2}$$

(5-118b)

Equations 5-117c and d can be simultaneously solved for S_{21} and S_{22} using Eqs. 5-11 to obtain

$$S_{21} = -\frac{6EI}{L^2}$$

(5-119a)

$$S_{22} = \frac{4EI}{L}$$

(5-119b)

The results of Eqs. 5-118 and 119 can be used to assemble the stiffness matrix as

$$\mathbf{S} = \begin{bmatrix} \frac{12EI}{L^3} & -\frac{6EI}{L^2} \\ -\frac{6EI}{L^2} & \frac{4EI}{L} \end{bmatrix}$$

(5-120)

The stiffness matrix is also symmetric, similar to the flexibility matrix. The stiffness matrix can be used to obtain the forces (e.g., reactions) due to known deformations (e.g., boundary conditions) according to Eq. 5-115.

In practice, the analysis of real indeterminate structures can result in flexibility and stiffness matrices of relatively large sizes requiring numerical evaluations rather than analytical ones, as was performed in Eqs. 5-118 and 5-119. The concepts behind the development of these matrices is similar to what is covered herein; however, a complete development of a case study would require more advanced background in the stiffness analysis of structures.

5.9.2 Concrete Mix Design

A concrete mix was designed to have the weight proportions of [cement : fine aggregates : coarse aggregates] as [1 : 1.9 : 2.8] with a water–cement ratio of 7 gallons of water per sack of cement. A sack of cement weighs 94 lb. Table 5-6 shows the unit weights for the components of the mix. How much of each constituent is needed to produce 50 cubic yards of concrete?

Using the notations of C = cement weight (lb), F = weight of fine aggregates (lb), G = weight of coarse aggregates (lb), and W = water volume (gallons), the following conditions can be developed based on the proportionality requirements:

$$F = 1.9C \qquad (5\text{-}121\text{a})$$

$$G = 2.8C \qquad (5\text{-}121\text{b})$$

Using the water–cement ratio, the following condition can be used:

$$W = \frac{7C}{94} \qquad (5\text{-}121\text{c})$$

Using the conversion factor of 1 cubic yard = 27 ft^3, the requirement for the total amount of concrete can be expressed as

$$50(27) = \frac{C}{195} + \frac{F+G}{165} + \frac{W}{7.48} \qquad (5\text{-}121\text{d})$$

TABLE 5-6   Unit Weights for
the Components of the Mix

Component	Unit Weight
Cement	195 pcf
Fine aggregates	165 pcf
Coarse aggregates	165 pcf
Water	62.4 pcf

where the factor 7.48 is to convert gallons to cubic feet. Equations 5-121 can be written as

$$
\begin{bmatrix}
1.9 & -1 & 0 & 0 \\
2.8 & 0 & -1 & 0 \\
\frac{7}{94} & 0 & 0 & -1 \\
\frac{1}{195} & \frac{1}{165} & \frac{1}{165} & \frac{1}{7.48}
\end{bmatrix}
\begin{bmatrix}
C \\ F \\ G \\ W
\end{bmatrix}
=
\begin{bmatrix}
0 \\ 0 \\ 0 \\ 1350
\end{bmatrix}
\tag{5-122}
$$

The system of equations can be arranged such that nonzero elements are on the diagonal of the matrix, resulting in

$$
\begin{bmatrix}
\frac{1}{195} & \frac{1}{165} & \frac{1}{165} & \frac{1}{7.48} \\
1.9 & -1 & 0 & 0 \\
2.8 & 0 & -1 & 0 \\
\frac{7}{94} & 0 & 0 & -1
\end{bmatrix}
\begin{bmatrix}
W \\ C \\ F \\ G
\end{bmatrix}
=
\begin{bmatrix}
1350 \\ 0 \\ 0 \\ 0
\end{bmatrix}
\tag{5-123}
$$

The Gauss–Jordan elimination can be used on the following coefficient matrix:

$$
\begin{bmatrix}
\frac{1}{195} & \frac{1}{165} & \frac{1}{165} & \frac{1}{7.48} & 1350 \\
1.9 & -1 & 0 & 0 & 0 \\
2.8 & 0 & -1 & 0 & 0 \\
\frac{7}{94} & 0 & 0 & -1 & 0
\end{bmatrix}
\tag{5-124}
$$

The result of the Gauss–Jordan elimination is

$$
\begin{bmatrix}
1 & 0 & 0 & 0 & 2{,}307.4 \\
0 & 1 & 0 & 0 & 30{,}985.6 \\
0 & 0 & 1 & 0 & 58{,}872.6 \\
0 & 0 & 0 & 1 & 86{,}759.6
\end{bmatrix}
\tag{5-125}
$$

Therefore, the amounts of the components of the mix are

$$\text{Water} = 2307.4 \text{ gal} \tag{5-126a}$$

$$\text{Cement} = 30{,}985.6 \text{ lb}$$

$$= 330 \text{ sacks} \tag{5-126b}$$

$$\text{Fine aggregates} = 58{,}872.6 \text{ lb}$$

$$= 29.4 \text{ tons} \tag{5-126c}$$

$$\text{Coarse aggregates} = 86{,}759.6 \text{ lb}$$

$$= 43.4 \text{ tons} \tag{5-126d}$$

5.10 PROBLEMS

Find the solution to Problems 5-1 through 5-6 using Gaussian elimination.

5-1. $X_1 + 2X_2 - X_3 = 3.1$
$-3X_1 + X_2 + X_3 = 1.4$
$-X_1 - X_2 + 4X_3 = 7.3$

5-2. $-2X_1 + 3X_2 - X_3 = -0.6$
$3X_1 + X_2 - 2X_3 = -3.3$
$X_1 + 2X_2 + X_3 = 1.9$

5-3. $2X_1 - 3X_2 - X_3 = -3.7$
$-3X_1 + 3X_2 - X_3 = -1.0$
$4X_1 - 2X_2 + 3X_3 = 4.4$

5-4. $3X_1 + 2X_2 - 4X_3 = -5.9$
$7X_1 - 4X_2 + 3X_3 = -5.7$
$-2X_1 + 2X_2 - 3X_3 = 1.2$

5-5. $2X_1 - 3X_2 + 2X_3 - 4X_4 = 2.3$
$-X_1 - 2X_2 + 3X_3 + 2X_4 = -3.0$
$2X_1 + X_2 - 4X_3 - 2X_4 = 7.7$
$-3X_1 + 4X_2 - X_3 + X_4 = -7.0$

5-6. $4X_1 - 2X_2 + 6X_3 - 2X_4 = 11.0$
$-3X_1 + 2X_2 + 2X_3 - 4X_4 = -9.9$
$X_1 + 5X_2 - 2X_3 - 2X_4 = -3.3$
$-2X_1 + X_2 - 3X_3 + X_4 = -5.5$

Find the solution to Problems 5-7 through 5-12 using Gauss–Jordan elimination.

5-7. $3X_1 + 3X_2 - 2X_3 = 7.6$
$2X_1 - 4X_2 + X_3 = 1.4$
$-X_1 - 2X_2 + 5X_3 = -6.3$

5-8. $2X_1 - 4X_2 + 6X_3 = 1.4$
$3X_1 + X_2 - 3X_3 = -9.8$
$-X_1 + 2X_2 - 3X_3 = -0.7$

5-9. $5X_1 - 3X_2 - X_3 = 20.3$
$2X_1 + 2X_2 - 3X_3 = -4.0$
$-3X_1 - 4X_2 + 5X_3 = 9.6$

5-10. $2X_1 - 3X_2 - X_3 = 5.2$
$-3X_1 - X_2 + 4X_3 = 3.9$
$5X_1 + 3X_2 - 4X_3 = -11.3$

5-11. $2X_1 - 3X_2 - X_3 + 2X_4 = 11.7$
$-3X_1 + X_2 + 2X_3 - X_4 = -11.5$
$-X_1 - 2X_2 + 3X_3 + 5X_4 = -0.4$
$3X_1 + 3X_2 - X_3 + X_4 = 3.6$

5-12. $-X_1 + 2X_2 + 3X_3 - 4X_4 = 1.5$
$\quad\quad 2X_1 - 3X_2 - X_3 - X_4 = -12.0$
$\quad\quad -X_1 - 3X_2 - X_3 + 5X_4 = -1.5$
$\quad\quad 3X_1 + 4X_2 + X_3 - 2X_4 = 5.1$

Find the solution to Problems 5-13 through 5-16 using the **LU** decomposition method.

5-13. $\quad X_1 + X_2 = 3$
$\quad\quad 4X_1 - 2X_2 = 2$

5-14. $2X_1 + 5X_2 = 3$
$\quad\quad -3X_1 - X_2 = 2$

5-15. The set of equations in Problem 5-1.

5-16. The set of equations in Problem 5-2.

Find the solution to Problems 5-17 through 5-20 using the Jacobi iteration method.

5-17. $\quad 3X_1 - X_2 + 2X_3 = -3.7$
$\quad\quad 2X_1 + 3X_2 - 2X_3 = 17.9$
$\quad\quad -X_1 + 2X_2 + 4X_3 = -6.2$

5-18. $\quad 4X_1 - 3X_2 - X_3 = -2.1$
$\quad\quad 3X_1 - 2X_2 + 5X_3 = -12.3$
$\quad\quad -2X_1 + 2X_2 + 3X_3 = 12.9$

5-19. $\quad 2X_1 + X_2 - 3X_3 + 2X_4 = 3.7$
$\quad\quad -3X_1 - X_2 - 4X_3 + 5X_4 = 3.3$
$\quad\quad X_1 + 5X_2 - 3X_3 + 4X_4 = 4.1$
$\quad\quad 5X_1 - 3X_2 + 2X_3 - 2X_4 = 6.4$

5-20. $3X_1 - 3X_2 + 2X_3 - 4X_4 = 7.9$
$\quad\quad -2X_1 - X_2 + 3X_3 - X_4 = -12.5$
$\quad\quad 5X_1 - 2X_2 - 3X_3 + 2X_4 = 18.0$
$\quad\quad -2X_1 + 4X_2 + X_3 + 2X_4 = -8.1$

Find the solution to Problems 5-21 through 5-26 using the Gauss–Seidel iteration method.

5-21. $5X_1 - 2X_2 + 3X_3 = 1.3$
$\quad\quad -2X_1 + 4X_2 - X_3 = -1.4$
$\quad\quad 2X_1 - X_2 + 3X_3 = 1.3$

5-22. $-2X_1 + 3X_2 - X_3 = -5.7$
$\quad\quad 3X_1 - 2X_2 - 5X_3 = 2.1$
$\quad\quad X_1 + 4X_2 + 2X_3 = 17.1$

5-23. $3X_1 - 3X_2 + 6X_3 = -5.7$
$\quad\quad 2X_1 + 4X_2 - 2X_3 = 11.2$
$\quad\quad -X_1 + X_2 - 2X_3 = 1.9$

5-24. $\quad 5X_1 + 2X_2 - 3X_3 = -25.2$
$\quad\quad -3X_1 + 4X_2 + 4X_3 = 19.2$
$\quad\quad 2X_1 - 2X_2 + 3X_3 = 2.8$

5-25. $2X_1 - 3X_2 + 3X_3 - X_4 = -7.9$
$-5X_1 + X_2 + X_3 - 2X_4 = -8.1$
$2X_1 + 3X_2 + 4X_3 - 3X_4 = -9.2$
$-X_1 - 2X_2 + 2X_3 - 3X_4 = -11.8$

5-26. $3X_1 + 4X_2 - 2X_3 - X_4 = 29.9$
$4X_1 - 2X_2 + 2X_3 + 3X_4 = 2.4$
$-2X_1 - 3X_2 + 4X_3 - X_4 = -29.6$
$2X_1 + 2X_3 - 3X_4 = 2.6$

For Problems 5-27 through 5-30, find the solution using both the Jacobi and Gauss–Seidel iteration methods and compare the rates of convergence.

5-27. $2X_1 - 3X_2 = 6.5$
$5X_1 + 2X_2 = 9.6$

5-28. $6X_1 - 5X_2 = -36.3$
$3X_1 + 4X_2 = -4.5$

5-29. $3X_1 + X_2 - 3X_3 = -4.2$
$-2X_1 + 4X_2 - X_3 = -14.1$
$4X_1 - 2X_2 + 3X_3 = 15.7$

5-30. $4X_1 - 2X_2 + 3X_3 = -10.8$
$3X_1 + 5X_2 - 2X_3 = 11.9$
$-2X_1 - 2X_2 + 5X_3 = -0.2$

Find the solution to Problems 5-31 through 5-36 using the method of determinants.

5-31. $2X_1 + 3X_2 = -0.3$
$-3X_1 + 4X_2 = -12.3$

5-32. $6X_1 - 5X_2 = -19.6$
$3X_1 + 4X_2 = 11.0$

5-33. $3X_1 - 2X_2 + 4X_3 = 1.2$
$-2X_1 + 5X_2 = 10.7$
$2X_1 + 3X_2 - 4X_3 = 8.5$

5-34. $6X_1 - 5X_2 + 2X_3 = -20.3$
$3X_1 - 2X_2 + 7X_3 = 2.1$
$4X_2 - 3X_3 = 1.1$

5-35. $2X_1 - 3X_2 + X_3 - 2X_4 = -10.3$
$3X_1 + 4X_2 - 5X_3 + X_4 = 4.3$
$-2X_1 - 2X_2 + 4X_4 = -5.4$
$X_1 + X_2 - 3X_3 + X_4 = -1.8$

5-36. $5X_1 - X_2 + X_3 + 3X_4 = 11.9$
$-2X_1 + 3X_2 - X_3 + 2X_4 = -2.0$
$3X_1 + X_2 - 3X_3 + 5X_4 = 14.1$
$2X_1 - 4X_2 + X_3 + 2X_4 = 4.4$

For Problems 5-37 through 5-40, find the inverse of the coefficient matrix shown.

5-37. $\begin{bmatrix} 1 & -0.45 \\ -0.45 & 1 \end{bmatrix}$

5-38. $\begin{bmatrix} 2.1 & -1.4 \\ 0.7 & 1.8 \end{bmatrix}$

5-39. $\begin{bmatrix} 1 & -0.62 & 0.37 \\ -0.62 & 1 & -0.51 \\ 0.37 & -0.51 & 1 \end{bmatrix}$

5-40. $\begin{bmatrix} 7 & -4 & 3 \\ -2 & 2 & -3 \\ 3 & 2 & -4 \end{bmatrix}$

In the statistical fitting of a linear equation of the form $Y = a + bX$, values for the fitting coefficients a and b are obtained by solving the two simultaneous equations

$$an + b \sum_{i=1}^{n} X_i = \sum_{i=1}^{n} Y_i$$

$$a \sum_{i=1}^{n} X_i + b \sum_{i=1}^{n} X_i^2 = \sum_{i=1}^{n} X_i Y_i$$

where n is the number of pairs of values of the two variables X and Y, and each summation is made over the n values. For Problems 5-41 through 5-43, solve for the values of a and b using any of the numerical algorithms to solve simultaneous equations.

5-41.

i	X_i	Y_i
1	1	2
2	2	1
3	3	1
4	3	3
5	6	4

5-42.

i	X_i	Y_i
1	0	5
2	1	3
3	2	4
4	4	3
5	6	1
6	7	2

5-43.

i	X_i	Y_i
1	4	− 2
2	9	− 3
3	13	2
4	21	1
5	24	4
6	28	7

Formulate a set of linear equations that represent the relationship between voltage, current, and resistances for the circuits shown in Problems 5-44 through 5-47. Solve the simultaneous equations using any of the methods discussed in this chapter.

5-44.

Figure P5.44

5-45.

Figure P5.45

5-46.

Figure P5.46

5-47.

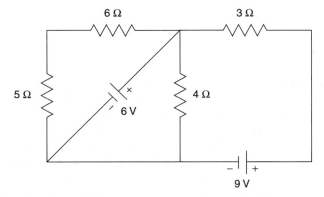

9 V   **Figure P5.47**

5-48. Write a computer program for the Gauss elimination method as outlined in Section 5.4.3 (in Eqs. 5-45). Test the program using Examples 5-3, 5-4, and 5-5.

5-49. Write a computer program for the Gauss–Jordan elimination method as outlined in Eqs. 5-45 in Section 5.4.3. Test the program using Example 5-6.

6

Numerical Interpolation

6.1 INTRODUCTION

Problems requiring interpolation between individual data points occur frequently in science and engineering. For example, the design of a computerized energy control system for a building may require as input data the typical temperature variation occurring in the building each day. Sample temperature values would be measured in the building at discrete time points. However, the computer program for the energy control system may require temperature at times other than those at which the sample measurements were taken, such as on an hourly increment. One way to overcome this problem is to have the program fit a curve to the measured temperature points and interpolate for values between the sample measurements.

Interpolation is required in many engineering applications that use tabular data as input. The basis of all interpolation algorithms is the fitting of some type of curve or function to a subset of the tabular data; linear interpolation uses a straight line. Interpolation algorithms differ in the form of their interpolation functions.

6.2 METHOD OF UNDETERMINED COEFFICIENTS

The method of undetermined coefficients is conceptually the simplest interpolation algorithm, and it illustrates many of the key points that hold for all interpolation schemes. In this method, an nth-order polynomial is used as the interpolation function, $f(x)$:

$$f(x) = b_0 + b_1 x + b_2 x^2 + b_3 x^3 + \cdots + b_n x^n \qquad (6\text{-}1)$$

The constants in Eq. 6-1, $b_0, b_1, b_2, b_3, \ldots, b_n$, are determined using the measured data points.

As an example of the method of undetermined coefficients, consider the easiest case, linear interpolation; that is, the interpolation function consists of a straight line. Table 6-1 contains numbers and their cubes for the purpose of illustration.

The method of undetermined coefficients is used to estimate the cube of 2.2. Of

TABLE 6-1   Data for Linear Interpolation

x	x^3
0	0
1	1
2	8
3	27
4	64

course, in this case, the answer can be calculated exactly, if we desired; however, linear interpolation is used instead to illustrate the general concepts. By truncating all except the first two terms of Eq. 6-1, the linear interpolation function has the form

$$f(x) = b_0 + b_1 x \tag{6-2}$$

Since there are two unknown coefficients, b_0 and b_1, in Eq. 6-2, two data points are necessary to obtain values for the coefficients. Any two data points could be used, but common sense dictates that the best approximation is obtained when the two data points closest to the value we are trying to determine are used. As shown in Fig. 6-1, the data points corresponding to $x = 2$ and $x = 3$ are used to estimate the value for $x = 2.2$; our interpolation function $f(x)$ is then the straight line connecting these two points from the table. Using concepts from algebra, the equation of the linear interpolation function is

$$f(x) = f(x_i) + \frac{x - x_i}{x_{i+1} - x_i} [f(x_{i+1}) - f(x_i)] \tag{6-3}$$

The ratio $[f(x_{i+1}) - f(x_i)]/(x_{i+1} - x_i)$ is the slope of the line between the two points. The difference $(x - x_i)$ represents a deviation from point x_i so that the product of the deviation and slope represents a correction to the base point value $f(x_i)$. Substituting values from Table 6-1 yields

$$f(x) = 8 + \frac{x - 2}{3 - 2} (27 - 8) \tag{6-4a}$$

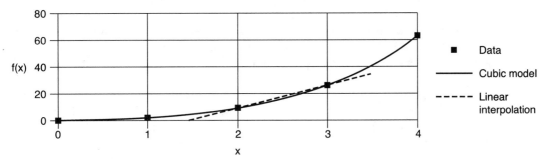

Figure 6-1   Linear Interpolation

Rearranging yields

$$f(x) = -30 + 19x \qquad (6\text{-}4b)$$

At the two observation points, the values for $f(x)$ from Eq. 6-4b exactly equal the cubes of the x values; at values for x other than the observation points, the values for $f(x)$ from Eq. 6-4b only approximate the true values for the cubes of x. From Fig. 6-1, it is evident that the accuracy of the approximation varies with the value of x, with the poorest accuracy within the range $2 < x < 3$ near the center of the interval.

Equation 6-3 is an algebraic form for estimating $f(x)$. Algebraic manipulation of Eq. 6-3 yields the following, which has the form of Eq. 6-2:

$$f(x) = \frac{x_{i+1}f(x_i) - x_i f(x_{i+1})}{x_{i+1} - x_i} + \frac{f(x_{i+1}) - f(x_i)}{x_{i+1} - x_i}x \qquad (6\text{-}5)$$

It can be noted that the manipulation produced the slope in the second term of Eq. 6-5. Using the above values of x and x^3 with Eq. 6-5 yields

$$f(x) = -30 + 19x \qquad (6\text{-}6)$$

This is the same interpolation function derived with Eq. 6-3.

Although we could write the interpolation function in Eq. 6-4 by inspection for this simple case, we can alternatively approach the problem in a more methodical fashion. Since the interpolation passed through the two data points spanning $x = 2.2$, Eq. 6-2 can be used with the values of the two points nearest to $x = 2.2$ to write the following two equations:

$$8 = b_0 + 2b_1 \qquad (6\text{-}7a)$$

$$27 = b_0 + 3b_1 \qquad (6\text{-}7b)$$

Equations 6-7 represent a system of two equations for the two unknown coefficients. These equations can be solved for b_0 and b_1 using the usual methods from algebra or the equation-solving algorithms covered in Chapter 5 to yield the solution $b_0 = -30$ and $b_1 = 19$. These values can be substituted into Eq. 6-2, producing the same interpolation function as we obtained in Eq. 6-6.

The linear interpolation function predicts a value of 11.8 for the cube of $x = 2.2$. Since the actual value for the cube of 2.2 is 10.648, the discrepancy or error in the interpolation is 1.152, or 10.8% of the true value at this point. As shown in Fig. 6-1, the reason for this discrepancy is the fact that the interpolation function only approximately matches the actual function between the two end points; in this case, the interpolation function always lies above the actual function. The error in the interpolation will always be zero at the observation *end* points and a maximum at some location between the end points.

The interpolation function could also be used to extrapolate values for the cube of x outside the range of our two observation points. However, as shown in Fig. 6-1, the errors in the approximation quickly grow very large. The extrapolated value from Eq. 6-6 for the cube of 4.0 is 46, which differs from the true value of 64 by over 28%. Extrapolation, in general, gives larger errors than interpolation.

Using a function that is more flexible than a straight line, the accuracy of an estimate can be improved. For example, a nonlinear function that has more coefficients than the linear interpolation function should provide greater accuracy. However, since a nonlinear function may contain more unknown coefficients than the straight line, it may be necessary to use more of the data points to estimate the coefficients. For example, by truncating the terms of Eq. 6-1 beyond the x^2 term, the following second-order polynomial could be used as an interpolation function:

$$f(x) = b_0 + b_1 x + b_2 x^2 \tag{6-8}$$

Determining the three unknown coefficients of Eq. 6-8 requires three pairs of values of x and its cube. To estimate the value of x^3 for $x = 2.2$ using the data of Table 6-1, coefficients for Eq. 6-8 can be fitted using x values of either 1, 2, and 3 or 2, 3, and 4. Since 2.2 is closer to 2 than to 3, 2 will be used as the center point. Evaluating Eq. 6-8 at the three data points yields the following set of simultaneous equations:

$$1 = b_0 + b_1 + b_2 \tag{6-9a}$$

$$8 = b_0 + 2b_1 + 4b_2 \tag{6-9b}$$

$$27 = b_0 + 3b_1 + 9b_2 \tag{6-9c}$$

Solving these three simultaneous equations yields the following values for the coefficients: $b_0 = 6$, $b_1 = -11$, and $b_2 = 6$. Therefore, the interpolation equation is

$$f(x) = 6 - 11x + 6x^2 \tag{6-9d}$$

The estimated value of $(2.2)^3$ from Eq. 6-9d is 10.840, which differs from the true value (10.648) by 1.8%. Thus, using a second-order function of Eq. 6-9d instead of the first-order function of Eq. 6-6 reduced the interpolation error by nearly an order of magnitude in this case. The cost of this increased precision is the additional effort required to determine the coefficients of the second-order equation, that is, Eq. 6-8 rather than Eq. 6-2.

The method of undetermined coefficients can be generalized to interpolation polynomials of any order. The general nth-order polynomial contains $n + 1$ undetermined coefficients and therefore requires $n + 1$ distinct data points for its solution. Each of the $n + 1$ data points is substituted into the interpolation polynomial in turn, generating $n + 1$ simultaneous equations that are solved to obtain the polynomial coefficients.

6.3 GREGORY–NEWTON INTERPOLATION METHOD

The previous analysis indicated that the second-order polynomial provides a significant improvement in accuracy when compared with the linear, or first-order, equation. Thus it may seem reasonable to try an interpolation polynomial of higher order. However, we need to recognize that the solution procedure becomes more complex as the order of the polynomial increases when the solution procedure based on solving a set of simultaneous equations is used. The Gregory–Newton method provides a means of developing an nth-order interpolation polynomial without requiring the solution of a set of simultaneous equations. The nth-order Gregory–Newton interpolation formula has the form

$$f(x) = a_1 + a_2(x - x_1) + a_3(x - x_1)(x - x_2) + a_4(x - x_1)(x - x_2)(x - x_3) + \cdots$$
$$+ a_n(x - x_1)(x - x_2) \ldots (x - x_{n-1}) + a_{n+1}(x - x_1)(x - x_2) \ldots (x - x_n) \tag{6-10}$$

in which x is the independent variable; x_i for $i = 1, 2, \ldots, n$ are n known values of the independent variable; a_i for $i = 1, 2, \ldots, (n + 1)$ are the unknown coefficients; and $f(x)$ is the value of the dependent variable for a value x of the independent variable. The $(n + 1)$ coefficients of Eq. 6-10 can be evaluated from the set of $(n + 1)$ values of the dependent and independent variables. Assuming that the $(n + 1)$ values of the independent variable are placed in increasing order (that is, $x_j < x_{j+1}$), the value of a_1 can be computed from Eq. 6-10 using $f(x_1)$ and x_1. For this pair of values, every term in Eq. 6-10 except the constant a_1 is zero because $(x - x_1)$ equals $(x_1 - x_1)$, which equals zero. Thus the value of a_1 equals $f(x_1)$. Given the just-computed value of a_1 and the pair of values x_2 and $f(x_2)$, we can compute a_2. For this case, all terms in Eq. 6-10 that contain $(x - x_2)$ are zero because $(x_2 - x_2)$ is zero. Thus a_2 is derived from

$$f(x_2) = a_1 + a_2(x_2 - x_1) \tag{6-11}$$

which can be solved for a_2:

$$a_2 = \frac{f(x_2) - a_1}{x_2 - x_1} = \frac{f(x_2) - f(x_1)}{x_2 - x_1} \tag{6-12}$$

The coefficient a_3 can be computed using the values of a_1 and a_2, and the pair $[x_3, f(x_3)]$. In this case, all terms containing $(x_3 - x_3)$ are zero, and therefore

$$f(x_3) = a_1 + a_2(x_3 - x_1) + a_3(x_3 - x_1)(x_3 - x_2) \tag{6-13}$$

Solving for a_3 yields

$$a_3 = \frac{f(x_3) - a_1 - a_2(x_3 - x_1)}{(x_3 - x_1)(x_3 - x_2)} \tag{6-14}$$

The process can be repeated to solve for the remaining coefficients. At each step a new pair of points $[x_i$ and $f(x_i)]$ is used with the values of a_i computed in previous steps. The Gregory–Newton method yields the same polynomial as the solution of the simultaneous equations used for the method of undetermined coefficients. Thus it will provide the same interpolated value and have the same accuracy.

Example 6-1 Cubic Function

The Gregory–Newton method can be demonstrated using the three pairs of values of x and x^3 from Table 6-1. Using the pair (1, 1), we get $f(x_1) = a_1 = 1$. Using the second pair (2, 8) and Eq. 6-12, the value of a_2 can be computed as

$$a_2 = \frac{8 - 1}{2 - 1} = 7 \tag{6-15}$$

The third data pair (3, 27) and Eq. 6-14 can be used to compute the value of a_3:

$$a_3 = \frac{27 - 1 - 7(3 - 1)}{(3 - 1)(3 - 2)} = 6 \tag{6-16}$$

This produces the following interpolation polynomial, which has the form of Eq. 6-10:

$$f(x) = 1 + 7(x - 1) + 6(x - 1)(x - 2) \tag{6-17}$$

Thus, for $x = 2.2$ we get

$$f(2.2) = 1 + 7(2.2 - 1) + 6(2.2 - 1)(2.2 - 2) = 10.84 \tag{6-18}$$

which is the same interpolated value as was computed before using Eqs. 6-9. Equation 6-17 can be rearranged to produce

$$f(x) = 6 - 11x + 6x^2 \tag{6-19}$$

which is the same interpolation function derived previously using the method of undetermined coefficients (Eq. 6-9d).

Example 6-2 Vapor Pressure of Water

The vapor pressure of water is an important element in many engineering problems, including chemical reactions involving steam, evapotranspiration estimation problems in agricultural engineering design, and cavitation problems in hydraulic engineering. Let's assume that the vapor pressure of water between temperatures of 40° and 72°C are of interest. Table 6-2 provides five pairs of values at 8° increments.

TABLE 6-2   Vapor Pressure Data

T (C°)	P (mm Hg)
40	55.3
48	83.7
56	123.8
64	179.2
72	254.5

The interpolation polynomial can be formed by applying Eq. 6-10 for each pair of values in increasing order of T. The calculations in this example are summarized to four significant figures in Table 6-3. This yields the following interpolating polynomial:

$$
\begin{aligned}
P(T) = {} & 55.3 + 3.55(T - 40) + 0.0914063(T - 40)(T - 48) \\
& + 0.001172(T - 40)(T - 48)(T - 56) \\
& + 0.00001017(T - 40)(T - 48)(T - 56)(T - 64)
\end{aligned} \tag{6-20}
$$

TABLE 6-3   Interpolation of Vapor Pressure

i	T (°C)	P (mm Hg)	Eq. 6-10	a_i
1	40	55.3	$55.3 = a_1$	55.3
2	48	83.7	$83.7 = a_1 + a_2(48 - 40)$	3.55
3	56	123.8	$123.8 = a_1 + a_2(56 - 40)$ $+ a_3(56 - 40)(56 - 48)$	0.0914063
4	64	179.2	$179.2 = a_1 + a_2(64 - 40)$ $+ a_3(64 - 40)(64 - 48)$ $+ a_4(64 - 40)(64 - 48)(64 - 56)$	0.001172
5	72	254.5	$254.5 = a_1 + a_2(72 - 40)$ $+ a_3(72 - 40)(72 - 48)$ $+ a_4(72 - 40)(72 - 48)(72 - 56)$ $+ a_5(72 - 40)(72 - 48)(72 - 56)(72 - 64)$	0.00001017

The interpolating polynomial of Eq. 6-20 is used to estimate the vapor pressure for a temperature of 52°C:

$$P(52) = 55.3 + 3.55(52 - 40) + 0.0914063(52 - 40)(52 - 48)$$
$$+ 0.001172(52 - 40)(52 - 48)(52 - 56) \qquad (6\text{-}21)$$
$$+ 0.00001017(52 - 40)(52 - 48)(52 - 56)(52 - 64) = 102.0 \text{ mm Hg}$$

The true value of P for a T of 52°C is 102.1; thus the interpolated value is in error by approximately 0.1, which is about 0.1% of the true value.

6.4 FINITE-DIFFERENCE INTERPOLATION

A finite-difference scheme can be used to develop an interpolation polynomial when the known values of the independent variable are equally spaced. If Δx is the increment on which the values of the independent variable are recorded, then the first finite difference of the dependent variable $y = f(x)$ is

$$\Delta f = f(x + \Delta x) - f(x) \qquad (6\text{-}22)$$

in which $f(x)$ is the value of the dependent variable for the value of the independent variable x. The second finite difference is the following difference of the first finite difference:

$$\Delta^2 f = \Delta[\Delta f] = [f(x + 2\Delta x) - f(x + \Delta x)] - [f(x + \Delta x) - f(x)] \qquad (6\text{-}23)$$

Continuing for successive differences, it can be shown that the nth finite difference is the following first difference of the $(n - 1)$th difference:

$$\Delta^n f = \Delta[\Delta^{n-1} f] \qquad (6\text{-}24)$$

The calculation of the interpolating polynomial is based on the following rule: The ith finite difference, where i is less than or equal to n, of an nth-degree polynomial is a polynomial of degree $(n - 1)$. This rule implies that the nth difference would be a constant. Assume the following nth-order polynomial:

$$f(x) = b_0 + b_1 x + b_2 x^2 + \cdots + b_{n-1} x^{n-1} + b_n x^n \tag{6-25}$$

in which b_n does not equal zero. For $(x + \Delta x)$, the polynomial of Eq. 6-25 becomes

$$f(x + \Delta x) = b_0 + b_1(x + \Delta x) + b_2(x + \Delta x)^2 + \cdots + b_n(x + \Delta x)^n \tag{6-26}$$

which can be rearranged into the form

$$f(x + \Delta x) = b_n x^n + (b_{n-1} + n\Delta x b_n)^{n-1} + \cdots \tag{6-27}$$

The first finite difference can be formed by subtracting Eq. 6-25 from Eq. 6-27 as follows:

$$\Delta f = f(x + \Delta x) - f(x) \tag{6-28a}$$

or

$$\Delta f = (b_{n-1} + n\Delta x b_n)x^{n-1} + \cdots \tag{6-28b}$$

where Eq. 6-28b is a polynomial of order $(n - 1)$. For $(x + 2\Delta x)$, Eq. 6-26 becomes

$$f(x + 2\Delta x) = b_n(x + 2\Delta x)^n + b_{n-1}(x + 2\Delta x)^{n-1} + \cdots \tag{6-29}$$

Equations 6-29, 6-27, and 6-25 can be substituted into Eq. 6-23 to compute the second finite difference, with the following result:

$$\Delta^2 f = n(n - 1)(\Delta x)^2 b_n x^{n-2} + \cdots \tag{6-30}$$

which is a polynomial of order $(n - 2)$. By induction, the ith finite difference is given by

$$\Delta^i f = n(n - 1)(n - 2) \cdots (n - i + 1)(\Delta x)^i b_n x^{n-i} + \cdots \tag{6-31}$$

which is a polynomial of order $(n - i)$. The nth difference equals the following constant:

$$\Delta^n f = n(n - 1)(n - 2) \ldots (1)(\Delta x)^n b_n = n! \, (\Delta x)^n b_n \tag{6-32}$$

For the case where n equals 2, Eq. 6-25 becomes

$$f(x) = b_2 x^2 + b_1 x + b_0 \tag{6-33}$$

For $(x + \Delta x)$ and $(x + 2\Delta x)$, we can define the following:

$$f(x + \Delta x) = b_2(x + \Delta x)^2 + b_1(x + \Delta x) + b_0 \tag{6-34a}$$

$$= b_2(x^2 + 2x\Delta x + (\Delta x)^2) + b_1(x + \Delta x) + b_0 \tag{6-34b}$$

$$= b_2 x^2 + (2\Delta x b_2 + b_1)x + (b_2(\Delta x)^2 + b_1\Delta x + b_0) \tag{6-34c}$$

$$f(x + 2\Delta x) = b_2(x + 2\Delta x)^2 + b_1(x + 2\Delta x) + b_0 \tag{6-35a}$$

$$= b_2(x^2 + 4x\Delta x + (2\Delta x)^2) + b_1(x + 2\Delta x) + b_0 \tag{6-35b}$$

$$= b_2 x^2 + (4b_2 \Delta x + b_1)x + (b_2(2\Delta x)^2 + 2b_1\Delta x + b_0) \qquad \text{(6-35c)}$$

The first difference is computed using Eqs. 6-22, 6-33, and 6-34 as follows:

$$\Delta f = f(x + \Delta x) - f(x) \qquad \text{(6-36a)}$$

$$= 2b_2(\Delta x)x + (b_2(\Delta x)^2 + b_1\Delta x) \qquad \text{(6-36b)}$$

which is a first-order polynomial. The second finite difference is computed for Eqs. 6-34c and 6-33 with Eq. 6-23 as

$$\Delta^2 f = [f(x + 2\Delta x) - f(x + \Delta x)] - [f(x + \Delta x) - f(x)] \qquad \text{(6-37a)}$$

$$= 2b_2(\Delta x)^2 \qquad \text{(6-37b)}$$

which is a constant.

Example 6-3 Finite-difference Interpolation Polynomial

The equations for the second-order polynomial (Eqs. 6-33 through 6-37) can be used to develop an interpolation polynomial for the data in Table 6-4. The first finite difference from Eq. 6-22 is

$$\Delta f = f(x + \Delta x) - f(x) = 5 - 3 = 2 \qquad \text{(6-38)}$$

TABLE 6-4   Data for Finite-difference Interpolation

x	$f(x)$
1	3
2	5
3	8

This can be equated to Eq. 6-36b with $\Delta x = 1$ to obtain the following:

$$5 - 3 = 2b_2(\Delta x)x + b_2(\Delta x)^2 + b_1\Delta x \qquad \text{(6-39a)}$$

$$2 = 2b_2(1)(1) + b_2(1)^2 - b_1(1) \qquad \text{(6-39b)}$$

$$2 = 3b_2 + b_1 \qquad \text{(6-39c)}$$

The second finite difference from Eq. 6-23 is

$$\Delta^2 f = [8 - 5] - [5 - 3] = 1 \qquad \text{(6-40)}$$

This can be equated to Eq. 6-37b to produce the following:

$$1 = 2b_2(\Delta x)^2 = 2b_2 \qquad \text{(6-41)}$$

Solving Eq. 6-41 yields $b_2 = 0.5$. Substituting this result into Eq. 6-39c yields $b_1 = 0.5$. Using these values of b_2 and b_1 and any one of the three points, a value of b_0 can be computed from Eq. 6-33. Using the point $f(1) = 3$, we get the following equation:

$$f(x) = 3 = b_2(1)^2 + b_1(1) + b_0 = 1 + b_0 \qquad \text{(6-42)}$$

Therefore, b_0 equals 2. Thus the interpolating polynomial is

$$f(x) = 0.5x^2 + 0.5x + 2 \qquad (6\text{-}43)$$

For any of the three points used to derive the interpolating polynomial, Eq. 6-43 yields an exact value. The equation can be used to interpolate a value of $f(x)$ for any x in the range of the data. For example, the value of $f(x)$ at $x = 2.3$ can be computed as

$$f(2.3) = 0.5(2.3)^2 + 0.5(2.3) + 2 = 5.795$$

6.4.1 Finite-difference Table

The finite-differences computed above can be organized into a table, as shown in Table 6-5. The finite-difference table can be used to derive an interpolating polynomial, as will be shown in the next section. The finite-difference table of Table 6-6 shows the differences when integers from 10 to 14 are the independent variable x and their cubes are the dependent variable $f(x)$.

TABLE 6-5   Finite-difference Table

x	$f(x)$	Δf	$\Delta^2 f$	$\Delta^3 f$	$\ldots$	$\Delta^n f$
x	$f(x)$					
		$\Delta f(x)$				
$x + \Delta x$	$f(x + \Delta x)$		$\Delta^2 f(x)$			
		$\Delta f(x + \Delta x)$		$\Delta^3 f(x)$		
$x + 2\Delta x$	$f(x + 2\Delta x)$		$\Delta^2 f(x + \Delta x)$			
		$\Delta f(x + 2\Delta x)$		$\Delta^3 f(x + \Delta x)$		
$x + 3\Delta x$	$f(x + 3\Delta x)$		$\Delta^2 f(x + 2\Delta x)$			
		$\Delta f(x + 3\Delta x)$		$\Delta^3 f(x + 2\Delta x)$		
			$\Delta^2 f(x + 3\Delta x)$			
$\vdots$	$\vdots$	$\vdots$	$\vdots$	$\vdots$	$\vdots\vdots\vdots$	$\Delta^n f(x)$
$x + (n-2)\Delta x$	$f[x + (n-2)\Delta x]$		$\Delta^2 f[x + (n-3)\Delta x]$			
		$\Delta f[x + (n-2)\Delta x]$		$\Delta^3 f[x + (n-3)\Delta x]$		
$x + (n-1)\Delta x$	$f[x + (n-1)\Delta x]$		$\Delta^2 f[x + (n-2)\Delta x]$			
		$\Delta f[x + (n-1)\Delta x]$				
$x + n\Delta x$	$f(x + n\Delta x)$					

TABLE 6-6   Finite-difference Table for Cubes

x	$f(x)$	Δf	$\Delta^2 f$	$\Delta^3 f$	$\Delta^4 f$
10	1000				
		331			
11	1331		66		
		397		6	
12	1728		72		0
		469		6	
13	2197		78		
		547			
14	2744				

Example 6-4 Interpolation of the Standard Normal Distribution Table

It is frequently necessary to interpolate values from probabilities tables, for example, the standard normal distribution. The finite-difference scheme provides a convenient method of developing an interpolating polynomial. Consider the values of the standard normal deviate z and the cumulative probability, $f(z)$, which are given in the finite-difference table of Table 6-7. To fit the coefficients of the second-order polynomial, Eqs. 6-36b, 6-37b, and 6-33 can be applied with the top diagonal row of the finite-difference table. For computing b_2, $\Delta z = -1.6 - (-1.5) = -0.1$ and Eq. 6-37b can be stated as

$$\Delta^2 f = 2b_2(\Delta z)^2 \tag{6-44a}$$

$$0.0018 = 2b_2(-0.1)^2 \tag{6-44b}$$

$$b_2 = 0.09 \tag{6-44c}$$

TABLE 6-7   Finite-difference Table for Standard Normal Distribution

z	$f(z)$	Δf	$\Delta^2 f$
−1.5	0.0668		
		−0.0120	
−1.6	0.0548		0.0018
		−0.0102	
−1.7	0.0446		

For computing b_1, Eq. 6-36b produces

$$\Delta f = 2b_2(\Delta z)\, z + b_2(\Delta z)^2 + b_1(\Delta z) \tag{6-45a}$$

$$-0.0120 = 2(0.09)(-0.1)(-1.5) + 0.09(-0.1)^2 + b_1(-0.1) \tag{6-45b}$$

$$b_1 = 0.399 \tag{6-45c}$$

For computing b_0, Eq. 6-33 produces

$$f(z) = b_2 z^2 + b_1 z + b_0 \tag{6-46a}$$

$$0.0668 = 0.09\,(-1.5)^2 + 0.399\,(-1.5) + b_0 \tag{6-46b}$$

$$b_0 = 0.4628 \tag{6-46c}$$

Thus the interpolating polynomial is

$$f(z) = 0.09\, z^2 + 0.399z + 0.4628 \tag{6-46d}$$

Substituting the values for z of -1.5, -1.6, and -1.7 yields the exact values of $f(z)$ shown in the finite-difference table. The interpolating polynomial can be used to estimate the probability f for any z. For example, given $z = -1.55$, the polynomial yields 0.0606, which equals the tabulated value in probability tables. The polynomial can be used to extrapolate; for example, for $z_0 = -1.8$, the polynomial yields 0.0362, which is about 1% greater than the actual value of 0.0359.

6.5 NEWTON'S METHOD

The finite-difference table of Table 6-5 provides a means for deriving an interpolation function. It can be shown that the finite differences of the first diagonal row are given by

$$f(x + \Delta x) = f(x) - \Delta f(x) \tag{6-47a}$$

$$\Delta f(x + \Delta x) = \Delta f(x) + \Delta^2 f(x) \tag{6-47b}$$

$$\Delta^2 f(x + \Delta x) = \Delta^2 f(x) + \Delta^3 f(x) \tag{6-47c}$$

$$\vdots$$

$$\Delta^{n-1} f(x + \Delta x) = \Delta^{n-1} f(x) + \Delta^n f(x) \tag{6-47d}$$

The same procedure could be used for any diagonal row. For example, the equations for the next diagonal row are

$$f(x + 2\Delta x) = f(x) + \Delta f(x) + \Delta f(x + \Delta x) \tag{6-48a}$$
$$= f(x) + \Delta f(x) + \Delta f(x) + \Delta^2 f(x)$$

$$= f(x) + 2\Delta f(x) + \Delta^2 f(x) \tag{6-48b}$$

$$\Delta f(x + 2\Delta x) = \Delta f(x) + 2\Delta^2 f(x) + \Delta^3 f(x) \tag{6-48c}$$

$$\Delta^2 f(x + 2\Delta x) = \Delta^2 f(x) + 2\Delta^3 f(x) + \Delta^4 f(x) \tag{6-48d}$$

$$\vdots$$

$$\Delta^{n-3}f(x + 2\Delta x) = \Delta^{n-3}f(x) + 2\Delta^{n-2}f(x) + \Delta^{n-1}f(x) \qquad (6\text{-}48e)$$

All these expansions have the form of a binomial expansion; thus they can be rewritten as

$$\Delta f(x + m\Delta x) = \sum_{i=0}^{m} b_i \Delta^i f(x) \qquad (6\text{-}49a)$$

$$\Delta f(x + m\Delta x) = f(x) + m\Delta f(x) + \frac{m(m-1)}{2}\Delta^2 f(x) + \cdots + \Delta^m f(x) \qquad (6\text{-}49b)$$

where b_i = constant as defined by Eqs. 6-49a and b. In general,

$$\Delta^r f(x + m\Delta x) = \sum_{i=0}^{m} b_i \Delta^{r+1} f(x) \qquad (6\text{-}50a)$$

$$\Delta^r f(x + m\Delta x) = \Delta^r f(x) + m\Delta^{r+1} f(x) + \cdots + \Delta^{r+m} f(x) \qquad (6\text{-}50b)$$

These equations yield the following Newton's interpolation formula:

$$\begin{aligned} f(x) = f(x_0) + n\Delta f(x_0) &+ \frac{n(n-1)}{2!}\Delta^2 f(x_0) + \cdots \\ &+ \frac{n(n-1)\cdots(n-m+1)}{m!}\Delta^m f(x_0) \end{aligned} \qquad (6\text{-}51)$$

in which

$$x = x_0 + n\Delta x \qquad (6\text{-}52)$$

or

$$n = \frac{x - x_0}{\Delta x} \qquad (6\text{-}53)$$

Thus, given the finite-difference table of Table 6-5, Newton's formula can be used to derive an interpolation polynomial.

Example 6-5 Tangents of Angles

Assume that we are interested in the values of the tangent of angles (θ) between 30° and 40° and a table with an increment of 2° is available. The accuracy of estimated values can be increased by developing a nonlinear interpolation polynomial, rather than using linear interpolation. The data are given in Table 6-8. Using the first diagonal row, an interpolation formula can be formed as follows:

$$\begin{aligned} \tan(\theta) = 0.5774 + n(0.0475) &+ \frac{n(n-1)}{2!}(0.0021) + \frac{n(n-1)(n-2)}{3!}(0.0003) \\ + \frac{n(n-1)(n-2)(n-3)}{4!}(0.0001) &+ \frac{n(n-1)(n-2)(n-3)(n-4)}{5!}(-0.0003) \end{aligned} \qquad (6\text{-}54)$$

For example, the tangent for θ equal to $36.5°$ can be computed using Eqs. 6-53 as

$$n = \frac{36.5 - 30}{2} = 3.25 \tag{6-55}$$

Thus Eq. 6-54 yields

$$\tan(36.5) = 0.5774 + 3.25(0.0475) + \frac{3.25(2.25)}{2}(0.0021)$$

$$+ \frac{3.25(2.25)(1.25)}{6}(0.0003) + \frac{3.25(2.25)(1.25)(0.25)}{24}(0.0001) \tag{6-56}$$

$$+ \frac{3.25(2.25)(1.25)(0.25)(-0.75)}{120}(-0.0003) = 0.73992$$

which is in error by 0.00004 since the true value is 0.73996. Given that the original values were recorded with four significant digits, the error is not significantly different from zero. Linear interpolation gives a value of 0.8402, which is in error by 0.00024 or approximately six times greater than the error resulting from the fifth-order polynomial of Eq. 6-56.

TABLE 6-8   Finite-difference Table for $\tan \theta$ Where $30° \le \theta \le 40°$

θ	$\tan \theta$	$\Delta \tan \theta$	$\Delta^2 \tan \theta$	$\Delta^3 \tan \theta$	$\Delta^4 \tan \theta$	$\Delta^5 \tan \theta$
30°	0.5774					
		0.0475				
32°	0.6249		0.0021			
		0.0496		0.0003		
34°	0.6745		0.0024		0.0001	
		0.0520		0.0004		−0.0003
36°	0.7265		0.0028		−0.0002	
		0.0548		0.0002		
38°	0.7813		0.0030			
		0.0578				
40°	0.8391					

Example 6-6 Dissolved Oxygen Concentration

The dissolved oxygen concentration is an important element in some water-quality analyses. The saturation values are a function of temperature. Table 6-9 gives the saturated values (D) for temperatures (T) from $0°$ to $25°C$ in increments of $5°$. Table 6-9 also shows the finite-difference table, with D as the dependent variable. The following is Newton's interpolation polynomial for the data of Table 6-9:

TABLE 6-9   Finite-difference Table for Dissolved Oxygen Concentrations (D)

$T\ (°C)$	D (mg/L)	ΔD	$\Delta^2 D$	$\Delta^3 D$	$\Delta^4 D$	$\Delta^5 D$
0	14.62					
		-1.82				
5	12.80		0.35			
		-1.47		-0.06		
10	11.33		0.29		-0.03	
		-1.18		-0.09		0.11
15	10.15		0.20		0.08	
		-0.98		-0.01		
20	9.17		0.19			
		-0.79				
25	8.38					

$$D = 14.62 + n(-1.82) + \frac{n(n-1)}{2}(0.35) + \frac{n(n-1)(n-2)}{6}(-0.06)$$
$$+ \frac{n(n-1)(n-2)(n-3)}{24}(-0.03) + \frac{n(n-1)(n-2)(n-3)(n-4)}{120}(0.11) \tag{6-57}$$

Since the increment of the independent variable is 5°, then Eq. 6-53 becomes

$$n = \frac{T-0}{5} = \frac{T}{5} \tag{6-58}$$

Thus, if it is necessary to predict the saturation concentration of D at 13°C, n equals 2.6 and Eq. 6-57 results in

$$D = 14.62 + 2.6(-1.82) + \frac{2.6(1.6)}{2}(0.35) + \frac{2.6(1.6)(0.6)}{6}(-0.06)$$
$$+ \frac{2.6(1.6)(0.6)(-0.4)}{24}(-0.03) + \frac{2.6(1.6)(0.6)(-0.4)(-1.4)}{120}(0.11) \tag{6-59a}$$

Therefore,

$$D = 10.593569 \text{ mg/L} \tag{6-59b}$$

The true value is 10.60 mg/L, which yields an error of 0.006 mg/L. Linear interpolation yields an estimated value of 10.622 mg/L, which yields an error of 0.022 mg/L.

The interpolating polynomial of Eq. 6-59 can also be used for extrapolation. For example, for T equal to 30°, and n equals 6, the computed value of D would be 7.41 mg/L. This represents an error of 0.22 mg/L since the true value is 7.63 mg/L, or 2.9% of the true value.

6.6 LAGRANGE POLYNOMIALS

For many problems in engineering analysis, the data collection cannot be controlled, and the resulting data are for measurements taken at unequal intervals. In this case, data are collected for one variable $f(x)$, which is called the dependent variable, as a function of a second variable x, which is called the independent variable. For the previously discussed methods of interpolation, the x variable was assumed to be measured at a constant interval, Δx. A method that can be used with data for which the independent variable is measured on a variable interval Δx is introduced here. The method is referred to as Lagrangian interpolation.

Before introducing the method, it is necessary to provide the notations that are used in this section. The data sample is assumed to consist of n pairs of values measured on x and $f(x)$, with x_i being the ith measured value of the independent variable. The method provides an estimate of the value of $f(x)$ at a specific value of x, which is denoted as x_0; the estimated value is denoted as $f(x_0)$. A measured value of the dependent variable is denoted as $f(x_i)$, where $i = 1, 2, \ldots, n$. The method involves a weighting function, with the weight given to the ith value of f for x_0, denoted as $w_i(x_0)$.

The Lagrangian interpolating equation for estimating the value of $f(x_0)$ is

$$f(x_0) = \sum_{i=1}^{n} w_i(x_0)f(x_i) \tag{6-60}$$

The values of $f(x_j)$ are the measured values of the dependent variable, and the weights are a function of x_0 and the n values of the independent variable x. The weights are given by

$$w_i(x_0) = \frac{\prod_{\substack{j=1 \\ j \neq i}}^{n}(x_0 - x_j)}{\prod_{\substack{j=1 \\ j \neq i}}^{n}(x_i - x_j)} \tag{6-61}$$

in which the products ($\prod$) include $(n-1)$ terms, with the term where $j = i$ omitted. Equations 6-60 and 6-61 can be illustrated using the case where $n = 3$. In this case, Eq. 6-60 becomes

$$f(x_0) = w_1(x_0)f(x_1) + w_2(x_0)f(x_2) + w_3(x_0)f(x_3) \tag{6-62}$$

and the three weights are given by:

$$w_1(x_0) = \frac{(x_0 - x_2)(x_0 - x_3)}{(x_1 - x_2)(x_1 - x_3)} \qquad \text{(6-63a)}$$

$$w_2(x_0) = \frac{(x_0 - x_1)(x_0 - x_3)}{(x_2 - x_1)(x_2 - x_3)} \qquad \text{(6-63b)}$$

$$w_3(x_0) = \frac{(x_0 - x_1)(x_0 - x_2)}{(x_3 - x_1)(x_3 - x_2)} \qquad \text{(6-63c)}$$

Example 6-7 Sine Function

This method is illustrated using the sine as the dependent variable $f(x)$ and the angle x in degrees as the independent variable; that is, $f(x) = sin(x)$. The data in Table 6-10 are applied with Eqs. 6-62 and 6-63. If we wish to estimate the sine of 12°, then, using the Eqs. 6-63, the weights are

$$w_1(12) = \frac{(12 - 11)(12 - 13)}{(10 - 11)(10 - 13)} = -\frac{1}{3} \qquad \text{(6-64a)}$$

$$w_2(12) = \frac{(12 - 10)(12 - 13)}{(11 - 10)(11 - 13)} = 1.0 \qquad \text{(6-64b)}$$

$$w_3(12) = \frac{(12 - 10)(12 - 11)}{(13 - 10)(13 - 11)} = \frac{1}{3} \qquad \text{(6-64c)}$$

TABLE 6-10 Sine Function Data

i	x_i	$f(x_i)$
1	10	0.17365
2	11	0.19081
3	13	0.22495

Substituting the weights and the tabulated values of the sine of x into Eq. 6-62 yields

$$f(12) = w_1(12)f(x_1) + w_2(12)f(x_2) + w_3(12)f(x3) \qquad \text{(6-65a)}$$

$$f(12) = -\frac{1}{3}(0.17365) + 1.0(0.19081) + \frac{1}{3}(0.22495) = 0.20791 \qquad \text{(6-65b)}$$

The computed value equals the true value to five significant digits.

Example 6-8 Lagrange Polynomial for Well-water Elevation

A well pumping at 250 gallons per minute has observation wells located at 15, 42, 128, 317, and 433 feet away along a straight line from the well. After 3 hours of pumping, the following drawdowns in the five wells were observed: 14.6, 10.7, 4.8, 1.7, and 0.3 feet, respectively. If an estimate of the drawdown (f) at a distance of 25 feet is needed, the interpolating polynomial is

$$f(25) = 14.6w_1 + 10.7w_2 + 4.8w_3 + 1.7w_4 + 0.3w_5 \qquad (6\text{-}66)$$

with the weights given by:

$$w_1(25) = \frac{(25-42)(25-128)(25-317)(25-433)}{(15-42)(15-128)(15-317)(15-433)} = 0.5416 \qquad (6\text{-}67a)$$

$$w_2(25) = \frac{(25-15)(25-128)(25-317)(25-433)}{(42-15)(42-128)(42-317)(42-433)} = 0.4915 \qquad (6\text{-}67b)$$

$$w_3(25) = \frac{(25-15)(25-42)(25-317)(25-433)}{(128-15)(128-42)(128-317)(128-433)} = -0.0362 \quad (6\text{-}67c)$$

$$w_4(25) = \frac{(25-15)(25-42)(25-128)(25-433)}{(317-15)(317-42)(317-128)(317-433)} = 0.0039 \qquad (6\text{-}67d)$$

$$w_5(25) = \frac{(25-15)(25-42)(25-128)(25-317)}{(433-15)(433-42)(433-128)(433-317)} = -0.0008 \quad (6\text{-}67e)$$

Thus, substituting the weights of Eqs. 6-67 into Eq. 6-66 yields $f(25) = 13.0$ feet.

6.7 INTERPOLATION USING SPLINES

For a given data set of the type $(x_i, f(x_i))$, $i = 1, 2, \ldots, n$, the highest-order interpolating polynomial that can be used is of the $(n-1)$ order. Generally, the accuracy of interpolation can be improved by using higher-order polynomials. However, sometimes the interpolation accuracy can decrease, especially in data sets that include local abrupt changes in the values of $f(x)$ for a steady change in the value of x. The resulting high-order interpolating polynomial includes oscillations around the abrupt changes. Lower-order polynomials in this case can be more accurate than higher-order polynomials. The lower-order polynomials are called *splines*. The splines can be classified depending on the order of these polynomials. The most commonly used splines are linear, quadratic, and cubic. In this section these three types are described. Other higher-order splines can be similarly developed with a large increase in computational difficulty.

6.7.1 Linear Splines

A linear spline used for interpolation with a data set of the type $(x_i, f(x_i))$, $i = 1, 2, \ldots, n$, results in connecting adjacent data points with straight-line segments. The resulting spline is sometimes called a first-order spline. The interpolation function can be expressed as

$$f_1(x) = f(x_1) + \frac{f(x_2) - f(x_1)}{x_2 - x_1}(x - x_1) \qquad \text{for } x_1 \le x \le x_2 \qquad (6\text{-}68a)$$

$$f_2(x) = f(x_2) + \frac{f(x_3) - f(x_2)}{x_3 - x_2}(x - x_2) \qquad \text{for } x_2 \le x \le x_3 \qquad (6\text{-}68b)$$

$$f_i(x) = f(x_i) + \frac{f(x_{i+1}) - f(x_i)}{x_{i+1} - x_i}(x - x_i) \qquad \text{for } x_i \le x \le x_{i+1} \qquad (6\text{-}68c)$$

$$f_{n-1}(x) = f(x_{n-1}) + \frac{f(x_n) - f(x_{n-1})}{x_n - x_{n-1}}(x - x_{n-1}) \qquad \text{for } x_{n-1} \le x \le x_n \qquad (6\text{-}68d)$$

These equations result in linear interpolation between the points. The resulting overall interpolation function satisfies the following condition:

$$f_i(x_i) = f_{i+1}(x_i) \qquad \text{for } i = 1, 2, \ldots, n - 1 \qquad (6\text{-}69)$$

However, the function is discontinuous at the data points, which might make it visually or physically unacceptable.

Example 6-9 Linear Spline for Well-water Elevation

The data set that was previously used in Example 6-8 can be utilized to illustrate the linear spline. Using Eqs. 6-68, Table 6-11 can be constructed. The resulting linear spline is shown in Fig. 6-2. The figure shows that the linear spline results in exact predictions at the data points, but has discontinuities of slope at the data points.

TABLE 6-11   Linear Spline for Well-water Elevation

i	x_i (ft)	$f(x_i)$ (ft)	Linear Spline $f_i(x)$
1	15	14.6	
			$14.6 + \dfrac{10.7 - 14.6}{42 - 15}(x - 15) = 14.6 - 0.1444(x - 15)$
2	42	10.7	
			$10.7 - 0.0686046\,(x - 42)$
3	128	4.8	
			$4.8 - 0.0164021\,(x - 128)$
4	317	1.7	
			$1.7 - 0.0120689\,(x - 317)$
5	433	0.3	

6.7.2 Quadratic Splines

The quadratic splines provide a quadratic equation connecting any two adjacent data points within a data set of the type $(x_i, f(x_i))$, $i = 1, 2, \ldots, n$. The general form of a quadratic equation between point $(x_i, f(x_i))$ and $(x_{i+1} f(x_{i+1}))$ is

$$f_i(x) = a_i x^2 + b_i x + c_i \qquad \text{for } i = 1, 2, \ldots, n-1 \tag{6-70}$$

Therefore, every two adjacent data points have an interpolation equation as given by Eq. 6-70, with three constants: a_i, b_i, and c_i. As a result, there are $3(n-1)$ unknowns that need to be determined using the given data set, requiring $3(n-1)$ conditions. The conditions in this case are given by the following:

1. The splines must pass through the data points. For the ith spline (where $i = 1, 2, \ldots, n-1$), this condition can be expressed as

$$f_i(x_i) = a_i x_i^2 + b_i x_i + c_i = f(x_i) \qquad \text{for } i = 1, 2, 3, \ldots, n-1 \tag{6-71a}$$

$$f_i(x_{i+1}) = a_i x_{i+1}^2 + b_i x_{i+1} + c_i = f(x_{i+1}) \qquad \text{for } i = 1, 2, 3, \ldots, n-1 \tag{6-71b}$$

2. The splines must be continuous at the interior data points. This condition can be expressed using first derivatives of the quadratic splines as

$$2a_i x_i + b_i = 2a_{i+1} x_i + b_{i+1} \qquad \text{for } i = 1, 2, 3, \ldots, n-2 \tag{6-72}$$

3. The last needed condition can be arbitrarily set as the second derivative for the spline between the first two data points to be zero. Since the second derivative for the first spline is $2a_1$, this condition can be expressed as

$$a_1 = 0 \tag{6-73}$$

Equations 6-71 through 6-73 provide the needed $3(n-1)$ conditions to solve for the $3(n-1)$ unknowns, a_i, b_i, and c_i, $i = 1, 2, \ldots, n-1$. The resulting quadratic splines provide the desired continuity while passing through the data points.

Example 6-10 Quadratic Spline for Well-water Elevation

The data set that was previously used in Examples 6-8 and 6-9 can be utilized to illustrate the quadratic spline. Using Eqs. 6-71a, the following equations can be developed:

$$225a_1 + 15b_1 + c_1 = 14.6 \tag{6-74a}$$

$$1764a_2 + 42b_2 + c_2 = 10.7 \tag{6-74b}$$

$$16{,}384a_3 + 128b_3 + c_3 = 4.8 \tag{6-74c}$$

$$100{,}489a_4 + 317b_4 + c_4 = 1.7 \tag{6-74d}$$

Equations 6-71b results in the following conditions:

$$1764a_1 + 42b_1 + c_1 = 10.7 \tag{6-75a}$$

$$16{,}384a_2 + 128b_2 + c_2 = 4.8 \tag{6-75b}$$

$$100{,}489a_3 + 317b_3 + c_3 = 1.7 \tag{6-75c}$$

$$187{,}489a_4 + 433b_4 + c_4 = 0.3 \tag{6-75d}$$

Equations 6-72 results in the following conditions:

$$2a_1(42) + b_1 = 2a_2(42) + b_2 \tag{6-76a}$$

$$2a_2(128) + b_2 = 2a_3(128) + b_3 \tag{6-76b}$$

$$2a_3(317) + b_3 = 2a_4(317) + b_4 \tag{6-76c}$$

The last condition comes from Eq. 6-73 as

$$a_1 = 0 \tag{6-77}$$

Since $a_1 = 0$, the resulting system of 11 equations can be summarized in a matrix format as

$$
\begin{bmatrix}
15 & 1 & 0 & 0 & 0 & 0 & 0 & 0 & 0 & 0 & 0 \\
0 & 0 & 1{,}764 & 42 & 1 & 0 & 0 & 0 & 0 & 0 & 0 \\
0 & 0 & 0 & 0 & 0 & 16{,}384 & 128 & 1 & 0 & 0 & 0 \\
0 & 0 & 0 & 0 & 0 & 0 & 0 & 0 & 100{,}489 & 317 & 1 \\
42 & 1 & 0 & 0 & 0 & 0 & 0 & 0 & 0 & 0 & 0 \\
0 & 0 & 16{,}384 & 128 & 1 & 0 & 0 & 0 & 0 & 0 & 0 \\
0 & 0 & 0 & 0 & 0 & 100{,}489 & 317 & 1 & 0 & 0 & 0 \\
0 & 0 & 0 & 0 & 0 & 0 & 0 & 0 & 187{,}489 & 433 & 1 \\
1 & 0 & -84 & -1 & 0 & 0 & 0 & 0 & 0 & 0 & 0 \\
0 & 0 & 256 & 1 & 0 & -256 & -1 & 0 & 0 & 0 & 0 \\
0 & 0 & 0 & 0 & 0 & 634 & 1 & 0 & -634 & -1 & 0
\end{bmatrix}
\begin{bmatrix}
b_1 \\ c_1 \\ a_2 \\ b_2 \\ c_2 \\ a_3 \\ b_3 \\ c_3 \\ a_4 \\ b_4 \\ c_4
\end{bmatrix}
=
\begin{bmatrix}
14.6 \\ 10.7 \\ 4.8 \\ 1.7 \\ 10.7 \\ 4.8 \\ 1.7 \\ 0.3 \\ 0 \\ 0 \\ 0
\end{bmatrix}
\tag{6-7}
$$

The solution of this system of equations can be determined numerically to be

$$
\begin{bmatrix}
b_1 \\
c_1 \\
a_2 \\
b_2 \\
c_2 \\
a_3 \\
b_3 \\
c_3 \\
a_4 \\
b_4 \\
c_4
\end{bmatrix}
=
\begin{bmatrix}
-0.144444 \\
16.766667 \\
8.818583x10^{-4} \\
-0.2185206 \\
18.32227 \\
-1.2506x10^{-4} \\
3.925179x10^{-2} \\
1.824836 \\
2.411243x10^{-4} \\
-0.1929122 \\
38.62283
\end{bmatrix}
\tag{6-79}
$$

The resulting quadratic splines are shown in Fig. 6-3. For example, the first spline is $f(x) = (0)x^2 + (-0.14444)x + (16.766667)$, which is valid for $15 \le x \le 42$. The first spline is shown in the figure over its range of applicability (i.e., $15 \le x \le 42$). Other splines can be similarly expressed. Plotting each spline over its respective range of applicability produces Fig. 6-3. The figure shows that the quadratic spline results in predicting the data points with the appearance of continuity in both magnitude and slope at the measured data points.

6.7.3 Cubic Splines and Other Higher-order Splines

The cubic splines provide a cubic equation connecting any two adjacent data points within a data set of the type $(x_i, f(x_i))$, $i = 1, 2, \ldots, n$. The general form of a cubic equation between point $(x_i, f(x_i))$ and $(x_{i+1}, f(x_{i+1}))$ is

$$
f_i(x) = a_i x^3 + b_i x^2 + c_i x + d_i \qquad \text{for } i = 1, 2, \ldots, n - 1 \tag{6-80}
$$

Therefore, every two adjacent data points have an interpolation equation as given by Eq. 6-80, with four constants: a_i, b_i, c_i, and d_i. As a result, there are $4(n - 1)$ unknowns that

Figure 6-3   Quadratic Spline for Well Pump

need to be determined using the given data set, requiring $4(n - 1)$ conditions. The conditions in this case are given by the following:

1. The splines must pass through the data points. This condition can be expressed as

$$f_i(x_i) = a_i x_i^3 + b_i x_i^2 + c_i x_i + d_i = f_i(x_i)$$

$$\text{for } i = 1, 2, 3, \ldots, n - 1 \tag{6-81a}$$

$$f_i(x_{i+1}) = a_i x_{i+1}^3 + b_i x_{i+1}^2 + c_i x_{i+1} + d_i = f_i(x_{i+1})$$

$$\text{for } i = 1, 2, 3, \ldots, n - 1 \tag{6-81b}$$

2. The splines must be continuous at the interior data points. This condition can be expressed using first derivatives of the cubic splines as

$$3a_i x_{i+1}^2 + b_i x_{i+1} + c_i = 3a_{i+1} x_{i+1}^2 + b_{i+1} x_{i+1} + c_{i+1}$$

$$\text{for } i = 1, 2, 3, \ldots, n - 2 \tag{6-82}$$

3. The splines need to satisfy the second derivative continuity at the interior points. This condition can be expressed as

$$6a_i x_{i+1} + b_i = 6a_{i+1} x_{i+1} + b_{i+1}$$

$$\text{for } i = 1, 2, 3, \ldots, n - 2 \tag{6-83}$$

4. The last needed two conditions can be set as the second derivative at the first and last data points to be zero. This condition can be expressed as

$$6a_1 x_i + b_1 = 0 \tag{6-84a}$$

$$6a_{n-1} x_n + b_{n-1} = 0 \tag{6-84b}$$

Equations 6-81 through 6-84 provide the needed $4(n - 1)$ conditions to solve for the $4(n - 1)$ unknowns, a_i, b_i, c_i, and d_i, $i = 1, 2, \ldots, n - 1$. The resulting cubic splines provide the desirable continuity while passing through the data points.

The cubic splines are sometimes referred to as the third-order splines. Higher-order splines can be developed using similar conditions on higher-order derivatives. However, the computational difficulty greatly increases by needing to solve larger systems of simultaneous equations. The gained visual or physical benefits might be too negligible to justify their use.

6.8 GUIDELINES FOR CHOICE OF INTERPOLATION METHOD

The best interpolation method to use for any problem often depends on the details of the particular problem. However, the following several general guidelines can be offered:

1. The method of undetermined coefficients is perhaps the easiest method to understand conceptually. However, the method does require the solution of a set of $n + 1$
   simultaneous equations for an nth-order interpolation polynomial. One advantage
   of this method is that the interpolation polynomial can be *customized* for the particular problem at hand; for example, some of the terms of the general nth-order interpolation polynomial can be dropped, if desired. This is not possible in most of the
   other interpolation methods.

2. Both the Gregory–Newton and Lagrange polynomial interpolation methods eliminate the need to solve a set of simultaneous equations. Of the two methods, the
   Lagrange polynomial scheme is perhaps the easier to program. Although both of
   these methods (as well as the method of undetermined coefficients) work with un-
   equally spaced data points, the best accuracy is usually achieved when the differences in the spacing of the data points are minimal.

3. The finite-difference interpolation scheme requires equally spaced data points for
   its application.

4. Interpolation using splines can be more accurate than other methods, especially in
   cases where the data show large sudden changes.

All the interpolation schemes can also be used for extrapolation. Always remember,
though, that extrapolation of data usually produces less accurate estimates than interpolated values within the range of the data, and the accuracy gets worse as the distance between the extrapolated point and the bounds of the data points increases.

6.9 MULTIDIMENSIONAL INTERPOLATION

So far the discussion has been limited to one independent variable x. In this section, the
objective is to generalize the previous concepts to several independent variables. In general, the function of interest can be of the following form:

$$f(x_1, x_2, x_3, \ldots, x_n) \tag{6-85}$$

where $x_1, x_2, x_3, \ldots, x_n$ are n independent variables. The objective herein is similar to the
objective in previous sections for one-dimensional interpolation. The objective can be
stated as follows: we are given data in the form of the function $f(x_1, x_2, x_3, \ldots, x_n)$ value
at specified sets of $x_1, x_2, x_3, \ldots, x_n$ values, and the value of the function at some other
values of $x_1, x_2, x_3, \ldots, x_n$ is needed.

The concepts behind multidimensional interpolation can be developed using two-
dimensional interpolation. Once understanding is established for the two-dimensional
case, the concepts can be extended to other higher dimensions. In this section, the discussion is limited to two-dimensional interpolation. Also, the discussion is limited to one interpolation method, which is linear interpolation. The development of other methods is
similar, but with an increased computational difficulty. In most engineering applications
that require multidimensional interpolation, linear interpolation is adequate.

6.9.1 Linear Interpolation in Two Dimensions

Two-dimensional interpolation involves one dependent variable and two independent variables. Thus a given data set consists of the triads x_1, x_2, and $f(x_1, x_2)$. The data set can have the following form for two independent variables x_1 and x_2:

$$
\begin{array}{c|ccccc}
 & x_{11} & x_{12} & x_{1i} & \cdots & x_{1n} \\
\hline
x_{21} & f(x_{11}, x_{21}) & f(x_{12}, x_{21}) & f(x_{1i}, x_{21}) & \cdots & f(x_{1n}, x_{21}) \\
x_{22} & f(x_{11}, x_{22}) & f(x_{12}, x_{22}) & f(x_{1i}, x_{22}) & \cdots & f(x_{1n}, x_{22}) \\
x_{2j} & f(x_{11}, x_{2j}) & f(x_{12}, x_{2j}) & f(x_{1i}, x_{2j}) & \cdots & f(x_{1n}, x_{2j}) \\
\vdots & \vdots & \vdots & \vdots & \vdots \\
x_{2m} & f(x_{11}, x_{2m}) & f(x_{12}, x_{2m}) & f(x_{1i}, x_{2m}) & \cdots & f(x_{1n}, x_{2m})
\end{array}
\tag{6-86}
$$

For given values of x_1 and x_2, the value of $f(x_1, x_2)$ needs to be determined. Assume that the values of x_1 and x_2 fall between x_{1i} and $x_{1(i+1)}$ and x_{2j} and $x_{2(j+1)}$, respectively, as follows:

$$
\begin{array}{c|ccc}
 & x_{1i} & x_1 & x_{1(i+1)} \\
\hline
x_2 & f(x_{1i}, x_{2j}) & f(x_1, x_{2j}) & f(x_{1(i+1)}, x_{2j}) \\
x_2 & f(x_{1i}, x_2) & f(x_1, x_2) & f(x_{1(i+1)}, x_2) \\
x_{2(j+1)} & f(x_{1i}, x_{2(j+1)}) & f(x_1, x_{2(j+1)}) & f(x_{1(i+1)}, x_{2(j+1)})
\end{array}
\tag{6-87}
$$

All the values in Eq. 6-87 are known except $f(x_1, x_{2j})$, $f(x_{1i}, x_2)$, $f(x_{1(i+1)}, x_2)$, $f(x_1, x_{2(j+1)})$, and $f(x_1, x_2)$. The objective of the interpolation is to determine $f(x_1, x_2)$ by performing the following one-dimensional interpolations:

1. A one-dimensional interpolation is performed for x_1 at x_{2j} to obtain $f(x_1, x_{2j})$ as follows:

$$
f(x_1, x_{2j}) = f(x_{1i}, x_{2j}) + \frac{f(x_{1(i+1)}, x_{2j}) - f(x_{1i}, x_{2j})}{x_{1(i+1)} - x_{1i}}(x_1 - x_{1i})
\tag{6-88}
$$

2. A one-dimensional interpolation is performed for x_1 at $x_{2(j+1)}$ to obtain $f(x_1, x_{2(j+1)})$ as follows:

$$
f(x_1, x_{2(j+1)}) = f(x_{1i}, x_{2(j+1)}) + \frac{f(x_{1(i+1)}, x_{2(j+1)}) - f(x_{1i}, x_{2(j+1)})}{x_{1(i+1)} - x_{1i}}(x_1 - x_{1i})
\tag{6-89}
$$

3. Using the results of the previous one-dimensional interpolations, a one-dimensional interpolation is performed for x_2 at x_1 to obtain $f(x_1, x_2)$ as follows:

$$
f(x_1, x_2) = f(x_{1i}, x_{2j}) + \frac{f(x_1, x_{2(j+1)}) - f(x_1, x_{2j})}{x_{2(j+1)} - x_{2j}}(x_2 - x_{2j})
\tag{6-90}
$$

Alternatively, the two-dimensional interpolation can be performed using one-dimensional interpolation over x_2 first, followed by interpolation over x_1. The computational procedure is similar to Eqs. 6-88 through 6-90.

Example 6-11 Gravitational Acceleration as a Function of Latitude and Elevation

The gravitational acceleration (g) in feet per second squared is a function of latitude (L) in degrees and elevation (E) in feet. This relationship is commonly tabulated with L and E as the independent variables and g as the dependent variable. For the purpose of illustration, a problem might require the value of g at $L = 12.5°$ and $E = 500$. The relevant portion of the table that defines this relationship is given by

	$E = 0$ ft	$E = 1000$ ft	$E = 2000$ ft	
$L = 0°$	32.0877	32.0847	32.0816	
$L = 5°$	32.0890	32.0859	32.0829	
$L = 10°$	32.0928	32.0897	32.0867	(6-91)
$L = 15°$	32.0991	32.0960	32.0929	
$L = 20°$	32.1075	32.1044	32.1013	

The two-dimensional interpolation can be performed as follows:

1. A one-dimensional interpolation is performed for E at $L = 10°$ to obtain $g(10°, 500$ ft$)$ as follows:

$$g(10,500) = 32.0928 + \frac{32.0897 - 32.0928}{1000 - 0}(500 - 0) = 32.09125 \text{ ft/sec}^2 \quad (6\text{-}92)$$

2. A one-dimensional interpolation is performed for E at $L = 15°$ to obtain $g(15°, 500$ ft$)$ as follows:

$$g(15,500) = 32.0991 + \frac{32.0960 - 32.0991}{1000 - 0}(500 - 0) = 32.09755 \text{ ft/sec}^2 \quad (6\text{-}93)$$

3. Using the results of the previous one-dimensional interpolations, a one-dimensional interpolation is performed for L at $E = 500$ ft to obtain $g(12.5°, 500$ ft$)$ as follows:

$$g(12.5,500) = 32.09125 + \frac{32.09755 - 32.09125}{15 - 10}(12.5 - 10) = 32.0944 \text{ ft/sec}^2 \quad (6\text{-}94)$$

Therefore, the value of g at $L = 12.5°$ and $E = 500$ ft is 32.0944 ft/sec^2.

6.10 APPLICATIONS

6.10.1 Probability of Wind Loading

A structural engineer is designing a temporary structure that will be in place for one year. To account for wind load in the design, the engineer obtains past records for wind speeds in the area. Using the 34-year record of the annual maximum wind speeds, the mean value ($\bar{Y}$) and standard deviation (S_y) are 90 and 20 km/hr, respectively. The engineer decides to use a wind speed of 125 km/hr for the design wind loading. The 34-year record also suggests that the annual maximum wind speed is normally distributed, so the engineer estimates the probability that the design loading will exceed 125 km/hr using the normal distribution. The standard normal deviate can be defined as

$$z = \frac{125 - \bar{Y}}{S_y} = \frac{125 - 90}{20} = 1.75 \qquad (6\text{-}95)$$

The table of normal probabilities that the engineer has available includes probabilities for z values to the nearest tenth. Thus an interpolating polynomial to estimate the probability must be developed.

For linear interpolation, values of z of 1.7 and 1.8 will be used, as follows:

$$\begin{aligned}
f(z = 1.75) &= f(1.7) + \frac{1.75 - 1.70}{1.80 - 1.70} = [f(1.8) - f(1.7)] \\
&= 0.0446 + \frac{0.05}{0.10}(0.0359 - 0.0446) \\
&= 0.04025
\end{aligned} \qquad (6\text{-}96)$$

The true value is 0.0401, so the estimate is in error by 0.00015, or 0.4%.

To obtain a more accurate estimate, Newton's finite-difference equation is used. The finite-difference table is shown in Table 6.12. The finite-difference interpolation formula is

$$\begin{aligned}
f(z) &= 0.0668 - 0.0120n + 0.0018\frac{n(n-1)}{2} - 0.0003\frac{n(n-1)(n-2)}{6} \\
&\quad + 0.0003\frac{n(n-1)(n-2)(n-3)}{24} - 0.0005\frac{n(n-1)(n-2)(n-3)(n-4)}{120}
\end{aligned} \qquad (6\text{-}97)$$

For $z = 1.75$, the value of n is

$$n = \frac{1.75 - 1.50}{0.1} = 2.5 \qquad (6\text{-}98)$$

Thus the estimate is

$$\begin{aligned}
f(1.75) &= 0.0668 - 0.0120(2.5) + 0.0018\frac{2.5(1.5)}{2} - 0.0003\frac{2.5(1.5)(0.5)}{6} \\
&\quad + 0.0003\frac{2.5(1.5)(0.5)(-0.5)}{24} \\
&\quad - 0.0005\frac{2.5(1.5)(0.5)(-0.5)(-1.5)}{120} \\
&= 0.0401
\end{aligned} \qquad (6\text{-}99)$$

Thus the Newton method gives an exact value.

TABLE 6-12   Finite-difference Table for Normal Probabilities

z	$f(z)$	Δf	$\Delta^2 f$	$\Delta^3 f$	$\Delta^4 f$	$\Delta^5 f$
1.5	0.0668					
		−0.0120				
1.6	0.0548		0.0018			
		−0.0102		−0.0003		
1.7	0.0446		0.0015		0.0003	
		−0.0087		0.0000		−0.0005
1.8	0.0359		0.0015		−0.0002	
		−0.0072		−0.0002		
1.9	0.0287		0.0013			
		−0.0059				
2.0	0.0228					

6.10.2 Shear Stress of Oil between Two Parallel Plates

In the development of a new piece of equipment, it is necessary to measure the shear stress (τ) between two parallel plates, where the lower plate is stationary and the upper plate moves at a velocity (V) of 2 m/sec. The plates are 0.025 m apart (h). The shear stress τ is estimated by

$$\tau = \mu \frac{V}{h} = \mu \frac{2 \text{ m/s}}{0.025 \text{ m}} = 80 \, \mu \tag{6-100}$$

The viscosity (μ) in N-sec/m^2 is a function of temperature. Laboratory measurements have determined μ at the temperatures shown in Table 6-13. Since the measurements are at unequal intervals, the method of Lagrange will be used to develop the interpolating polynomial. For a temperature of 38°C, the weights for the values of μ are

$$w_1(38) = \frac{(38 - 20)(38 - 30)(38 - 50)(38 - 55)}{(5 - 20)(5 - 30)(5 - 50)(5 - 55)} = 0.03482 \tag{6-101a}$$

$$w_2(38) = \frac{(38 - 5)(38 - 30)(38 - 50)(38 - 55)}{(20 - 5)(20 - 30)(20 - 50)(20 - 55)} = -0.34194 \tag{6-101b}$$

$$w_3(38) = \frac{(38 - 5)(38 - 20)(38 - 50)(38 - 55)}{(30 - 5)(30 - 20)(30 - 50)(30 - 55)} = 0.96941 \tag{6-101c}$$

TABLE 6-13 Viscosity
Coefficient Data

T (°C)	μ(N-sec/m²)
5	0.08
20	0.015
30	0.009
50	0.006
55	0.0055

$$w_4(38) = \frac{(38 - 5)(38 - 20)(38 - 30)(38 - 55)}{(50 - 5)(50 - 20)(50 - 30)(5 - 55)} = 0.59840 \qquad \text{(6-101d)}$$

$$w_5(38) = \frac{(38 - 5)(38 - 20)(38 - 30)(38 - 50)}{(55 - 5)(55 - 20)(55 - 30)(55 - 50)} = -0.26068 \qquad \text{(6-101e)}$$

Based on Eq. 6-60, the interpolating polynomial is

$$\mu(38) = 0.03482(0.08) - 0.34194(0.015) + 0.96941(0.009) - 0.59840(0.006)$$

$$+ \ 0.26068(0.0055) \qquad \text{(6-102)}$$

$$= 0.00854$$

Thus the estimated shear stress is

$$t = 0.00854 \ \frac{2}{0.025} = 0.6832 \ \text{N/m}^2 \qquad \text{(6-103)}$$

6.11 PROBLEMS

Use the following data set for Problems 6-1 through 6-5:

x	0.4	0.5	0.6	0.7
$f(x)$	0.064	0.125	0.216	0.343

Derive an interpolating polynomial using the method of undetermined coefficients for the function indicated and find the interpolated value of $f(x)$ for the given value of x.

Problem	Function, $f(x)$	x
6-1.	$a + bx$	0.47
6-2.	$a + bx$	0.62
6-3.	$a + bx + cx^2$	0.47
6-4.	$a + bx + cx^2$	0.58
6-5.	$a + bx + cx^2 + dx^3$	0.58

6-6. When an interpolating polynomial is of the same order as the function used to derive the data, show that the polynomial will provide errorless interpolated values.

6-7. Using the values in the following table, derive second-order interpolating polynomials for estimating the snow-covered area (A) in percent as a function of the accumulated degree-days (D) in °C. Use the method of undetermined coefficients. Estimate the value of A for D of 50° and 450°C. If the true values are 100% and 6%, respectively, compute the percentage of error.

A	100	91	44	20	9	4
D	0	100	200	300	400	500

6-8. The following table gives the values of $\sin(x)$ for selected angles (x) in degrees.

x (degrees)	10	12	14	16
$\sin(x)$	0.17365	0.20791	0.24192	0.27564

Using the method of undetermined coefficients, derive linear, quadratic, and cubic interpolating polynomials for the interval $12° \leq x \leq 14°$ and estimate the value of $\sin(x)$ for angles of 12.5° and 13°. Compare the accuracy achieved given the true values of 0.21644 and 0.22495, respectively.

6-9. Given the following data set:

x	0.50	0.55	0.60	0.65
$f(x)$	0.4621	0.5005	0.5370	0.5717

derive an interpolating polynomial using the Gregory–Newton method. Estimate the value of $f(x)$ for $x = 0.475$ and $x = 0.525$. Compare the accuracy of the estimated values if the true values are 0.4422 and 0.4815, respectively.

6-10. Use the Gregory–Newton method with the data of Problem 6-7 to derive an interpolation polynomial. Establish the value of A for D of 50° and 450°C. If the true values are 100% and 6%, respectively, compute the relative error (%).

6-11. The following table gives the specific enthalpy (h) for superheated steam as a function of temperature (t) and at a constant pressure of 2500 lb/in.² Using the Gregory–Newton method, derive an interpolating polynomial and estimate the enthalpy at 1100°F.

h (Btu/lb)	1303.6	1458.4	1585.3	1706.1	1826.2
t (°F)	800	1000	1200	1400	1600

6-12. For the data given in Problem 6-7, derive a fifth-order interpolating polynomial using the Gregory–Newton method. Evaluate the error of the estimated value of A for $D = 50$°C.

6-13. The following table gives the sine for angles at an interval of 5°:

Sine	0.000	0.087	0.174	0.259	0.342	0.423	0.500
Angle θ (degrees)	0	5	10	15	20	25	30

   a. Using the Gregory–Newton method, derive a fourth-order interpolating polynomial.

   b. Estimate the sine of 7°. If the true value is 0.122, compute the error.

   c. Derive an interpolating polynomial using an increment of 10° and estimate the sine of 7°. Compare the error with the error for part (b).

6-14. Examine the sensitivity of the interpolating polynomial of Problem 6-13 if there is an error of 5% of the sine of 10°.

6-15. Using the data of Problem 6-13, show the change in error for an interpolated value of the sine of 2.5° as the order of the polynomial is increased. Use successively more pairs of values with an increment of 5° when increasing the order of the polynomial.

6-16. Using the data of Problem 6-7, derive both a finite-difference table and an interpolating polynomial based on the table. Evaluate the error of the estimated value of A for $D = 50$°C.

6-17. Using the data of Problem 6-13, derive both a finite-difference table and an interpolating polynomial based on the table. Evaluate the error of the estimated value of $\sin(\theta)$ for 7° (the true value is 0.122).

6-18. The data in the following table are the short-wave radiation flux (R) at the outer limit of the atmosphere in gram-calories per square centimeter per day for the month of September:

R	891	856	719	494	219
Latitude (°N)	0°	20°	40°	60°	80°

   a. Compute the finite-difference table and derive an interpolating polynomial using Newton's method.

   b. Estimate R at a latitude of 35°.

6-19. Using the data of Problem 6-11 compute a finite-difference table and derive an interpolating polynomial using Newton's method. Estimate the enthalpy at a temperature of 1100°F.

6-20. The following table gives the times of sunrise and sunset in hours and minutes at 40° latitude for 4 days, with an interval of 28 days:

Date	Sunrise	Sunset
May 1	4 hr 59 m	18 hr 53 m
May 29	4 hr 34 m	19 hr 19 m
June 26	4 hr 32 m	19 hr 32 m
July 24	4 hr 50 m	19 hr 21 m

a. Compute finite-difference tables and derive interpolating polynomials using Newton's method for the time of both sunrise and sunset.
b. Estimate the time of sunrise on May 12.
c. Estimate the time of sunset on July 16.
d. Find the date and time of the earliest sunrise.
e. Find the date and time of the latest sunset.
f. Find the date and length of the longest period of daylight.

6-21. Repeat Problem 6-20 using the following data for 20° latitude:

Date	Sunrise	Sunset
May 1	5 hr 30 m	18 hr 32 m
May 29	5 hr 19m	18 hr 34 m
June 26	5 hr 22 m	18 hr 42 m
July 24	5 hr 32 m	18 hr 40 m

6-22. The following table give the times of sunrise and sunset in hours and minutes at 60° latitude. The data are given on an unequal time scale.

Date	Sunrise	Sunset
Nov. 1	7 hr 21 m	16 hr 04 m
Dec. 1	8 hr 34 m	15 hr 02 m
Jan. 1	9 hr 02 m	15 hr 04 m
Feb. 1	8 hr 13 m	16 hr 13 m
Mar. 1	6 hr 58 m	17 hr 26 m

a. Derive Lagrangian interpolation polynomials for the time of sunrise and sunset.
b. Estimate the time of sunrise on February 20.
c. Estimate the time of sunset on November 11.
d. Find the date and time of the latest sunrise.
e. Find the date and time of the earliest sunset.
f. Find the date and length of the shortest period of daylight.

6-23. The following table gives the current density (D) in amperes per square millimeter (A/mm^2) of insulated copper conductors for selected cross-sectional area (A) in square millimeters:

A	0.75	1.0	1.5	2.5	4.0	6.0	10.0	16.0
D	17.4	15.0	13.3	10.8	9.0	7.7	6.8	5.7

Derive a Lagrangian interpolating polynomial for predicting the density D. Estimate D for cross-sectional areas of 1.2 and 2.0 mm^2.

6-24. The following tables gives the percentage of acetic acid ionized (P) at different concentrations (C) in moles per liter.

C	0.1	0.05	0.02	0.0098	0.0059	0.00103
P	1.35	1.91	2.99	4.22	5.40	12.38

Derive a Lagrangian interpolating polynomial for predicting the percentage ionized. Estimate P for concentrations of 0.015 and 0.003 mole/liter.

6-25. The percentage of light (L) passing through the sea surface that penetrates to specific depths (D) in clean ocean water is as follows:

D (m)	0	1	2	10	50	100
L (%)	100	45	39	22	5	0.5

Find the value of L at a depth of 1.6 m using a Lagrangian interpolation polynomial.

6-26. Laboratory tests on a flocculant suspension in a settling column give the following values of the percentage of solids removed (S) as a function of the sampling time (T):

T (min)	5	11	23	33	49	60
S (%)	9	31	57	64	68	71

Find the percent removal for detention times of 15 and 30 min using a Lagrangian interpolation polynomial.

6-27. Using the data for the well pumping example (Eq. 6-66), estimate the drawdown at a distance of 225 ft from the well.

6-28. Using the data of Problem 6-24, develop linear splines and quadratic splines for the purpose of interpolation. Plot your results.

6-29. Using the data of Problem 6-25, develop linear splines and quadratic splines for the purpose of interpolation. Plot your results.

6-30. Using the data of Problem 6-26, develop linear splines and quadratic splines for the purpose of interpolation. Plot your results.

6-31. The following data provide the value of a function at any selected pairs of x and y:

	$y = 10$	$y = 20$	$y = 30$
$x = 1$	0.1940	0.4610	0.6781
$x = 2$	0.3125	0.8123	0.9610
$x = 3$	0.8190	2.1670	2.8901
$x = 4$	1.8231	3.1621	4.0070

a. Determine the value of the function using linear interpolation at $x = 2.3$ and $y = 22.3$.

b. Determine the value of the function using linear interpolation at $x = 3.4$ and $y = 10.6$.

7

Differentiation and Integration

7.1 NUMERICAL DIFFERENTIATION

A number of engineering problems require a numerically derived estimate of a derivative of a function $f(x)$. There are two general approaches to the problem. First, if the function is known but the derivative cannot be computed analytically, the derivative can be estimated by computing the function for two values of the independent variable(s) separated by distance Δx and dividing the difference by Δx as follows:

$$\frac{df(x)}{dx} \approx \frac{f(x + \Delta x) - f(x)}{\Delta x} \tag{7-1}$$

in which $f(x)$ and x are the dependent and independent variables, respectively, and $f(x)$ denotes the value of the dependent variable when x is the value of the independent variable.

The second approach to numerical differentiation is to fit a function to a set of points that describes the relationship between the dependent and independent variables and then differentiate the fitted function. Specifically, an interpolation polynomial of order n could be fit to the data and the derivative of the polynomial used as the estimate of the derivative. The selection of one of the two methods depends on the form in which the data are presented and the desired level of accuracy.

7.1.1 Finite-difference Differentiation

Equation 7-1 is just one possible difference method; it is called the *forward difference* equation because it uses the base point x and the value of x that is greater than x by an amount Δx. The following *backward difference* formula could also be used:

$$\frac{df(x)}{dx} \approx \frac{f(x) - f(x - \Delta x)}{\Delta x} \tag{7-2}$$

For Eq. 7-2, the value of x is incremented by $-\Delta x$ and the value of the function is determined for both x and $(x - \Delta x)$; the difference between the values of the two functions is divided by Δx to approximate the derivative. The *two-step method*, which is an alternative to Eqs. 7-1 and 7-2, computes the value of the function for values of the independent variable of $(x - \Delta x)$ and $(x + \Delta x)$, which provides an estimate of the derivative by

$$\frac{df(x)}{dx} \approx \frac{f(x + \Delta x) - f(x - \Delta x)}{2\Delta X} \tag{7-3}$$

For many highly variable functions, the two-step method may provide slightly more accurate estimates than either the forward or backward finite-difference approximations.

The finite-difference approach can be used both when the data are given as tabular values at some finite increment Δx or when a functional form is given, but the functional form cannot be differentiated analytically. Examples 7-1 and 7-2 are used to illustrate the first and second cases, respectively.

Example 7-1 Evaporation Rates

A design engineer must make estimates of evaporation rates when the amount of needed water to meet irrigation demands is required. One input to a frequently used formula for estimating evaporation rates is the slope of the saturation vapor pressure curve at air temperature T. A small part of the table that is used to make such estimates is given in Table 7-1. If it is necessary to make design estimates at a temperature of 22°C, then the slope of the saturation vapor pressure curve at 22°C could be estimated using, as follows, the forward (Eq. 7-1), backward (Eq. 7-2), or two-step (Eq. 7-3) method:

$$\frac{de_s}{dT} = \frac{e_s(23) - e_s(22)}{23 - 22} = \frac{21.05 - 19.82}{1} = 1.23 \text{ mm Hg/°C} \tag{7-4}$$

$$\frac{de_s}{dT} = \frac{e_s(22) - e_s(21)}{22 - 21} = \frac{19.82 - 18.65}{1} = 1.17 \text{ mm Hg/°C} \tag{7-5}$$

$$\frac{de_s}{dT} = \frac{3_s(23) - e_s(21)}{23 - 21} = \frac{21.05 - 18.65}{2} = 1.20 \text{ mm Hg/°C} \tag{7-6}$$

TABLE 7-1 Saturation Vapor Pressure (e_s) in mm Hg as a Function of Temperature (T) in °C

T (°C)	e_s (mm Hg)
20	17.53
21	18.65
22	19.82
23	21.05
24	22.37
25	23.75

When applying the two-step method in Eq. 7-6, a full increment was actually taken in each direction because this was the smallest increment for which data were available. The true value is 1.20 mm Hg/°C, so the two-step method provided the most accurate estimate.

Example 7-2 Model Fitting

A common problem in engineering analysis is fitting models to engineering data. Simple model forms, such as a straight line or a plane, can be fit using the methods of Chapter 10; these methods require that some objective function be minimized. The minimization of the sum of the squares of the errors is the most frequently used objective function in engineering analysis. The objective function is given by

$$F = \min \sum_{i=1}^{n} e_i^2 = \min \sum_{i=1}^{n} (\hat{Y}_i - Y_i)^2 \tag{7-7}$$

in which F is the value of the objective function; n is the number of observations, or sample size; e_i is the error, or residual, for observation i; and $\hat{Y}_i$ and Y_i are the predicted and measured values of the criterion variable, respectively. If the model used to predict $f(x)$ is relatively simple in structure, Eq. 7-7 can be differentiated analytically with respect to the coefficients (i.e., unknowns) of the model. If the model is more complex, then an analytical analysis is not possible, and a numerical solution of the derivatives is necessary to find the value or the values of the coefficients that provide the minimum value of F.

Let's examine the problem using a model employed by the environmental engineer in predicting the dissolved oxygen deficit (ppm) in a stream as a function of time:

$$\hat{Y} = \frac{K_1}{1 + e^{-K_2(t - t_0)}} - K_3 \tag{7-8}$$

in which t is the number of days from the time when a point waste-load is discharged into a stream; K_1, K_2, K_3, and t_0 are unknowns that must be found from measured data; and $\hat{Y}$ is the predicted dissolved oxygen deficit. Substituting Eq. 7-8 into Eq. 7-7 yields

$$F = \sum_{i=1}^{n} \left(\frac{K_1}{1 + e^{-K_2(t - t_0)}} - K_3 - Y_i \right)^2 \tag{7-9}$$

The subscript i on t_i, which is the time at which the dissolved oxygen deficit y_i is measured, and n refers to the number of the observation in the data set. For purposes of illustration, assume that the data shown in Table 7-2 were measured at a point on a stream approximately 500 ft below the point where an industrial plant discharges waste into the stream.

To find the values of the coefficients, it would be necessary to take the derivative of F with respect to each of the four unknowns using analytical calculus. This would not be a simple task, so the solution can be found using a numerical analysis. The numerical solution

TABLE 7-2 Dissolved Oxygen Data

i	1	2	3	4	5
t (days)	0	1	2	3	4
y_i (ppm)	0	0.62	1.82	3.03	3.64

requires initial estimates of the unknowns; for example, the following values are assumed: $K_1 = 4.5$, $K_2 = 1.0$, $K_3 = 0.1$, and $t_0 = 2$. Using these values and the time in days shown in Table 7-2, the following predicted values of the dissolved oxygen deficit are obtained using Eq. 7-8: 0.44, 1.11, 2.15, 3.19, and 3.86 ppm. These values can be used to compute the value of F according to Eq. 7-7.

$$F = \sum_{i=1}^{5} (\hat{Y}_i - Y_i)^2$$

$$= (0.44 - 0)^2 + (1.11 - 0.62)^2 + (2.15 - 1.82)^2 \tag{7-10}$$
$$+ (3.19 - 3.03)^2 + (3.86 - 3.64)^2$$

$$= 0.6166$$

To evaluate the derivative of F with respect to K_1, we could increase K_1 by $\Delta K_1 = 0.1$, re-compute the values of $\hat{Y}$ using Eq. 7-8, and compute a new value of F using Eq. 7-7, which is denoted as F_1. Then an estimate of the derivative is given by

$$\frac{\Delta F}{\Delta K_1} = \frac{F_1 - F}{\Delta K_1} \tag{7-11}$$

which is the forward difference formula. For $K_1 = 4.6$, predicted dissolved oxygen concentrations of 0.45, 1.14, 2.20, 3.26, and 3.95 ppm are obtained. These values yield a value for F_1 of

$$F_1 = (0.45 - 0)^2 + (1.14 - 0.62)^2 + (2.20 - 1.82)^2$$
$$+ (3.26 - 3.03)^2 + (3.95 - 3.64)^2 \tag{7-12}$$

$$= 0.7663$$

Thus the estimate of the derivative is

$$\frac{\Delta F}{\Delta K_1} = \frac{0.7663 - 0.6166}{0.1} = 1.497 \tag{7-13}$$

Obviously, if we wish to find the minimum value of F, then the value of K_1 needs to be decreased, because F increased when K_1 was increased from 4.5 to 4.6.

The derivative of F with respect to K_3 is easily computed numerically. If we increase the initial value of K_3 from 0.1 to 0.15, then the predicted dissolved oxygen deficits would be 0.39, 1.06, 2.10, 3.14, and 3.81 ppm, respectively. This would yield a value for F of

$$F_2 = (0.39 - 0)^2 + (1.06 - 0.62)^2 + (2.10 - 1.82)^2$$
$$+ (3.14 - 3.03)^2 + (3.81 - 3.64)^2 \tag{7-14}$$

$$= 0.4651$$

Thus the numerical estimate of the derivative would be

$$\frac{\Delta F}{\Delta K_3} = \frac{F_2 - F}{\Delta K_3} = \frac{0.4651 - 0.6166}{0.05} = -3.03 \tag{7-15}$$

Since increasing the value of K_3 from 0.1 to 0.15 caused the value of F to decrease, we would want to increase the value of K_3 further to find the minimum value of the objective function.

The objective of this example was to show how the derivative of a complex functional form can be evaluated numerically. The example also illustrates that the numerical evaluation of derivatives can be used to optimize complex engineering models when it is not possible to evaluate the derivatives analytically.

7.1.2 Differentiation Using a Finite-difference Table

The finite-difference table introduced in Chapter 6 can also be used to approximate the derivatives of a function that is expressed in incremental form. Consider the problem of estimating the cube of a number. The finite-difference table is given in Table 6-6. The values in the Δf column can be used to make estimates of the first derivative; values in the $\Delta^i f$ column can be used to estimate the ith derivative.

For estimating the first derivative at any value of x, the Δf value in the diagonal row below the row for the value of x can be used to estimate the forward finite difference. The value of Δf in the diagonal row above the row for the value of x can be used to estimate the backward finite difference. The average of these two Δf values can be used to estimate the first derivative with the two-step method. Specifically, for a value of x equal to 11, the estimates of the first derivative by the forward, backward, and two-step methods based on Table 6-6 on page 156 are, respectively,

$$\frac{df}{dx} = \frac{\Delta f_{i+1}}{\Delta x} = \frac{397}{1} = 397 \qquad (7\text{-}16)$$

$$\frac{df}{dx} = \frac{\Delta f_i}{\Delta x} = \frac{331}{1} = 331 \qquad (7\text{-}17)$$

$$\frac{df}{dx} = \frac{\Delta f_i + \Delta f_{i+1}}{2\Delta x} = \frac{331 + 397}{2} = 364 \qquad (7\text{-}18)$$

The true value can be found by differentiating $f(x) = x^3$ with respect to x, which is $3x^2$. For an x of 11, the true value would be 363. Thus the two-step method provides the most accurate estimate. The large error for the forward and backward difference equations is the result of the large incremental value of the independent variable, that is, Δx.

A value for the second derivative is obtained from the value of $\Delta^2 f$ of Table 6-6. In this case, the estimate equals the value of $\Delta^2 f$ divided by Δx, which for the example would be 66. This agrees with the true value of $6x = 6(11) = 66$.

Example 7-3 Specific Heat Capacity

The specific heat capacity is an important element in thermodynamic processes. For a process in which the pressure is constant, the specific heat capacity (c_p) equals the slope of the relationship between the specific enthalpy (h) and the temperature (T):

$$c_p = \frac{dh}{dT} \qquad (7\text{-}19)$$

Table 7-3 gives the finite-difference table for data that defines the relationship between temperature and the enthalpy of a superheated steam. Using the first finite difference (Δh), the specific heat capacity can be estimated. For example, at a temperature of 1200°F, the forward finite-difference equation would yield a specific heat capacity of

$$c_p = \frac{\Delta h}{\Delta T} = \frac{1705 - 1585}{1400 - 1200} = \frac{120}{200} = 0.6 \text{ Btu/lb/°F} \qquad (7\text{-}20)$$

The backward difference formula would yield the following estimate:

$$c_p = \frac{\Delta h}{\Delta T} = \frac{1585 - 1460}{1200 - 1000} = \frac{125}{200} = 0.625 \text{ Btu/lb/°F} \qquad (7\text{-}21)$$

Using a full increment on either side of the base point, the two-step method would lead to the following estimate:

$$c_p = \frac{\Delta h}{\Delta T} = \frac{1705 - 1460}{1400 - 1000} = \frac{245}{400} = 0.6125 \text{ Btu/lb/°F} \qquad (7\text{-}22)$$

The rate of change of c_p with respect to T can be approximated by the second derivative, which is computed from the second finite difference. For the data of Table 7-3, the rate of change of c_p with respect to T at $T = 1200$°F is

$$\frac{\Delta^2 h}{\Delta T^2} = \frac{-5}{200} = -0.025 \text{ Btu/lb/(°F)}^2 \qquad (7\text{-}23)$$

At a temperature of 1000°F the rate of change would be -0.15 Btu/lb/(°F)^2.

TABLE 7-3   Finite-difference Table for Specific Enthalpy (h) in Btu/lb and Temperature (T) in °F

T	h	Δh	$\Delta^2 h$	$\Delta^3 h$	$\Delta^4 h$
800	1305				
		155			
1000	1460		−30		
		125		25	
1200	1585		−5		−20
		120		5	
1400	1705		0		
		120			
1600	1825				

7.1.3 Differentiating an Interpolating Polynomial

Given data in a tabular form, an interpolating polynomial of the dependent variable as a function of the independent variable can be derived and differentiated. The interpolation methods discussed in Chapter 6 can be used for this purpose. If the interpolating polynomial is expressed as an nth-order polynomial, then the derivative is a polynomial of order $(n - 1)$. In general, a polynomial can be expressed as

$$f(x) = b_n x^n + b_{n-1} x^{n-1} + b_{n-2} x^{n-2} + \cdots + b_1 x + b_0 \tag{7-24}$$

The derivative of the polynomial is

$$\frac{df(x)}{dx} = n b_n x^{n-1} + (n-1) b_{n-1} x^{n-2} + \cdots + b_1 \tag{7-25}$$

When the interpolating polynomial is expressed in the form of the Gregory–Newton formula (Eq. 6-10), the derivative is somewhat more difficult to evaluate directly. For example, differentiating the first four terms of Eq. 6-10 with respect to x yields

$$\begin{aligned}\frac{df(x)}{dx} &= a_2 + a_3[(x - x_2) + (x - x_1)] + a_4[(x - x_2)(x - x_3) \\ &\quad + (x - x_1)(x - x_3) + (x - x_1)(x - x_2)]\end{aligned} \tag{7-26}$$

This is obviously more difficult to evaluate than the $(n - 1)$th-order derivative of Eq. 7-25.

Example 7-4 Specific Heat Capacity

Using the data of Example 6-2 that are given in Table 6-2, the Gregory–Newton method was used to develop the interpolation equation between vapor pressure of water (P) and temperature (T) as given by Eq. 6-20. The resulting a_i, where $i = 1, 2, 3, 4,$ and 5, can be used in Eq. 7-26 to evaluate the derivative of P with respect to T. Several methods of estimating evaporation rates from open water surfaces require the slope of the vapor pressure curve at some design temperature. The derivative of Eq. 6-20 is given by

$$\begin{aligned}\frac{dP(T)}{dT} &= 3.55 + 0.0914063[(T - 48) + (T - 40)] \\ &\quad + 0.001172[(T - 48)(T - 56) + (T - 40)(T - 56) + (T - 40)(T - 48)] \\ &\quad + 0.00001017[(T - 48)(T - 56)(T - 64) + (T - 40)(T - 56)(T - 64) \\ &\quad + (T - 40)(T - 48)(T - 64) + (T - 40)(T - 48)(T - 56)]\end{aligned} \tag{7-27}$$

The slope of the vapor pressure curve at a temperature of 70°C is computed by substituting 70°C for T and yields $dP(T)/dT = 10.1084$ mm Hg/°C, which is accurate to three significant digits.

7.1.4 Differentiation Using Taylor Series Expansion

The basic finite-difference equation for differentiation as given by Eq. 7-1 results from the following forward Taylor series expansion:

$$f(x + \Delta x) = f(x) + \frac{df(x)}{dx} \Delta x + \frac{df^2(x)}{dx^2} \frac{(\Delta x)^2}{2!} + \frac{df^3(x)}{dx^3} \frac{(\Delta x)^3}{3!} + \cdots \quad (7\text{-}28)$$

Truncating terms with the second derivative and higher derivatives results in the following equation:

$$f(x + \Delta x) \approx f(x) + \frac{df(x)}{dx} \Delta x \quad (7\text{-}29)$$

Rearranging the terms to solve for $f'(x) = df(x)/dx$ results in the forward finite-difference differentiation equation of Eq. 7-1. Therefore, this equation provides a first-order approximation of the first derivative of $f(x)$.

The second-order approximation of the first derivative of $f(x)$ can be obtained by including the second-order term from the Taylor series expansion in Eq. 7-29. The resulting second-order approximation, after solving for $f'(x)$, is

$$\frac{df(x)}{dx} \approx \frac{f(x + \Delta x) - f(x)}{\Delta x} - \frac{df^2(x)}{dx^2} \frac{\Delta x}{2!} \quad (7\text{-}30)$$

The second-order approximation requires knowledge of the second derivative of $f(x)$. The finite-difference approximation can also be used to compute higher-order derivatives. For example, if we let $f'(x)$ be the first derivative of $f(x)$ with respect to x, then the forward difference approximation of the second derivative is given by

$$\frac{d^2 f(x)}{dx^2} \approx \frac{f'(x + \Delta x) - f'(x)}{\Delta x} \quad (7\text{-}31)$$

Substituting the first-order approximation of $f'(x)$ in Eq. 7-31 produces the following equation for the second derivative:

$$\frac{d^2 f(x)}{dx^2} \approx \frac{f(x + 2\Delta x) - 2f(x + \Delta x) + f(x)}{(\Delta x)^2} \quad (7\text{-}32)$$

Equation 7-32 is a first-order approximation of the second derivative. Now, by substituting Eq. 7-32 in Eq. 7-30, the second-order approximation of the first derivative is obtained as

$$\frac{df(x)}{dx} \approx \frac{-f(x + 2\Delta x) + 4f(x + \Delta x) - 3f(x)}{2\Delta x} \quad (7\text{-}33)$$

The resulting Eq. 7-33 does not require any higher-order derivatives; it can be evaluated using the function $f(x)$. Similarly, the first-order estimate of the second derivative as given by the Eqs. 7-31 and 7-32 can be revised by including a second-order term from the Taylor series expansion to obtain its second-order approximation as

$$\frac{d^2 f(x)}{dx^2} \approx \frac{-f(x + 3\Delta x) + 4f(x + 2\Delta x) - 5f(x + \Delta x) + 2f(x)}{(\Delta x)^2} \tag{7-34}$$

Using backward difference in the Taylor series expansion, the following first- and second-order estimates, respectively, for the first derivative of $f(x)$ can be obtained:

$$\frac{df(x)}{dx} \approx \frac{f(x) - f(x - \Delta x)}{\Delta x} \tag{7-35}$$

$$\frac{df(x)}{dx} \approx \frac{3f(x) - 4f(x - \Delta x) + f(x - 2\Delta x)}{2\Delta x} \tag{7-36}$$

Using backward difference in the Taylor series expansion, the following first- and second-order estimates, respectively, for the second derivative of $f(x)$ can be obtained:

$$\frac{d^2 f(x)}{dx^2} \approx \frac{f(x) - 2f(x - \Delta x) + f(x - 2\Delta x)}{(\Delta x)^2} \tag{7-37}$$

$$\frac{d^2 f(x)}{dx^2} \approx \frac{2f(x) - 5f(x - \Delta x) + 4f(x - 2\Delta x) - f(x - 3\Delta x)}{(\Delta x)^2} \tag{7-38}$$

Using the two-step method in the Taylor series expansion, the following first- and second-order estimates, respectively, for the first derivative of $f(x)$ can be obtained:

$$\frac{df(x)}{dx} \approx \frac{f(x + \Delta x) - f(x - \Delta x)}{2\Delta x} \tag{7-39}$$

$$\frac{df(x)}{dx} \approx \frac{-f(x + 2\Delta x) + 8f(x + \Delta x) - 8f(x - \Delta x) + f(x - 2\Delta x)}{12\Delta x} \tag{7-40}$$

Using the two-step method in the Taylor series expansion, the following first- and second-order estimates, respectively, for the second derivative of $f(x)$ can be obtained:

$$\frac{d^2 f(x)}{dx^2} \approx \frac{f(x + \Delta x) - 2f(x) + f(x - \Delta x)}{(\Delta x)^2} \tag{7-41}$$

$$\frac{d^2 f(x)}{dx^2} \approx \frac{-f(x + 2\Delta x) + 16f(x + \Delta x) - 30f(x) - 16f(x - \Delta x) - f(x - 2\Delta x)}{12(\Delta x)^2} \tag{7-42}$$

Example 7-5 Evaporation Rates

Example 7-1 was used to illustrate the first-order finite-difference equations for computing the slope of the saturation vapor pressure curve based on the data in Table 7-1. Using the second-order equations (Eqs. 7-33, 7-36, and 7-40) for the forward, backward, and the two-step methods produces, respectively, the following results:

$$\frac{df(x)}{dx} \approx \frac{-e_s(24) + 4e_s(23) - 3e_s(22)}{2(1)} \tag{7-43}$$

$$\approx \frac{-22.37 + 4(21.05) - 3(19.82)}{2(1)} = 1.185 \text{ mm Hg/°C}$$

$$\frac{df(x)}{dx} \approx \frac{3e_s(22) - 4e_s(21) + e_s(20)}{2(1)} \tag{7-44}$$

$$\approx \frac{3(19.82) - 4(18.65) + 17.53}{2(1)} = 1.195 \text{ mm Hg/°C}$$

$$\frac{df(x)}{dx} \approx \frac{-e_s(24) + 8e_s(23) - 8e_s(21) + e_s(20)}{12(1)} \tag{7-45}$$

$$\approx \frac{-(22.37) + 8(21.05) - 8(18.65) + (17.53)}{12(1)} = 1.19667 \text{ mm Hg/°C}$$

The accuracy of the three methods is improved by using the second-order approximation. The two-step method still provides the best estimate, since the true value at $T = 22°C$ is 1.2 mm Hg/°C.

7.2 NUMERICAL INTEGRATION

Integration is introduced in analytical calculus using examples in which the function describes a continuous curve over some interval. For example, we may be interested in the area under the curve $f(x) = x^2$ between the lower and upper limits of a and b, respectively, as shown in Fig. 7-1. The integration is expressed as

$$\text{integral} = \int_a^b f(x)\, dx \tag{7-46}$$

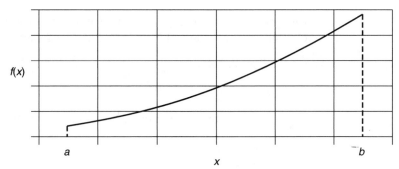

Figure 7-1 Analytical Integration

The result of the integral is the area under the $f(x)$ curve between a and b. In engineering, sometimes the function $f(x)$ does not have a known analytical solution. Therefore, a numerical evaluation of the integral is needed.

In many engineering problems, the data are presented as a set of discrete values for some dependent variable. For example, the volume rate of flow (Q) of water in a channel or through a pipe is the integral of the velocity (V) and the incremental area (dA):

$$Q = \int V \, dA \tag{7-47}$$

In practical applications for a natural channel that is irregular in shape, velocity measurements are made at a number of elevations and the volume flow rate computed for each subarea, with the total flow in the channel being the sum of the flow rates computed for the subareas. Thus the integral of Eq. 7-47 is not computed through analytical integration but by a numerical approximation.

The methods presented in this section for numerical integration can be used for both classes of problems, that is, problems with a known $f(x)$ that cannot be integrated analytically and problems with discrete data. There are two general approaches to numerical integration. First, a polynomial could be numerically fitted to the discrete points and then the polynomial can be integrated analytically. For example, either the Gregory–Newton or the finite-difference method could be used to derive an interpolating polynomial to a set of points, and then the polynomial could be integrated using analytical calculus. Second, the area can be approximated using simple geometric shapes defined by adjacent points. It can be shown that the two methods are computationally equivalent, although they differ in physical appearance.

Other methods for numerical integration are discussed in Section 8.11. These methods are based on converting the integration problem of the type given in Eq. 7-46 to a differential equation problem; hence numerical methods for solving differential equations can be used.

7.2.1 Interpolation Formula Approach

The Gregory–Newton method for deriving an interpolation formula was introduced in Chapter 6. The end result was a polynomial expressed in the following form:

$$f(x) = a_1 + a_2(x - x_1) + a_3(x - x_1)(x - x_2) + \cdots + a_{n+1}(x - x_1)(x - x_2) \ldots (x - x_n) \tag{7-48}$$

The terms could be rearranged to form an nth-order polynomial of the form

$$f(x) = b_1 x^n + b_2 x^{n-1} + \cdots + b_{n+1} \tag{7-49}$$

in which the coefficients b_i, $i = 1, 2, 3, \ldots, n + 1$, are a function of the values of x_i and the a_i. After the coefficients of Eq. 7-48 are derived and the polynomial of Eq. 7-49 is formed, Eq. 7-49 could be integrated analytically as follows:

$$\int f(x)dx = \frac{b_1}{n+1}x^{n+1} + \frac{b_2}{n}x^n + \cdots + b_{n+1}x \qquad (7\text{-}50)$$

The value of Eq. 7-50 can be computed for any specified limits of integration.

Example 7-6 Traffic Safety Engineer

The traffic safety engineer is interested in the time gap between successive cars on a freeway during rush hour. In an attempt to develop a relationship between the probability of accidents and the mean gap on a section of a highway, the engineer collects the data given in Table 7-4 for a single location on the highway. The engineer can use the data on the frequency of occurrence to compute the probability for any gap length. For example, if the engineer believes that a gap of 4 seconds is the gap below which accidents are most likely to occur, the following equation can be used to approximate the desired probability:

$$P(\text{gap} \le 4 \text{ sec}) = \frac{\displaystyle\int_0^4 f(x)dx}{\displaystyle\int_0^{12} f(x)dx} \qquad (7\text{-}51)$$

The Gregory–Newton formula was used to derive the following interpolating polynomial based on mid-intervals (x) as shown in Table 7-4:

$$\begin{aligned}
f(x) = {} & 12 + 7(x - x_1) - \frac{11}{4}(x - x_1)(x - x_2) + \frac{13}{24}(x - x_1)(x - x_2)(x - x_3) \\
& - \frac{1}{12}(x - x_1)(x - x_2)(x - x_3)(x - x_4) \\
& + \frac{11}{960}(x - x_1)(x - x_2)(x - x_3)(x - x_4)(x - x_5)
\end{aligned} \qquad (7\text{-}52)$$

TABLE 7-4 Traffic Gap Data

Gap Length (seconds)	Mid-gap Length, x	Frequency of Occurrence
0–2	1	12
2–4	3	26
4–6	5	18
6–8	7	14
8–10	9	8
10–12	11	6
12 or more	NA	0

This interpolating polynomial can be rearranged to form the following fifth-order polynomial:

$$f(x) = \frac{11}{960}x^5 - \frac{71}{192}x^4 + \frac{433}{96}x^3 - \frac{2465}{96}x^2 + \frac{20{,}633}{320}x - \frac{1981}{64} \qquad (7\text{-}53)$$

The integrals of Eq. 7-51 can be computed using the polynomial of Eq. 7-53:

$$\int f(x)dx = \frac{11}{5760}x^6 - \frac{71}{960}x^5 + \frac{433}{384}x^4 - \frac{2465}{288}x^3 + \frac{20{,}633}{640}x^2 - \frac{1981}{64}x \qquad (7\text{-}54)$$

By evaluating Eq. 7-54 between the limits for the two integrals of Eq. 7-51, the following probability was estimated:

$$P(\text{gap} \le 4 \text{ sec}) = \frac{64.99}{162.19} = 0.40 \qquad (7\text{-}55)$$

If similar estimates of the probability are computed for other locations on the highway, the traffic safety engineer can develop a relationship between the accident rate and the probability of cars that follow each other with a small time gap.

7.2.2 Trapezoidal Rule

The trapezoidal rule approximates the area of a function defined by a set of discrete points by fitting a trapezoid to each pair of adjacent points that defines the dependent variable and summing the individual areas, as shown in Fig. 7-2a. That is, assuming that the function defining the dependent variable $f(x)$ is quantified by an array of points at selected values of the dependent variable x, the area under the function can be approximated by (see Fig. 7-2b)

$$\int_{x_1}^{x_n} f(x)dx \approx \sum_{i=1}^{n-1} (x_{i+1} - x_i)\frac{f(x_{i+1}) + f(x_i)}{2} \qquad (7\text{-}56)$$

The first term of Eq. 7-56 is the base width of the trapezoid, while the second term is the average height of the end points. The accuracy of the estimate can be improved by reducing the width of the interval; however, this increases the computational effort. A comparison of Fig. 7-2a and Fig. 7-2c shows the change in accuracy, where the difference between the curve $f(x)$ and the linear approximations reflects the error. Therefore, Eq. 7-56 can be easily applied to a function defined by a set of discrete values.

The trapezoidal rule, Eq. 7-56, can be derived by fitting a linear interpolating polynomial to each pair of points. Using the Gregory–Newton formula (Eq. 6-10) with values of the independent variable of x_{i+1} and x_i, the following equation is obtained:

$$f(x) = f(x_i) + \frac{f(x_{i+1}) - f(x_i)}{x_{i+1} - x_i}(x - x_i) \qquad (7\text{-}57)$$

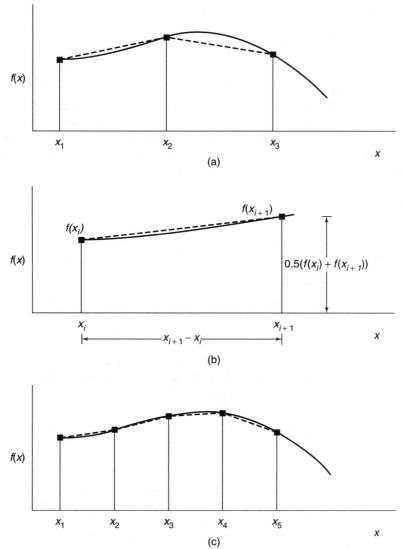

Figure 7-2   Trapezoidal Rule for Integration. **(a)** Wide-interval Approximation; **(b)** Terms of Eq. 7-56; **(c)** Narrow-interval Approximation

Integrating Eq. 7-57 between two points, say *a* and *b,* produces

$$\int_a^b f(x)dx \approx \frac{b-a}{x_{i+1}-x_i}\left[\frac{(b+a)}{2}(f(x_{i+1})-f(x_i)) + x_{i+1}f(x_i) - x_i f(x_{i+1})\right] \quad (7\text{-}58)$$

Letting $b = x_{i+1}$ and $a = x_i$, results in

$$\int_{x_i}^{x_{i+1}} f(x)dx \approx \frac{x_{i+1} - x_i}{2} [f(x_{i+1}) + f(x_i)] \tag{7-59}$$

which is the rule given by Eq. 7-56.

Numerical integration can result in errors, where a numerical error is defined as the difference between the true value and the estimated value. The true value is unknown; otherwise, it would not be necessary to use the trapezoidal rule. An approximate upper bound on the error can be determined based on the second term of a Taylor series expansion:

$$\frac{x_{i+1} - x_i}{2} f''(x) \tag{7-60}$$

in which $f''(x)$ is the value of the second derivative evaluated at the value of x that maximizes the value of $f''(x)$ between x_i and x_{i+1}.

Example 7-7   Trapezoidal Rule for Integration

For example, the data shown in Table 7-5 include three intervals; that is, $n = 4$. The discrete $f(x_i)$ values are for the following true (but supposedly unknown) function:

$$f(x) = x^3 - 5x^2 + 4x + 3 \tag{7-61}$$

The points and true function are shown in Fig. 7-3, with the true area being the area between the line $f(x) = 0$ and the solid line. Using Eq. 7-56, the trapezoidal rule provides the following estimate of the integral of Eq. 7-61:

$$\int f(x)dx \approx \frac{2-1}{2}[3 + (-1)] + \frac{3-2}{2}[-1 + (-3)] + \frac{4-3}{2}[3 + (-3)] \tag{7-62}$$

$$= 1 + (-2) + 0 = -1$$

The trapezoidal approximation is shown in Fig. 7-3 as a dashed line. The true value, which is obtained by integrating Eq. 7-61 between the limits of 1 and 4, is -2.25, which indicates that the trapezoidal estimate of Eq. 7-62 is in error by 1.25 (absolute value). The error is the algebraic difference between the two curves of Fig. 7-3. The error could be reduced by decreasing the grid spacing.

TABLE 7-5   Data for Trapezoidal Rule

i	1	2	3	4
x_i	1	2	3	4
$f(x_i)$	3	−1	−3	3

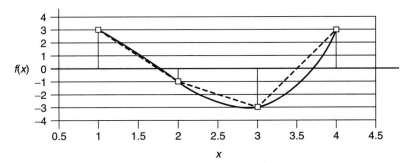

Figure 7-3   Trapezoidal Rule for Integration

Example 7-8 Flow Rate

The flow rate (Q) of an incompressible fluid is given by the integral:

$$Q = \int V \, dA \tag{7-63}$$

in which V is the velocity and A is the area. For a circular pipe of radius r, the incremental area dA is equal to $2\pi r \, dr$, and the quantity must be integrated over the cross section of the pipe. The incremental distance dr is measured from the center of the pipe as shown in Fig. 7-4; that is, the expression of Eq. 7-63 is integrated from $r = 0$ to $r = r_0$, where r_0 is the radius of the pipe. Thus Eq. 7-63 becomes

$$Q = \int_0^{r_0} V(2\pi r \, dr) \tag{7-64}$$

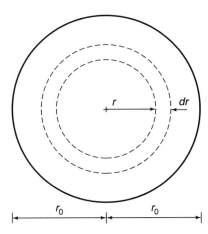

Figure 7-4   Flow Rate for a Pipe

If an analytical expression exists for the velocity V as a function of r, Eq. 7-64 can be solved analytically. However, if only measurements of V are available at selected distances from the center of the pipe, the integral must be solved numerically.

Table 7-6 gives values of the velocity (V) measured at selected distances from the center of the pipe for a 1-ft-diameter pipe. The trapezoidal rule can be used to estimate the value of the integral in Eq. 7-64 as follows:

$$Q = \frac{\Delta r}{2} \sum_{i=1}^{6} [V_i(2\pi r_i) + V_{i+1}(2\pi r_{i+1})] \tag{7-65}$$

in which Δr is the increment of r in which values of V are given in Table 7-6 $\left(\text{i.e., } \Delta r = \frac{1}{12} \text{ ft.}\right)$, and V_i is the value of V at distance r_i. For a pipe with a diameter of 1 ft, Eq. 7-65 becomes

$$Q = \frac{2\pi\left(\frac{1}{12}\right)}{2} \sum_{i=1}^{6} [V_i r_i + V_{i+1} r_{i+1}] \tag{7-66a}$$

$$Q = \frac{\pi}{12} \left\{ \left[10(0) + 9.772\left(\frac{1}{12}\right) \right] + \left[9.722\left(\frac{1}{12}\right) + 8.889\left(\frac{1}{6}\right) \right] + \left[8.889\left(\frac{1}{6}\right) + 7.5\left(\frac{1}{4}\right) \right] \right.$$

$$+ \left[7.5\left(\frac{1}{4}\right) + 5.556\left(\frac{1}{3}\right) \right] + \left[5.556\left(\frac{1}{3}\right) + 3.056\left(\frac{5}{12}\right) \right] \tag{7-66b}$$

$$\left. \left[3.056\left(\frac{5}{12}\right) + 0\left(\frac{1}{2}\right) \right] \right\} = 3.818 \text{ ft}^3/\text{sec}$$

While the true value is usually not known, the data of Table 7-6 were derived from the following relationship:

$$V = 10(1 - 4r^2) \tag{7-67}$$

TABLE 7-6 Application of Trapezoidal Rule for Estimating the Flow Rate of a Fluid in a Circular Pipe

i	r_i (ft)	V_i (fps)
1	0	10.000
2	$\frac{1}{12}$	9.722
3	$\frac{1}{6}$	8.889
4	$\frac{1}{4}$	7.500
5	$\frac{1}{3}$	5.556
6	$\frac{5}{12}$	3.056
7	$\frac{1}{2}$	0.

Therefore, Eq. 7-67 could be solved analytically. The resulting Q is 3.927 ft³/sec, which means that the estimate provided by the trapezoidal rule (Eqs. 7-66) is in error by 0.109 ft³/sec, or an error of 2.8% of the true value.

7.2.3 Simpson's Rule

The trapezoidal rule is based on a linear interpolating polynomial. The accuracy in an estimate of an integral can usually be improved by using a higher-order polynomial as the interpolation formula. Simpson's rule is a numerical method of estimating the value of an integral when a second-order polynomial is used as the interpolating formula. Figure 7-5 shows both a second-order polynomial (Simpson's rule) and the linear equations for the trapezoidal rule in comparison with a higher-order function (the solid line). Following the procedure for integrating the linear interpolating polynomial for the trapezoidal rule, it can be shown that Simpson's rule has the following form:

$$\int_a^b f(x)\, dx = \frac{\Delta x}{3}\,[\,f(a) + 4f(a + \Delta x) + f(b)\,] \tag{7-68}$$

in which a and b are the lower and upper bounds on the interval, and the interval width is equal to $2\Delta x$. Thus the interval from $x = a$ to $x = b$ consists of two subintervals, each separated by a distance of $\Delta x = (b - a)/2$. Simpson's rule can only be applied when there are an even number of subintervals, that is, an odd number of data pairs, x and $f(x)$. For $n - 1$ intervals of equal size, the Simpson's rule can be expressed as

$$\int_{x_1}^{x_n} f(x)\, dx \approx \sum_{i=1,3,5}^{n-2} \frac{x_{i+1} - x_i}{3}\,[\,f(x_i) + 4f(x_{i+1}) + f(x_{i+2})\,] \tag{7-69}$$

Given that Simpson's rule is based on a second-order interpolating polynomial, it

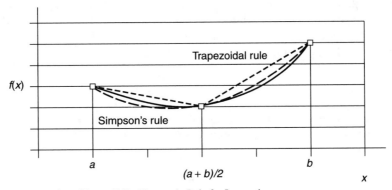

Figure 7-5   Simpson's Rule for Integration

can be shown that the error bound is a function of the fourth derivative. Specifically, the absolute value of the upper bound on the error is estimated by

$$\text{error} = \frac{(x_{i+1} - x_i)^5}{90} \frac{d^4 f(x_i)}{dx^4} \tag{7-70}$$

in which $d^4 f(x_i)/dx^4$ is the fourth derivative of the relationship between the dependent and independent variables, with the derivative being evaluated at the value of the independent variable (x_j) that maximizes the error between x_i and x_{i+1}. The error given by Eq. 7-70 is for the specific interval being evaluated, and the bound on the total error equals the sum over all the intervals of the absolute values of the error limits defined by Eq. 7-70.

Example 7-9 Flow Rate Problem

Applying Simpson's rule to the data of Table 7-6, the following flow rate (Q) can be obtained using Eq. 7-69:

$$Q = \int_0^{r_0} V 2\pi r \, dr = \frac{2\pi\left(\frac{1}{12}\right)}{3} \sum_{r=1,3}^{5} [V_i r_i + 4V_{i+1} r_{i+1} + V_{i+2} r_{i+2}]$$

$$= \frac{\pi}{18}\left[10(0) + 4(9.722)\left(\frac{1}{12}\right) + 8.889\left(\frac{1}{6}\right)\right] + \left[8.889\left(\frac{1}{6}\right) + 4(7.5)\left(\frac{1}{4}\right)\right. \tag{7-71}$$

$$\left. + 5.556\left(\frac{1}{3}\right)\right] + \left[5.556\left(\frac{1}{3}\right) + 4(3.056)\left(\frac{5}{12}\right) + 0\left(\frac{1}{2}\right)\right] = 3.927 \text{ ft}^3/\text{sec}$$

The estimate from Simpson's rule equals the true value, which is obviously better than the error resulting from the trapezoidal rule. In this case, Simpson's rule gives an exact value because the true function is a second-order polynomial; Simpson's rule will always exactly integrate a quadratic function.

7.2.4 Romberg Integration

It was shown that Simpson's rule provided greater accuracy than the trapezoidal rule, but at the expense of a larger number of evaluations of the function $f(x)$. It then should seem reasonable that the accuracy should improve as the number of evaluations of the function increases. Denoting the trapezoidal estimate as I_{01},

$$I_{01} = \frac{b-a}{2}(f(a) + f(b)) \tag{7-72}$$

where a and b are the start and end of an interval. Then a second estimate of the integral can be made using an average of areas obtained with a center point; this estimate is denoted as I_{11}:

$$I_{11} = \frac{b-a}{2}\left(\frac{1}{2}f(a) + f(m) + \frac{1}{2}f(b)\right) \tag{7-73}$$

in which $f(m)$ is the value of the function at the midpoint of the interval (a, b). Equation 7-73 can be rewritten as

$$I_{11} = \frac{1}{2}\left[I_{01} + (b - a)f\left(a + \frac{b - a}{2}\right)\right] \tag{7-74}$$

A third estimate I_{21} of the integral can be obtained using three equally spaced intermediate points m_1, m_2, and m_3:

$$I_{21} = \frac{b - a}{2}\left[\frac{1}{4}f(a) + \frac{1}{2}f(m_1) + \frac{1}{2}f(m_2) + \frac{1}{2}f(m_3) + \frac{1}{4}f(b)\right] \tag{7-75}$$

Expressing Eq. 7-75 in terms of I_{11} yields

$$I_{21} = \frac{1}{2}\left[I_{11} + \frac{b - a}{2}\sum_{\substack{k = 1 \\ k \neq 2}}^{3} f\left(a + \frac{b - a}{4}k\right)\right] \tag{7-76}$$

Continuing this subdividing of the interval leads to the following recursive relationship:

$$I_{i1} = \frac{1}{2}\left[I_{i-1,1} + \frac{b - a}{2^{i-1}}\sum_{k = 1, 3, 5}^{2^i - 1} f\left(a + \frac{b - a}{2^i}k\right)\right] \quad \text{for } i = 1, 2, \ldots \tag{7-77}$$

As i increases, the accuracy of the integral I_{i1} is expected to increase. If the integration interval, $b - a$, is halved N times, then the subintervals are of length $h = (b - a)/2^N$ and we have a vector of $N + 1$ values of I_{i1}, $i = 0, 1, 2, \ldots, N$, where I_{01} is the estimate of the integral made with the trapezoidal rule.

Richardson provided a method of combining estimates of integrals to obtain a more accurate estimate. The general extrapolation formula in recursive form is

$$I_{ij} = \frac{4^{j-1}I_{i+1,j-1} - I_{i,j-1}}{4^{j-1} - 1} \tag{7-78}$$

where $i = 0, 1, \ldots, N - j + 1$ and $j = 1, 2, 3, \ldots, N$. For any value of j, the values I_{ij} form a vector. The values of I_{i2} are computed from the vector I_{i1}; the values of I_{i3} are computed from the vector I_{i2}. When $j = N$, the vector I_{iN} will contain the one value I_{0N}, which is the best estimate of the integral.

The values of I_{ij} are best presented in the following upper-triangular matrix form:

$$
\begin{matrix}
I_{01} & I_{02} & I_{03} & I_{04} & \cdots & I_{0,N-1} & I_{0,N} & I_{0,N+1} \\
I_{11} & I_{12} & I_{13} & \cdot & \cdots & I_{1,N-1} & I_{1,N} & \\
I_{21} & I_{22} & \cdot & \cdot & \cdots & I_{2,N-1} & & \\
I_{31} & \cdot & \cdot & \cdot & & & & \\
\cdot & \cdot & \cdot & & & & & \\
\cdot & & \cdot & I_{N-2,3} & & & & \\
\cdot & I_{N-1,2} & & & & & & \\
I_{N1} & & & & & & &
\end{matrix}
\tag{7-79}
$$

To compute the estimate of I_{ON}, Eq. 7-77 can be used to obtain the values of I_{i1}. Then the recursive relationship of Eq. 7-78 can be used to compute the vector I_{i2} and then successively the values of I_{i3}, I_{i4}, and so on. It is only necessary to store the elements of adjacent vectors I_{ij} and $I_{i,j+1}$. Once the vector $I_{i,j+1}$ has been computed, the vector I_{ij} is not necessary, so the elements can be used to store the values of $I_{i,j+2}$.

Example 7-10 Romberg Method for Integration

To illustrate the Romberg method, a function with a known integral will be used. For the function $f(u) = ue^{au}$, the integral is

$$\int ue^{au}\, du = \frac{e^{au}}{a^2}(au - 1) + c \tag{7-80}$$

If $a = 2$ and the function is integrated over the interval from 0 to 1, the solution of Eq. 7-80 yields an integral of 2.097264 (seven significant digits). The Romberg integration matrix is expressed using six significant digits in Table 7-7. The values in Table 7-7 were computed using Eqs. 7-77 and 7-78. The table is expressed according to Eq. 7-79. For example, the values in Table 7-7 that corresponds to $i = 0, 1, 2, 3, 4$ and 5 were computed as follows:

For $i = 0$, the trapezoidal rule is used to obtain I_{01}:

$$I_{01} = 0.5(b - a)(f(a) + f(b)) = 0.5(1 - 0)[(0)\exp(2(0)) + (1)\exp(2(1))] = 3.69453$$

For $i = 1$, Eq. 7-74 or 7-77 is used to obtain I_{11}:

$$I_{11} = 0.5\left[I_{01} + (b - a)f\left(a + \frac{b - a}{2}\right)\right] = 0.5[3.69453 + (1 - 0)(0.5\exp(2(0.5)))] = 2.52683$$

For $i = 2$, Eq. 7-76 or 7-77 is used to obtain I_{21}:

$$I_{21} = 0.5\left[I_{11} + 0.5(b - a)\left(f\left(a + \frac{(b - a)}{4}\right) + f\left(a + 3\frac{(b - a)}{4}\right)\right)\right]$$
$$= 0.5[2.52683 + 0.5(0.25\exp(2(0.25)) + 0.75\exp(2(0.75)))] = 2.20678$$

TABLE 7-7. Computations According to the Romberg Method

i	j = 1	2	3	4	5	6
0	3.69453	2.13760	2.09759	2.09727	2.09726	2.09726
1	2.52683	2.10009	2.09727	2.09726	2.09726	
2	2.20678	2.09745	2.09726	2.09726		
3	2.12478	2.09728	2.09726			
4	2.10415	2.09727				
5	2.09899					

For $i = 3$, Eq. 7-77 is used to obtain I_{31}:

$$I_{31} = 0.5\left[I_{21} + 0.25(b - a)\left(f\left(a + \frac{b - a}{8}\right) + f\left(a + \frac{3(b - a)}{8}\right)\right.\right.$$
$$\left.\left. + f\left(a + \frac{5(b - a)}{8}\right) + f\left(a + \frac{7(b - a)}{8}\right)\right)\right]$$
$$= 0.5[2.20678 + 0.25(0.125 \exp(2(0.125)) + 0.375 \exp(2(0.375))$$
$$+ 0.625 \exp(2(0.625)) + 0.875 \exp(2(0.875)))] = 2.12478$$

For $i = 4$, Eq. 7-77 is used to obtain I_{41}:

$$I_{41} = 0.5\left[I_{31} + 0.125(b - a)\left(f\left(a + \frac{b - a}{16}\right) + f\left(a + \frac{3(b - a)}{16}\right)\right.\right.$$
$$+ f\left(a + \frac{5(b - a)}{16}\right) + f\left(a + \frac{7(b - a)}{16}\right) + f\left(a + \frac{9(b - a)}{16}\right) + f\left(a + \frac{11(b - a)}{16}\right)$$
$$\left.\left. + f\left(a + \frac{13(b - a)}{16}\right) + f\left(a + \frac{15(b - a)}{16}\right)\right)\right]$$
$$= 0.5[2.12478 + 0.125(0.0625 \exp(2(0.0625)) + 0.1875 \exp(2(0.1875))$$
$$+ 0.3125 \exp(2(0.3125)) + 0.4375 \exp(2(0.4375)) + 0.5625 \exp(2(0.5625))$$
$$+ 0.6875 \exp(2(0.6875)) + 0.8125 \exp(2(0.8125)) + 0.9375 \exp(2(0.9375)))]$$
$$= 2.10415$$

Similarly for $i = 5$, Eq. 7-77 is used to obtain $I_{51} = 2.09899$.

The rest of the entries in Table 7-7 were computed using Eq. 7-78. For example, I_{02} is computed as follows:

$$I_{02} = \frac{4^{2-1}(2.52683) - 3.69453}{4^{2-1} - 1} = 2.13760$$

The value of I_{03} is computed as follows:

$$I_{03} = \frac{4^{3-1}(2.10009) - 2.13760}{4^{3-1} - 1} = 2.09759$$

Similar computations are performed for the other values in the table. The Romberg method yields an exact value to six significant digits for the integral of 2.09726.

7.3 APPLICATIONS

7.3.1 Estimating Areas

A land developer must clear a small wooded area on a site to be developed into a commercial shopping center. Figure 7-6 shows the plot. To estimate the time required to clear the wooded area, the developer needs an estimate of the area. This is obtained by assuming that the latitude is the independent variable and the longitude is the dependent variable. The wooded area is the difference between the area below the upper perimeter between latitudes of 400 and 2400 ft and the area below the lower perimeter between the same two points. Six points are used to delineate the two lines. Since the perimeters are

Figure 7-6 Plot for Commercial Shopping Center

defined by an even number of points, the trapezoidal rule is used. The area below the upper perimeter is computed using Eq. 7-56 as

$$A_u = \frac{400}{2}[1200 + 2(1700 + 1900 + 1800 + 1500) + 900]$$

$$= 3,180,000 \text{ ft}^2 = 73.00 \text{ acres} \tag{7-81}$$

The area below the lower perimeter is computed using Eq. 7-56 as

$$A_L = \frac{400}{2}[1200 + 2(600 + 900 + 300 + 600) + 900] \tag{7-82}$$

$$= 1,380,000 \text{ ft}^2 = 31.68 \text{ acres}$$

Therefore, the perimeter of the wooded area encloses 41.32 acres, which is the difference between 73.00 and 31.68 acres.

7.3.2 Pipe Flow Rate

Using an instrument for measuring flow velocity, velocity measurements were made over the cross section of a 3-ft-diameter pipe. Table 7-8 gives the velocity at the center line of the cross section and at distances (r) measured from the center line of the pipe (see Fig. 7-7). The flow rate is obtained using the following continuity equation:

$$Q = \int V\, dA = \int V_r\, 2\pi r\, dr = 2\pi \int_0^{1.5} V_r r\, dr \tag{7-83}$$

If an expression for V_r were available and could be integrated, an analytical solution could be used. In this case, measurements are available at discrete points. An estimate of Q can be made with the trapezoidal rule.

When applying the trapezoidal rule to this case, the difference $(x_{i+1} - x_i)$ in Eq.

TABLE 7-8 Calculation of Volume Flow Rates Using the
Trapezoidal Rule

r (in.)	Velocity (fps)	Incremental Area (ft^2)	Incremental Q (ft^3/sec)
0	5.0		
		0.196	0.970
3	4.9		
		0.589	2.739
6	4.4		
		0.982	4.026
9	3.8		
		1.374	4.534
12	2.8		
		1.767	3.799
15	1.5		
		2.160	1.620
18	0.0		

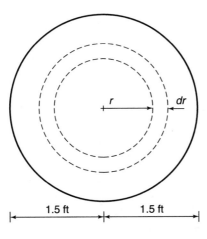

1.5 ft   1.5 ft

Figure 7-7 Flow Rate for a Pipe

7-56 must be replaced by the area associated with the points in the form of an increment area (ΔA_i) that is given by $\Delta A_i = \pi(r_{i+1}^2 - r_i^2)$. Thus the trapezoidal rule becomes

$$Q = \sum_{i=1}^{n-1} \pi(r_{i+1}^2 - r_i^2)\frac{V_{i+1} + V_i}{2} \tag{7-84}$$

which uses the product of the average velocity and the incremental area as the incremental flow rate ΔQ_i. Summing the ΔQ_i values gives the total flow rate Q. The calculations are shown in Table 7-8, with a flow rate of 17.7 cfs for the measured data. The flow rate was computed as the summation of the last column in Table 7-8.

7.3.3 Volume of Gabion Retaining Wall

A geotechnical engineer is planning to use a gabion retaining wall to protect a section of highway from debris and mud. A gabion is a wire-mesh container filled with rocks. The engineer must estimate the volume of rock required. The gabion wall will have a width of 6 ft, and the cross-sectional area of the retaining wall is shown in Fig. 7-8.

The estimate can be made with the trapezoidal rule using Eq. 7-56 as

$$\text{volume} = 6\left[\frac{40}{2}(0 + 2(5) + 2(7) + 2(10) + 2(8) + 0)\right]$$

$$= 7200 \text{ ft}^3 \tag{7-85}$$

In this case, each value of $f(x)$ is subtracted from 10 ft to compute the ordinates; in essence, the volume of the structure is computed using the top of the structure as the axis. The volume of rock required to fill the wire-mesh gabions would depend on the amount of void space within the baskets holding the rocks.

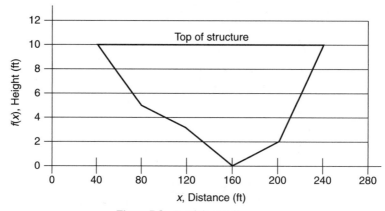

Figure 7-8   Retaining Wall

7.3.4 Sediment Loads in Rivers

Environmental engineers are concerned with the environmental effects of sediment in streams and rivers. Measurements of sediment transport rates are made during flood runoff events, which is the time when most sediment is transported. Large volumes of sediment that are included in runoff into a reservoir can reduce the storage volume of the reservoir, as well as reduce the quality of the water for fish and wildlife.

At a particular location the following 6-hr measurements of the sediment rate (q_s) were made during a storm event: 0, 8, 18, 23, 28, 26, 21, 18, 14, 11, 6, 2, and 2 lb/sec. Each measurement is assumed to reflect the average rate over the 6-hr period. A plot of the sediment rate as a function of time (t) is called a sediment graph, or sometimes a pollutograph.

To find the total sediment load (pounds) during the storm (Q_s), Simpson's rule can be used to integrate the function defined by the points:

$$Q_s = \int q_s \, dt \approx \frac{6 \, hr}{6} [0 + 4(8) + 2(18) + 4(23) + 2(28) + 4(26) \qquad (7\text{-}86)$$
$$+ \, 2(21) + 4(18) + 2(14) + 4(11) + 2(6) + 4(2) + 2] = 528 \text{ lb}$$

At a specific weight of 165 lb/ft^3, the sediment would have a volume of 11,520 ft^3. If the sediment entered a reservoir with a bottom area of 25 acres and settled uniformly over the bottom, it would settle to a depth of 0.0106 ft, or $\frac{1}{8}$ in. Following many such storms, the capacity of the reservoir to hold water would be greatly diminished because of sediment accumulation.

7.3.5 Probability Computations Using the Standard Normal Probability Distribution

The standard normal probability distribution is commonly used in engineering. The probability density function $f(x)$ is given by

$$f(x) = \frac{1}{\sqrt{2\pi}} \exp(-0.5x^2) \qquad (7\text{-}87)$$

Figure 7-9 shows this symmetric function with respect to $x = 0$. The following integral can be used to compute the probability that x is between a and b:

$$P(a \leq x \leq b) = \int_a^b \frac{1}{\sqrt{2\pi}} \exp(-0.5x^2) \, dx \qquad (7\text{-}88)$$

The integral of Eq. 7-88 cannot be analytically computed. Numerical integration can be used for this purpose. For example, the following probability can be determined using the trapezoidal rule:

$$P(-1 \leq x \leq 1) = \int_{-1}^1 \frac{1}{\sqrt{2\pi}} \exp(-0.5x^2) \, dx \qquad (7\text{-}89)$$

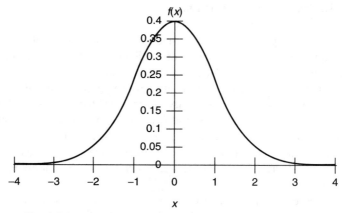

Figure 7-9   Standard Normal Probability Density Function

Using a constant interval of 0.2, the probability can be computed as

$$P(-1 \le x \le 1) = \frac{0.2}{2}\left(f(-1) + f(1) + 2\sum_{i=1}^{9} f(-1 + 0.2i)\right) = 0.6810741 \qquad (7\text{-}90)$$

The exact value can be computed from tables in probability textbooks; in this case the exact value is 0.68269. The error is 0.0016159 (0.2%). The error level depends on the selected area under the curve and the size of the selected interval.

7.4 PROBLEMS

7-1. The following table gives the water vapor capacity of air (V_c) in grains per cubic foot for selected temperatures (T) in degrees Fahrenheit. Determine the slope of the water vapor capacity curve at a temperature of 54°F using the forward, backward, and the two-step finite-difference approximations.

T	50	54	58	62	66
V_c	4.108	4.725	5.420	6.203	7.082

7-2. Given the following values of $f(x)$ and x, determine the first derivative of $f(x)$ with respect to x at $x = 3$ using the forward, backward, and two-step finite-difference approximation. Compare the accuracy if the true value of the first derivative is given by $6x^5 - 25x^4 + 48x^3 - 54x^2 + 70x + 80$.

x	1	2	3	4	5
$f(x)$	5	152	455	1676	6425

7-3. The table below gives the sine for selected angles (θ). Determine the first derivative of $\sin \theta$ at $\theta = 7°$ using the forward, backward, and two-step finite-difference approximation. Compare the accuracy with the true value. (*Note:* $1° = 0.017453$ radian.)

θ	4	5	6	7	8
$\sin \theta$	0.06976	0.08716	0.10453	0.12187	0.13917

7-4. Using the data in the following table, compute the first derivative at $x = 4$ using the forward, backward, and two-step finite-difference approximation. Compute the second derivative at $x = 4$ using the two-step approximation. Compare these estimates with the true values of 0.25 and -0.0625, respectively.

x	2	3	4	5	6
$f(x)$	0.69315	1.09861	1.38629	1.60944	1.79176

7-5. Using the data in the following table, compute the first derivative at $x = 1.4$ using the forward, backward, and two-step finite-difference approximations. Compute the second derivative at $x = 1.4$ using the two-step approximation.

x	1.0	1.2	1.4	1.6	1.8	2.0
$f(x)$	1.0	0.91817	0.88716	0.89352	0.93138	1.0

7-6. Using the data of Problem 6-9, estimate the first derivative at $x = 0.55$ using a finite-difference table.

7-7. The data in the following table are the distances (D) traveled by a car, which was stationary at time $t = 0$, for selected times. Using a finite-difference table, determine (a) the velocity at $t = 6$ sec, (b) the velocity at $t = 8.7$ sec, (c) the acceleration at $t = 1$ sec, and (d) the acceleration at $t = 6.3$ sec.

t (sec)	0	2	4	6	8	10
D (ft)	0	10	50	150	330	610

7-8. A car is traveling at a constant velocity. At time $t = 0$, the driver applies the brakes, with distances (D) traveled at specified times. Using a finite-difference table, estimate (a) the velocity at time $t = 0$ sec, (b) the velocity at time $t = 3.5$ sec, and (c) the rate for deceleration at time $t = 3.5$ sec.

t (sec)	0	1	2	3	4	5
D (ft)	0	90	160	200	220	225

7-9. Fit an interpolating polynomial to the data of Problem 7-1 and estimate the first derivative at (a) $T = 54°$ and (b) $T = 64°$.

7-10. Fit an interpolating polynomial to the data of Problem 7-2 and estimate the first derivative at (a) $x = 3$ and (b) $x = 5.5$. Compare the estimates with the true values.

7-11. Fit an interpolating polynomial to the data of Problem 7-3 and estimate the first derivative at (a) $\theta = 7°$ and (b) $\theta = 5.5°$. Compare the estimates with the true values.

7-12. Fit an interpolating polynomial to the data of Problem 7-4 and estimate the first and second derivatives at (a) $x = 4$ and (b) $x = 1.7$. For part (a), compare the estimated values with the true values.

7-13. Fit an interpolating polynomial to the data of Problem 7-5 and estimate the first and second derivatives at $x = 1.4$.

7-14. Two parallel plates are separated by a fluid having a viscosity (μ) of 10^{-3} lb-sec/ft². While the lower plate is stationary, the upper plate is movable and connected to a weight (W) by a cable, as shown in the figure. The shear stress across the space is given by $\tau = \mu \, dV/dy$. Velocity measurements were made using stop-action photography when the system was in steady state as shown in the figure. Determine the distribution of shear stress across the space between the plates using an interval of $0.25Y$, where $Y = 0.016$ ft.

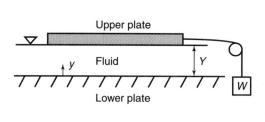

Y (ft)	V (ft/sec)
0	0.000
0.2Y	0.001
0.4Y	0.002
0.6Y	0.005
0.8Y	0.015
1.0Y	0.040

Figure P7.14

7-15. The shear stress (τ) between a rotating shaft and a stationary cylinder is given by $\tau = \mu \, dV/dr$, where μ is the viscosity of the fluid between the cylinder and the shaft, and dV/dr is the velocity gradient within the space. The viscosity of the fluid is 10^{-4} lb-sec/ft^2, and the radius of the shaft (R_1) and cylinder (R_2) are 3 in. and 3.20 in., respectively. Estimate the angular velocity at the outer surface of the shaft and find the distribution of the shear stress across the fluid.

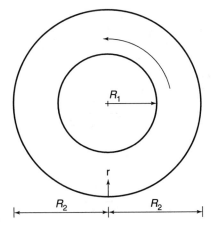

r (inch)	V (ft/sec)
0	0
0.04	1.7
0.08	2.9
0.12	4.8
0.16	7.8

Figure P7.15

7-16. Using the trapezoidal rule and the following data, find the area under the sine curve from 0° to 90°. Compare the result with the true integral.

θ	0°	15°	30°	45°	60°	75°	90°
$\sin \theta$	0	0.2588	0.5	0.70701	0.8660	0.9659	1.0

7-17. The system reliability function $R(t)$ is given by $R(t) = 1 - F(t)$, where $F(t)$ is the probability that the component will fail in the interval from 0 to t. $F(t)$ is determined from $F(t) = \int f(t)\,dt$ where $f(t)$ is the probability density of the time to failure of the system. If the following estimates of $f(t)$ are made from sample data, use the trapezoidal rule to estimate (a) the reliability function at different t values and (b) the probability that the system will provide reliable service for 5 days.

t (days)	0	1	2	3	4	5	6	7	8	9
$f(t)$	0.33	0.24	0.17	0.12	0.09	0.06	0.04	0.03	0.02	0.01

7-18. Based on the concepts of Problem 7-17, use the following data to estimate (a) the reliability function and (b) the probability that the system will survive for at least 2 hr but no more than 6 hr. Use the trapezoidal rule.

t (hours)	0	2	4	6	8	10	12	14	16
$f(t)$	0.02	0.05	0.06	0.07	0.08	0.07	0.06	0.05	0.02

7-19. Use the trapezoidal rule to estimate $\int_0^3 x^{1/3}\,dx$ for interval widths of 1, 0.5, and 0.25. Compare these values with the corresponding true values.

7-20. Ground surface elevations $f(x)$ are taken every 25 ft across the stream channel at a bridge opening. Using the trapezoidal rule, estimate the flow Q in the river when the water surface is at the bottom of the bridge (22 ft) (see Fig. P7.20). The flow Q is given by the integral $Q = \int V\,dA$, where V is the average velocity and equals a constant of 6 ft/sec.

x (ft)	0	25	50	75	100	125	150	175	200
$f(x)$ (ft)	22	15	11	0	3	8	11	16	22

Figure P7.20

7-21. For Problem 7-20, estimate the flow Q if the velocity is not constant but varies with the location according to the following:

x (ft)	0		25		50		75		100		125		150		175		200
V (ft/sec)		2		4		7		8		8		6		6		3	

7-22. Using the data of Problem 7-16 and Simpson's rule, compute the area under the sine curve and compare the estimated value with both the true value and the value estimated using the trapezoidal rule.

7-23. Using the data of Problem 7-17 and Simpson's rule, compute the probability that the system will provide reliable service for at least 6 days.

7-24. For Problem 7-20, estimate the flow of Q using Simpson's rule.

7-25. Using Simpson's rule, estimate $\int_0^3 x^{1/3}\, dx$ for interval widths of 1.5, 0.5, and 0.25. Compare these estimated values with the corresponding true values.

7-26. Using Simpson's rule, estimate $\int_0^2 \dfrac{dx}{(6 - x^2)^{3/2}}$ using 2, 4, 6, 8, and 10 intervals.

7-27. Using Simpson's rule, estimate $\int_1^4 \dfrac{dx}{3 + 2e^x}$ using 4, 8, and 10 intervals.

8

Differential Equations

8.1 INTRODUCTION

8.1.1 Definitions

An equation that defines a relationship between an unknown function and one or more of its derivatives is referred to as a *differential equation*. To begin, let's assume that the relationship involves two variables, the independent variable and the dependent variable. The derivatives in the differential equation are of the dependent variable with respect to the independent variable, and the unknown function is a function of either the independent variable, the dependent variable, or both. The order of the differential equation is equivalent to the order of the highest derivative in the equation; for example, an equation that includes only first and second derivatives would be a second-order differential equation.

To illustrate these items, a first-order differential equation with y as the dependent variable and x as the independent variable would be

$$\frac{dy}{dx} = f(x, y) \tag{8-1}$$

in which $f(x, y)$ is a function. A second-order differential equation would have the form

$$\frac{d^2y}{dx^2} = f\left(x, y, \frac{dy}{dx}\right) \tag{8-2}$$

The functions on the right sides of Eqs. 8-1 and 8-2 do not necessarily have to include all the variables shown. For example, the function of Eq. 8-1 could also be either $f(x)$ or $f(y)$, and for Eq. 8-2 the function could be $f(x, dy/dx)$, $f(y, dy/dx)$, $f(x, y)$, $f(x)$, $f(y)$, or $f(dy/dx)$. Thus Eqs. 8-1 and 8-2 are written in general terms.

An ordinary differential equation is one with a single independent variable. Thus, while Eqs. 8-1 and 8-2 are ordinary differential equations, the following is not:

$$\frac{dy}{dx_1} = f(x_1, x_2, y) \tag{8-3}$$

in which y is the dependent variable and x_1 and x_2 are independent variables.

Differential equations may be either linear or nonlinear. An nth-order differential equation is linear if the function on the right of the equation is linear in the independent variables, dependent variable, and their derivatives, that is, $f(x, y, dy/dx, d^2y/dx^2, \ldots, d^{n-1}y/dx^{n-1})$ is linear.

The solution of the differential equation is a function of the independent variable that, when substituted for the dependent variable, reduces the equation to an identity in the independent variable.

Boundary conditions are constraints placed on the solutions; that is, in addition to specifying the differential equation, there may be additional conditions placed on the solution. For example, the differential equation might be

$$\frac{dy}{dx} = 5x \tag{8-4}$$

A particular problem may specify that the desired solution should have $y = 2$ at $x = 1$; this constraint on the solution is called a *boundary condition*. The solution of a differential equation such as Eq. 8-4 may be plotted on a graph of y versus x; for the above boundary condition the plotted function passes through the point ($y = 2$, $x = 1$). For a different boundary condition, the solution would be different.

8.1.2 Origin of Differential Equations

Differential equations can originate from either geometric or physical problems. To illustrate a geometric origin, consider the case of a pair of variables (x and y) in which the slope of the relationship between y and x equals the product of a constant c and the difference between y and x. This geometric problem would define the following differential equation:

$$\frac{dy}{dx} = c(y - x) \tag{8-5}$$

The solution would be a relationship of the form $y = g(x)$. Equation 8-5 may be subject to one or more boundary conditions.

Physical problems can also be defined by differential equations. Simple problems in electrical circuits and heat transfer are commonly used to introduce differential equations. Simple motion problems can also be governed by differential equations. For example, consider a mass of m slugs that falls from rest in a medium in which the resistance is proportional to the square of the velocity (V) of the mass at any time t. The net force (F) on the mass equals the difference between the force exerted by gravity and the resistance of the medium. For a medium resistance defined as R (in mass units per second), the governing equation is

$$F = ma = m\frac{dV}{dt} \tag{8-6a}$$

The net force is also given by

$$F = mg_r - RV \tag{8-6b}$$

where a = acceleration and g_r = gravitational acceleration. The fact that the mass begins at rest defines a boundary condition that V (the dependent variable) equals zero at time (the independent variable) zero ($V = 0$ at $t = 0$). Dividing through by m yields

$$\frac{dV}{dt} = g_r - \frac{R}{m}V \qquad \text{such that } V = 0 \text{ at } t = 0 \tag{8-6c}$$

A solution that shows V as a function of t depends on the values of R and m.

8.2 TAYLOR SERIES EXPANSION

8.2.1 Fundamental Case

Assume that the problem is a first-order ordinary differential equation of the form

$$\frac{dy}{dx} = f(x) \qquad \text{subject to } y = y_0 \text{ at } x = x_0 \tag{8-7}$$

If we separate the variables and integrate both sides, we have

$$\int_{y_0}^{y} dy = \int_{x_0}^{x} f(x)\, dx \tag{8-8a}$$

or

$$y = y_0 + \int_{x_0}^{x} f(x)\, dx \tag{8-8b}$$

For a single independent variable x, a Taylor series expansion has the form

$$g(x) = g(x_0) + (x - x_0)g'(x_0) + \frac{(x - x_0)^2}{2!} g''(x_0) + \cdots \tag{8-9}$$

where x_0 is a value where the function $g(x_0)$ is known, and the derivatives $g'(x_0)$, $g''(x_0)$, $\ldots$, and so on, with respect to x can be computed; x is some other value of the independent variable for which the value of the dependent variable $y = g(x)$ is not known and needs to be determined. If we can compute the derivatives, the values of the function can be computed for any value of the independent variable.

Equations 8-8b and 8-9 can be used to solve a differential equation. The terms on the right side of Eq. 8-9 beginning with $(x - x_0)g'(x_0)$ can be used to represent the integral on the right side of Eq. 8-8b. Since $y_0 = g(x_0)$ and $g'(x_0) = f(x_0)$, the solution to the differential equation of Eq. 8-7 can be obtained using the Taylor series expansion of Eq. 8-9.

Example 8-1 First-order Differential Equation

Given the following differential equation,

$$\frac{dy}{dx} = 3x^2 \qquad \text{such that } y = 1 \text{ at } x = 1 \tag{8-10a}$$

the higher-order derivatives are easily computed:

$$\frac{d^2 y}{dx^2} = 6x \tag{8-10b}$$

$$\frac{d^3 y}{dx^3} = 6 \tag{8-10c}$$

$$\frac{d^n y}{dx^n} = 0 \qquad \text{for } n \geq 4 \tag{8-10d}$$

Based on Eqs. 8-10a and b, the Taylor series of Eq. 8-9 yields a function $y = g(x)$ that describes the dependent variable for any value of x:

$$g(x) = 1 + (x - 1)\frac{dy}{dx} + \frac{(x - 1)^2}{2!}\frac{d^2 y}{dx^2} + \frac{(x - 1)^3}{3!}\frac{d^3 y}{dx^3} \tag{8-10e}$$

$$= 1 + (x - 1)(3x_0^2) + \frac{(x - 1)^2}{2!}(6x_0) + \frac{(x - 1)^3}{3!}(6) \tag{8-10f}$$

$$= 1 + 3x_0^2(x - 1) + 3x_0(x - 1)^2 + (x - 1)^3 \tag{8-10g}$$

where $x_0 = 1$. Equation 8-10g can be used to compute the value of y [i.e., $g(x)$] for any value of x. The solution is given in Table 8-1 and shown graphically in Fig. 8-1 for values of x from 1 to 2. Note that the higher-order terms are small near the boundary condition of $x = 1$,

TABLE 8-1 Taylor Series Solution

x	One Term	Two Terms	Three Terms	Four Terms
1	1	1	1	1
1.1	1	1.3	1.33	1.331
1.2	1	1.6	0.72	1.728
1.3	1	1.9	2.17	2.197
1.4	1	2.2	2.68	2.744
1.5	1	2.5	3.25	3.375
1.6	1	2.8	3.88	4.096
1.7	1	3.1	4.57	4.913
1.8	1	3.4	5.32	5.832
1.9	1	3.7	6.13	6.859
2	1	4	7	8

but they become larger as x deviates farther from the boundary condition. For the case of Fig. 8-1, the derivatives beyond the third equal zero. Thus there is no error in the solution when all derivatives were used, as in Eqs. 8-10e, f, and g. If only the first and second derivatives were used, the solution would be in error by an amount equal to the term containing the third derivative. Thus the error would be the difference between the upper two lines of Fig. 8-1. At x equal to 2, the exact solution is $y = 8$. The error would be 12.5% of the true value of y if the term with the third derivative were omitted. If the terms with the second and third derivatives were omitted, then, at x equal to 2, the error would be 50% of the true value of y.

8.2.2 General Case

Assume that the problem is a first-order ordinary differential equation of the following general form:

$$\frac{dy}{dx} = f(x, y) \qquad \text{subject to } y = y_0 \text{ at } x = x_0 \tag{8-11}$$

For this general case, a Taylor series expansion has the form

$$g(x) = g(x_0, y_0) + (x - x_0)g'(x_0, y_0) + \frac{(x - x_0)^2}{2!} g''(x_0, y_0) + \cdots \tag{8-12}$$

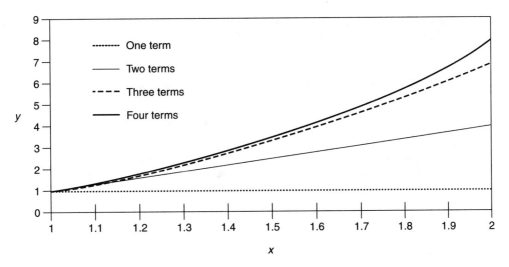

Figure 8-1   Graphical Representation of the Terms of a Taylor Series Expansion

where x_0 and y_0 are the values when the function $g(x_0, y_0)$ is known, and the derivatives $g'(x_0, y_0)$, $g''(x_0, y_0)$, . . . , and so on, with respect to x can be computed; x is some other value of the independent variable for which the value of the dependent variable $y = g(x)$ is not known and needs to be determined. If we can compute the derivatives, the values of the function can be computed for any value of the independent variable. The derivatives in this case are more difficult to evaluate. Implicit differentiation using the chain rule for differentiation might be necessary.

8.3 EULER'S METHOD

In some cases the derivatives are not easily computed. Therefore, the Taylor series of Eqs. 8-9 and 8-12 can be truncated so that only the term with the first derivative is used. Thus the value of the dependent variable $g(x)$ can be computed using

$$g(x) = g(x_0) + (x - x_0)\frac{dy}{dx} + e \qquad (8\text{-}13)$$

The derivative dy/dx in Eq. 8-13 is evaluated at the beginning of the interval, i.e., x_0. It should be evident from Fig. 8-1 that, as the distance or interval $(x - x_0)$ increases, the error increases because the nonlinear terms (i.e., second order and higher) of the Taylor series expansion become more important. Thus it is important to keep the distance $(x - x_0)$ small. In practice, Eq. 8-13 is applied iteratively over small distances in order to reduce the error. The computations according to this method are performed one step at a time. Therefore, the method is sometimes referred to as the one-step Euler's method. This is best illustrated using an example.

Example 8-2 First-order Differential Equation

Consider the differential equation

$$\frac{dy}{dx} = 4x^2, \qquad \text{such that } y = 1 \text{ at } x = 1 \tag{8-14}$$

This equation is easily solved analytically, and the analytical solution is useful here as a basis for assessing the accuracy of estimates obtained numerically. For example, for x equal to 1.1, the true value of y is

$$\int_1^y dy = \int_1^{1.1} 4x^2 \, dx \tag{8-15a}$$

$$y - 1 = \frac{4}{3} x^3 \Big|_1^{1.1} = 0.44133^+ \tag{8-15b}$$

Therefore, at $x = 1.1$, $y = 1.44133^+$.

Using Eq. 8-13 with a step size of $\Delta x = (x - x_0) = 0.1$, we get

$$g(1.1) = 1 + 0.1[4(1)^2] = 1.4 \tag{8-16}$$

Thus, when a step size of 0.1 is used, Euler's equation results in an error (in absolute value) of 0.04133^+. If, instead, we use a step size of 0.05 and apply Euler's equation twice (at $x = 1$ and at $x = 1.05$), we get a value of y at $x = 1.1$ of

$$g(1.05) = g(1) + (1.05 - 1.00)[4(1)^2] = 1 + 0.2 = 1.2 \tag{8-17a}$$

$$g(1.10) = g(1.05) + (1.10 - 1.05)[4(1.05)^2] = 1.4205 \tag{8-17b}$$

Thus, by decreasing the step size from 0.1 to 0.05, the error was decreased from 0.04133^+ to 0.020833^+, a reduction of 50%. For a step size of 0.02, the estimated value, after five steps, is 1.43296, which is in error by 0.008373^+.

8.3.1 Errors with Euler's Method

Errors in numerical solutions of differential equations can be classified as global or local errors. The local errors occur over one step size, whereas the global errors are cumulative over the range of solution. The global errors can be computed as the difference between the numerical and exact solutions. The local errors can be computed as the difference between a numerical solution at the end of the step using the exact value at the beginning of the step and the exact solution at the end of the step.

The error (ϵ) using Euler's method can be approximated using the second term of the Taylor series expansion as

$$\epsilon = \frac{(x - x_0)^2}{2!} \frac{d^2 y}{dx^2} \tag{8-18}$$

where the second derivative is evaluated at the point where it is maximum over the interval $(x - x_0)$. If the range is divided into n increments, then the error at the end of the range of interest for x would be $n\epsilon$.

Example 8-3 Analysis of Errors

Considering the differential equation in Example 8-2, the local and global errors can be computed as shown in Tables 8-2a, b, and c for step sizes of 0.1, 0.05, and 0.01, respectively. It can be observed from these tables that by decreasing the step size both the local and global errors can be decreased.

In this example, $d^2 y/dx^2$ is $8x$. Thus an approximate assessment of the error is given by

$$\epsilon = \frac{(x - x_0)^2}{2} (8x) = 4x(x - x_0)^2 \tag{8-19}$$

TABLE 8-2a   Local and Global Errors for Example 8-3 Using a Step Size of 0.1

x	Exact Solution	Numerical Solution	Local Error (%)	Global Error (%)
1	1	1	0	0
1.1	1.4413333	1.4	-2.8677151	-2.8677151
1.2	1.9706667	1.884	-2.300406	-4.3978349
1.3	2.596	2.46	-1.9003595	-5.238829
1.4	3.3253333	3.136	-1.6038492	-5.6936648
1.5	4.1666667	3.92	-1.376	-5.92
1.6	5.128	4.82	-1.1960478	-6.0062402
1.7	6.2173333	5.844	-1.0508256	-6.004718
1.8	7.4426667	7	-0.9315657	-5.947689
1.9	8.812	8.296	-0.8321985	-5.8556514
2	10.333333	9.74	-0.7483871	-5.7419355

TABLE 8-2b Local and Global Errors for Example 8-3 Using a Step Size of 0.05

x	Exact Solution	Numerical Solution	Local Error (%)	Global Error (%)
1	1	1	0	0
1.05	1.2101667	1.2	−0.8401047	−0.8401047
1.1	1.4413333	1.4205	−0.7400555	−1.4454209
1.15	1.6945	1.6625	−0.6589948	−1.8884627
1.2	1.9706667	1.927	−0.5920162	−2.2158322
1.25	2.2708333	2.215	−0.5357798	−2.4587156
1.3	2.596	2.5275	−0.4879301	−2.6386749
1.35	2.9471667	2.8655	−0.4467568	−2.771023
1.4	3.3253333	3.23	−0.4109864	−2.8668805
1.45	3.7315	3.622	−0.3796507	−2.9344768
1.5	4.1666667	4.0425	−0.352	−2.98
1.55	4.6318333	4.4925	−0.3274441	−3.0081681
1.6	5.128	4.973	−0.3055122	−3.0226209
1.65	5.6561667	5.485	−0.2858237	−3.0261956
1.7	6.2173333	6.0295	−0.2680678	−3.0211237
1.75	6.8125	6.6075	−0.2519878	−3.0091743
1.8	7.4426667	7.22	−0.2373701	−2.9917592
1.85	8.1088333	7.868	−0.2240355	−2.9700121
1.9	8.812	8.5525	−0.2118323	−2.9448479
1.95	9.5531667	9.2745	−0.2006316	−2.9170083
2	10.333333	10.035	−0.1903226	−2.8870968

For increments of 0.1, 0.05, and 0.02, the upper limits on the error would be, respectively,

$$\epsilon_{0.1} = 4(1.1)(0.1)^2 = 0.044 \tag{8-20a}$$

$$\epsilon_{0.05} = 2(4)(1.1)(0.05)^2 = 0.022 \tag{8-20b}$$

TABLE 8-2c   Local and Global Errors for Example 8-3 Using a Step Size of 0.02

x	Exact Solution	Numerical Solution	Local Error (%)	Global Error (%)
1	1	1	0	0
1.02	1.0816107	1.08	−0.1489137	−0.1489137
1.04	1.1664853	1.163232	−0.1408219	−0.2789005
1.06	1.254688	1.24976	−0.1334728	−0.392767
1.08	1.3462827	1.339648	−0.1267688	−0.4928138
1.1	1.4413333	1.43296	−0.120629	−0.5809436
1.2	1.9706667	1.95312	−0.0963464	−0.8903924
1.3	2.596	2.56848	−0.0793015	−1.0600924
1.4	3.3253333	3.28704	−0.0667201	−1.1515638
1.5	4.1666667	4.1168	−0.057088	−1.1968
1.6	5.128	5.06576	−0.049506	−1.2137285
1.7	6.2173333	6.14192	−0.0434055	−1.212953
1.8	7.4426667	7.35328	−0.0384092	−1.2010032
1.9	8.812	8.70784	−0.0342563	−1.1820245
2	10.333333	10.2136	−0.0307613	−1.1587097

$$\epsilon_{0.02} = 5(4)(1.1)(0.02)^2 = 0.0088 \qquad (8\text{-}20c)$$

The errors given by Eqs. 8-20 are greater than the actual errors.

8.4 MODIFIED EULER'S METHOD

To this point, two methods for solving differential equations were introduced. The use of a Taylor series provided accurate estimates, but it required estimates of the higher-order derivatives. Euler's method did not require higher-order derivatives, but it required a small step size and an iterative solution. A third method, the modified Euler's method, attempts to improve the accuracy of the estimated solution while using only the first derivative. It is structurally similar to the Euler's method of Section 8.3, but it uses an average slope, rather than the slope at the start of the interval (or step).

The general procedure for the most basic form is as follows:

a. Evaluate the slope (i.e., dy/dx) at the start of the interval.

b. Estimate the value of the dependent variable y at the end of the interval Δx (i.e., y at $x + \Delta x$) using the Euler's method.

c. Evaluate the slope at the end of the interval.

d. Find the average slope using the slopes of steps a and c.

e. Compute a revised value of the dependent variable at the end of the interval using the average slope of step d with Euler's method.

Example 8-4 Modified Euler's Method

The procedure is best illustrated using an example. Consider the differential equation

$$\frac{dy}{dx} = x\sqrt{y} \qquad \text{such that } y = 1 \text{ at } x = 1 \tag{8-21}$$

The five steps of the first iteration for $\Delta x = 0.1$ are as follows:

1a. $\left.\dfrac{dy}{dx}\right|_{1} = 1\sqrt{1} = 1$

1b. $g(1.1) = g(1.0) + (1.1 - 1.0)\left.\dfrac{dy}{dx}\right|_{1} = 1 + 0.1(1) = 1.1$

1c. $\left.\dfrac{dy}{dx}\right|_{1.1} = 1.1\sqrt{1.1} = 1.15369$

1d. $\left.\dfrac{dy}{dx}\right|_{a} = \dfrac{1}{2}(1 + 1.15369) = 1.07684$

1e. $g(1.1) = g(1.0) + (1.1 - 1.0)\left.\dfrac{dy}{dx}\right|_{a} = 1 + 0.1(1.07684) = 1.10768$

where, for example, dy/dx_1 means the first derivative of y with respect to x evaluated at $x = 1$ and the corresponding y value, and dy/dx_a means the average of the first derivative of y with respect to x evaluated at the start and end of an interval. It is especially important to note the following: (1) estimates of the dependent variable were made at both steps 1b and 1e, and (2) dy/dx_1 was used in step 1b, whereas dy/dx_a was used in step 1e. An estimate could be obtained at $x = 1.2$ by performing a second execution with the ends of the interval at $x = 1.1$ and $x = 1.2$. The steps for the second interval are as follows:

2a. $\left.\dfrac{dy}{dx}\right|_{1.1} = x\sqrt{y} = 1.1\sqrt{1.10768} = 1.15771$

2b. $g(1.2) = g(1.1) + (1.2 - 1.1)\, dy/dx_{1.1} = 1.10768 + 0.1(1.15771) = 1.22345$

2c. $\left.\dfrac{dy}{dx}\right|_{1.2} = 1.2\sqrt{1.22345} = 1.32732$

2d. $\left.\dfrac{dy}{dx}\right|_a = \dfrac{1}{2}\left(\left.\dfrac{dy}{dx}\right|_{1.1} + \left.\dfrac{dy}{dx}\right|_{1.2}\right) = 1.24251$

2e. $yg(1.2) = g(1.1) + (1.2 - 1.1)\,dy/dx_a = 1.23193$

The procedure could be continued for additional values of x.

8.5 RUNGE–KUTTA METHODS

The Euler's and modified Euler's methods are considered to be one-step methods since they use information from one interval in estimating the value of y at the end of the interval. A class of methods called Runge–Kutta methods, which include the Euler's and modified Euler's methods as special cases, is widely used. In these methods, the y value at the end of an interval is determined based on the value at the beginning of the interval, the step size, and some representative slope over the interval. For example, Euler's method uses the slope at the beginning of the interval as the representative slope over the interval.

The Runge–Kutta methods are classified according to the order that they represent in their solutions. However, it should be noted that these methods do not require higher-order derivatives for their solutions. In this section, example cases of the second- and third-order Runge–Kutta methods are provided. Only the fourth-order Runge–Kutta methods are described in detail. The use of second- and third-order Runge–Kutta methods is similar to fourth-order Runge–Kutta methods. Also the second- and third-order methods are used in the applications section of this chapter.

8.5.1 Second-order Runge–Kutta Methods

The modified Euler's method is considered a case of the second-order Runge–Kutta methods. It can be expressed as

$$y_{i+1} = y_i + 0.5[f(x_i, y_i) + f(x_i + h, y_i + hf(x_i, y_i))]h \tag{8-22}$$

where $y_i = g(x_i)$, $y_{i+1} = g(x_i + \Delta x)$, $x_i = x$ value at the start of an interval, and $x_{i+1} = x_i + \Delta x$; $h = \Delta x$. The computations according to this method can be summarized in the following steps:

1. Evaluate the slope at the start of an interval, that is, at (x_i, y_i). It is called slope S_1, that is,

$$S_1 = f(x_i, y_i) \tag{8-23a}$$

2. Evaluate the slope at the end of the interval (x_{i+1}, y_{i+1}), called slope S_2 as follows:

$$S_2 = f(x_i + h, y_i + hS_1) \tag{8-23b}$$

3. Evaluate y_{i+1} using the average slope of S_1 and S_2 as

$$y_{i+1} = y_i + 0.5(S_1 + S_2)h \tag{8-23c}$$

A second representation of the second-order Runge–Kutta methods is given by the following:

$$y_{i+1} = y_i + [f(x_i + 0.5h, y_i + 0.5hf(x_i, y_i))]h \tag{8-24}$$

The computational steps in this case are summarized as follows:

1. Evaluate the slope at (x_i, y_i). It is called slope S_1 and is given by Eq. 8-23a.
2. Evaluate the slope at mid-interval, S_2, as

$$S_2 = f(x_i + 0.5h, y_i + 0.5hS_1) \tag{8-25a}$$

3. Evaluate the quantity of interest y_{i+1} as

$$y_{i+1} = y_i + S_2h \tag{8-25b}$$

A third example of the second-order Runge–Kutta methods is given by the following predictor:

$$y_{i+1} = y_i + \left[\frac{f(x_i, y_i)}{3} + \frac{2f(x_i + 0.75h, y_i + 0.75hf(x_i, y_i))}{3}\right]h \tag{8-26}$$

The computational steps in this case are summarized as follows:

1. Evaluate the slope at (x_i, y_i), called S_1 using Eq. 8-23a.
2. Evaluate a second slope estimate at an intermediate point in the interval, called S_2, as

$$S_2 = f(x_i + 0.75h, y_i + 0.75hS_1) \tag{8-27a}$$

3. Evaluate the quantity of interest y_{i+1} as

$$y_{i+1} = y_i + \left(\frac{1}{3}S_1 + \frac{2}{3}S_2\right)h \tag{8-27b}$$

8.5.2 Third-order Runge–Kutta Methods

The following is an example of the third-order Runge–Kutta methods:

$$y_{i+1} = y_i + \frac{1}{6}[f(x_i, y_i) + 4f(x_i + 0.5h, y_i + 0.5hf(x_i, y_i))$$
$$+ f(x_i + h, y_i - hf(x_i, y_i) + 2hf(x_i + 0.5h, y_i + 0.5hf(x_i, y_i)))]h \tag{8-28}$$

The computational steps for this method are as follows:

1. Evaluate the slope at (x_i, y_i), called S_1, using Eq. 8-22a.
2. Evaluate a second slope estimate at the mid-point of the step, called S_2, as

$$S_2 = f(x_i + 0.5h, y_i + 0.5hS_1) \qquad (8\text{-}29a)$$

3. Evaluate a third slope S_3 as

$$S_3 = f(x_i + h, y_i - hS_1 + 2hS_2) \qquad (8\text{-}29b)$$

4. Estimate the quantity of interest y_{i+1} as

$$y_{i+1} = y_i + \frac{1}{6}[S_1 + 4S_2 + S_3]h \qquad (8\text{-}29c)$$

8.5.3 Fourth-order Runge–Kutta Methods

The fourth-order Runge–Kutta methods are a commonly used class of the Runge–Kutta methods and are introduced here in a greater detail than the previous Runge–Kutta methods. For the differential equation

$$\frac{dy}{dx} = f(x, y) \qquad \text{such that } y = y_0 \text{ at } x = x_0 \qquad (8\text{-}30)$$

the following procedure can be used to estimate the value of the dependent variable y at the end of the interval, denoted as y_{i+1}, based on values y_i and x_i at the beginning of the interval and the width of the interval $\Delta x = h$:

1. Compute the slope S_1 at (x_i, y_i) using $f(x, y)$ according to Eq. 8-23a:

$$S_1 = f(x_i, y_i)$$

2. Estimate y at the mid-point of the interval using $y_i, f(x_i, y_i)$, and $h/2$:

$$y_{i+1/2} = y_i + \frac{h}{2}f(x_i, y_i) \qquad (8\text{-}31a)$$

where the subscript $\left(i + \frac{1}{2}\right)$ is used to denote the center of the interval.

3. Estimate the slope at mid-interval S_2 as

$$S_2 = f(x_i + 0.5h, y_i + 0.5hS_1) \qquad (8\text{-}31b)$$

4. Revise the estimate of y at mid-interval $\left(i + \frac{1}{2}\right)$ using y_i, S_2, and $h/2$:

$$y_{i+1/2} = y_i + \frac{h}{2}S_2 \qquad (8\text{-}31c)$$

5. Compute a revised estimate of the slope at mid-interval S_3 using $y_{i+1/2}$ of step 4 by substituting in $f(x, y)$. The slope S_3 can be computed as

$$S_3 = f(x_i + 0.5h, y_i + 0.5hS_2) \tag{8-31d}$$

6. Estimate y at the end of the interval (y_{i+1}) using y_i, S_3, and h:

$$y_{i+1} = y_i + hS_3 \tag{8-31e}$$

7. Estimate the slope S_4 at the end of the interval using

$$S_4 = f(x_i + h, y_i + hS_3) \tag{8-31f}$$

8. Using S_1 of step 1, S_2 of step 3, S_3 of step 5, and S_4 of step 7, estimate y_{i+1} using

$$y_{i+1} = y_i + \frac{h}{6}(S_1 + 2S_2 + 2S_3 + S_4) \tag{8-32}$$

Equation 8-32 uses four estimates of the slope to estimate the value of the y_{i+1}.

Example 8-5 Fourth-order Runge–Kutta Method

The fourth-order Runge–Kutta method is illustrated using the following differential equation:

$$\frac{dy}{dx} = \frac{xy}{1+x^2} \qquad \text{such that } y = 1 \text{ at } x = 1 \tag{8-33}$$

Using an interval of $h = 0.1$, the following steps can be used to estimate y at $x = 1.1$, that is, $g(1.1)$:

1. The slope at $x = 1$ is computed by entering the values of the boundary condition into Eq. 8-33.

$$S_1 = \frac{1(1)}{1 + (1)^2} = 0.5$$

2. y at $x = 1.05$ is computed using Euler's method according to Eq. 8-31a:

$$g(1.05) = g(1.0) + \frac{h}{2}S_1 = 1 + \frac{0.10}{2}(0.5) = 1.025$$

3. The slope at $x = 1.05$ is computed using Eq. 8-31b as

$$S_2 = \frac{1.05(1.025)}{1 + (1.05)^2} = 0.51189$$

4. y at $x = 1.05$ is revised as

$$g(1.05) = g(1.0) + \frac{h}{2}S_2 = 1 + \frac{0.10}{2}(0.51189) = 1.02559$$

5. Revise the estimate of the slope at the mid-point of the interval:

$$S_3 = \frac{1.05(1.02559)}{1 + (1.05)^2} = 0.51219$$

6. Make an initial estimate of y at the end of the interval:

$$g(1.1) = g(1.0) + hS_3 = 1 + 0.1(0.51219) = 1.05122$$

7. Estimate the slope at $x = 1.1$:

$$S_4 = \frac{1.1(1.05122)}{1 + (1.1)^2} = 0.52323$$

8. Revise the estimate of y at $x = 1.1$ using Eq. 8-32:

$$g(1.1) = g(1.0) + \frac{h}{6}(S_1 + 2S_2 + 2S_3 + S_4)$$

$$= 1 + \frac{0.1}{6}[0.5 + 2(0.51189) + 2(0.51219) + 0.52323]$$

$$= 1.05119$$

The following is the exact solution to Eq. 8-33:

$$y = \left(\frac{1 + x^2}{2}\right)^{0.5} \tag{8-34}$$

Thus the exact value of y at $x = 1.1$ is 1.05119, which is the same as the fourth-order Runge–Kutta solution to five significant digits. Euler's method would provide the estimate of $y = 1.05000$, which is in error by 0.113%.

8.6 PREDICTOR–CORRECTOR METHODS

Unless the step sizes are small, procedures such as Euler's method and the Runge–Kutta methods may not yield precise solutions. A class of algorithms called predictor–corrector methods can be used to improve the accuracy of the solution. These methods make use of solutions from previous intervals to project to the end of the next interval. The disadvantage of predictor–corrector algorithms is that, for the initial intervals, values for previous intervals may not be available; in such cases, one-step methods, such as Euler's method, may have to be used to generate values for the first few steps. In addition to the multistep aspect of the solution, predictor–corrector algorithms iterate several times over the same interval until the solution converges to within an acceptable tolerance; this is the corrector aspect of the algorithm.

There are two parts to each method, the predictor and corrector parts. The predictor part is used to get an initial estimate of the dependent variable at the end of the interval;

this is used with the corrector algorithm to improve the estimate. The solution iterates with the corrector algorithm to improve the accuracy of the estimate.

8.6.1 Euler–trapezoidal Method

The Euler–trapezoidal method uses Euler's method as the predictor algorithm and the trapezoidal rule as the corrector equation. Using the first subscript to indicate the interval and the second subscript to indicate the trial number, Euler's formula is

$$y_{i+1,0} = y_{i,*} + h \frac{dy}{dx}\bigg|_{i,*} \tag{8-35}$$

where 0 and * as the second subscripts indicate the initial and final estimate, respectively. For the corrector equation, the trapezoidal rule is given in the form

$$y_{i+1,j} = y_{i,*} + \frac{h}{2}\left[\frac{dy}{dx}\bigg|_{i,*} + \frac{dy}{dx}\bigg|_{i+1,j-1}\right] \tag{8-36}$$

where j is an iteration counter for applying the corrector equation. The corrector equation can be applied as many times as necessary to get convergence.

Example 8-6 Euler–trapezoidal Method

The problem of Eq. 8-21,

$$\frac{dy}{dx} = x\sqrt{y} \qquad \text{such that } y = 1 \text{ at } x = 1$$

is used to illustrate the Euler–trapezoidal method. The initial (predictor) estimate for y at $x = 1.1$ is

$$\frac{dy}{dx}\bigg|_{0,0} = 1\sqrt{1} = 1$$

$$y_{1,0} = y_{0,*} + 0.1\left[\frac{dy}{dx}\bigg|_{0,0}\right]$$

$$y_{1,0} = 1 + 0.1(1) = 1.1$$

The corrector equation of Eq. 8-36 can be used to improve the estimate:

$$\frac{dy}{dx}\bigg|_{1,0} = 1.1\sqrt{1.1} = 1.15369$$

$$y_{1,1} = y_{0,*} + \frac{h}{2}\left[\frac{dy}{dx}\bigg|_{0,0} + \frac{dy}{dx}\bigg|_{1,0}\right] = 1 + \frac{0.1}{2}[1 + 1.15369] = 1.10768$$

$$\frac{dy}{dx}\bigg|_{1,2} = 1.1\sqrt{1.10768} = 1.15771$$

$$y_{1,2} = y_{0,*} + \frac{h}{2}\left[\frac{dy}{dx}\bigg|_{0,1} + \frac{dy}{dx}\bigg|_{1,1}\right] = 1 + \frac{0.1}{2}[1 + 1.15771] = 1.10789$$

$$\frac{dy}{dx}\bigg|_{1,3} = 1.1\sqrt{1.10789} = 1.15782$$

$$y_{1,3} = y_{0,*} + \frac{h}{2}\left[\frac{dy}{dx}\bigg|_{0,1} + \frac{dy}{dx}\bigg|_{1,2}\right] = 1 + \frac{0.1}{2}[1 + 1.15782] = 1.10789$$

Since $y_{1,3} = y_{1,2}$, then further iteration will not change the estimate of y at $x = 1.1$. Thus an estimate of y at $x = 1.2$ can be made using the predictor equation (Eq. 8-35) as

$$y_{2,0} = y_{1,3} + h\frac{dy}{dx}\bigg|_{1,*} = 1.10789 + 0.1(1.15782) = 1.22367$$

The corrector estimate using Eq. 8-36 is

$$\frac{dy}{dx}\bigg|_{2,1} = 1.2\sqrt{1.22367} = 1.32744$$

$$y_{2,1} = y_{1,3} + \frac{h}{2}\left[\frac{dy}{dx}\bigg|_{1,*} + \frac{dy}{dx}\bigg|_{2,1}\right] = 1.10789 + \frac{0.1}{2}[1.15782 + 1.32744] = 1.23215$$

$$\frac{dy}{dx}\bigg|_{2,2} = 1.2\sqrt{1.23215} = 1.33203$$

$$y_{2,2} = y_{1,3} + \frac{h}{2}\left[\frac{dy}{dx}\bigg|_{1,*} + \frac{dy}{dx}\bigg|_{2,2}\right] = 1.10789 + \frac{0.1}{2}[1.15782 + 1.33203] = 1.23238$$

$$\frac{dy}{dx}\bigg|_{2,3} = 1.2\sqrt{1.23238} = 1.33215$$

$$y_{2,3} = y_{1,3} + \frac{h}{2}\left[\frac{dy}{dx}\bigg|_{1,*} + \frac{dy}{dx}\bigg|_{2,3}\right] = 1.10789 + \frac{0.1}{2}[1.15782 + 1.33215] = 1.23239$$

Again, the corrector algorithm converged in three iterations.

The Euler–trapezoidal solution can be compared with the modified Euler's solution of Example 8-4. At $x = 1.1$, the modified Euler's method yielded an estimate of 1.10768; the predictor–corrector analysis yields 1.10789. At $x = 1.2$, the modified Euler's and Euler–trapezoidal methods yielded estimates of 1.23193 and 1.23239, respectively.

8.6.2 Milne–Simpson Method

The Milne–Simpson method uses Milne's equation as the predictor equation and Simpson's rule for integration as the corrector formula. The Milne equation uses values for three prior ordinates y and has the form

$$y_{i+1,0} = y_{i-3,*} + \frac{4h}{3}\left[2\frac{dy}{dx}\bigg|_{i,*} - \frac{dy}{dx}\bigg|_{i-1,*} + 2\frac{dy}{dx}\bigg|_{i-2,*}\right] \tag{8-37}$$

The corrector equation is

$$y_{i+1,j} = y_{i-1,*} + \frac{h}{3}\left[\frac{dy}{dx}\bigg|_{i+1,j-1} + 4\frac{dy}{dx}\bigg|_{i,*} + \frac{dy}{dx}\bigg|_{i-1,*} \right] \tag{8-38}$$

The predictor equation requires estimates at two previous sampling points. For the two initial sampling points, a one-step method such as Euler's equation (Eq. 8-13) can be used.

Example 8-7 Milne–Simpson Method

The problem of Example 8-6 can be used to illustrate the Milne–Simpson method. To use the predictor method, a value would be needed for $y_{i-3,*}$, which is y at $x = 0.7$. Since this value is not available, estimates for y at $x = 1.1$, 1.2, and 1.3 are needed before the predictor equation can be used. Therefore, the following values were obtained from the Euler–trapezoidal method in Example 8-6:

x	y	$\dfrac{dy}{dx}$
1	1	1
1.1	1.10789	1.15782
1.2	1.23239	1.33215

To compute the initial estimate of y at $x = 1.3$, Euler's method can be used:

$$y_{3,0} = y_{2,*} + h\frac{dy}{dx}\bigg|_{2,*} = 1.23239 + 0.1(1.33215) = 1.36560$$

$$\frac{dy}{dx}\bigg|_{3,0} = 1.3\sqrt{1.36560} = 1.51917$$

The corrector formula of Eq. 8-38 can now be used:

$$y_{3,1} = y_{1,*} + \frac{0.1}{3}\left[\frac{dy}{dx}\bigg|_{3,0} + 4\frac{dy}{dx}\bigg|_{2,*} + \frac{dy}{dx}\bigg|_{1,*} \right]$$

$$= 1.10789 + \frac{0.1}{3}[1.51917 + 4(1.33215) + 1.15782]$$

$$= 1.37474$$

$$\frac{dy}{dx}\bigg|_{3,1} = 1.3\sqrt{1.37474} = 1.52424$$

$$y_{3,2} = 1.10789 + \frac{0.1}{3}[1.52424 + 4(1.33215) + 1.15782]$$

$$= 1.37491$$

$$\frac{dy}{dx}\bigg|_{3,2} = 1.3\sqrt{1.37491} = 1.52434$$

$$y_{3,3} = 1.10789 + \frac{0.1}{3}[1.52434 + 4(1.33215) + 1.15782]$$

$$= 1.37492$$

Thus computations for $x = 1.3$ are complete. Now the Milne predictor equation (Eq. 8-37) can be used to estimate y at $x = 1.4$:

$$y_{4,0} = y_{0,*} + \frac{4h}{3}\left[2\frac{dy}{dx}\bigg|_{3,*} - \frac{dy}{dx}\bigg|_{2,*} + 2\frac{dy}{dx}\bigg|_{0,*}\right]$$

$$= 1 + \frac{4(0.1)}{3}[2(1.52434) - 1.33215 + 2(1.15782)] = 1.53762$$

$$\frac{dy}{dx}\bigg|_{4,1} = 1.4\sqrt{1.53762} = 1.73601$$

$$y_{4,1} = y + \frac{h}{3}\left[\frac{dy}{dx}\bigg|_{4,0} + 4\frac{dy}{dx}\bigg|_{3,*} + \frac{dy}{dx}\bigg|_{2,*}\right]$$

$$= 1.23239 + \frac{0.1}{3}[1.73601 + 4(1.52434) + 1.33215] = 1.53791$$

$$\frac{dy}{dx}\bigg|_{4,1} = 1.4\sqrt{1.53791} = 1.73617$$

$$y_{4,2} = 1.23239 + \frac{0.1}{3}[1.73617 + 4(1.52434) + 1.33215] = 1.53791$$

Additional intervals could be similarly evaluated.

8.7 LEAST-SQUARES METHOD

The methods detailed previously present the solution as a series of points using the step size or interval h. The least-squares method provides a function that can be used to approximate the value of the dependent variable $y = g(x)$ for any value of the independent variable x. The procedure for deriving the least-squares function is as follows:

1. For a given ordinary differential equation (Eq. 8-1), assume that the solution is a differentiable function. For example, an nth-order polynomial could be used:

$$\hat{y} = b_0 + b_1 x + b_2 x^2 + \cdots + b_n x^n \tag{8-39}$$

2. Use the boundary condition of the ordinary differential equation to evaluate one of the coefficients of the assumed function or to provide a condition toward the solution of the coefficients.

3. Define the least-squares objective function F, which is the integral of the errors (e) squared:

$$F = \int_x e^2 \, dx \qquad (8\text{-}40)$$

where the error is defined as the difference between the derivatives for the assumed function (Eq. 8-39) and the actual function (Eq. 8-1):

$$e = \frac{d\hat{y}}{dx} - \frac{dy}{dx} \qquad (8\text{-}41)$$

4. Find the minimum of F with respect to the unknowns ($b_1, b_2, b_3, \ldots, b_n$) of the assumed function of Eq. 8-39 by taking the derivatives of Eq. 8-40 with respect to the unknowns and setting the derivatives equal to zero:

$$\frac{\partial F}{\partial b_i} = \int_{\text{all } x} 2e \frac{\partial e}{\partial b_i} \, dx = 0 \qquad (8\text{-}42)$$

5. The integrals of Eq. 8-42 are called the normal equations; the solution of the normal equations yields values of the unknown coefficients.

Using the coefficients with Eq. 8-39, an estimate of y for any x can be made. The accuracy of the estimate depends on how well the function of Eq. 8-39 can approximate the actual solution and how far from the boundary condition the estimate is made.

The least-squares method has the advantage that the solution is represented by a function. It has the following disadvantages: (1) the computational effort is greater than for the step methods, and (2) it is more difficult to provide a generalized computer program that provides a solution.

Note that the objective function of Eq. 8-40 corresponds to the least-squares solution for a set of discrete points (see Eqs. 10-17 through 10-22). For solution of an ordinary differential equation, the error is defined in terms of the derivative (Eq. 8-41), rather than values of the dependent variable (y of Eq. 10-17).

Example 8-8 Least-squares Method

The least-squares method can be used to solve the following ordinary differential equation:

$$\frac{dy}{dx} = xy \qquad \text{such that } y = 1 \text{ at } x = 0 \qquad (8\text{-}43)$$

As a basis for comparison, Eq. 8-43 can be solved analytically:

$$y = e^{x^2/2} \qquad (8\text{-}44)$$

To illustrate the method, a solution can be obtained for the interval $0 \le x \le 1$.

As a first attempt, a linear model is used for the approximating function:

$$\hat{y} = b_0 + b_1 x \qquad (8\text{-}45)$$

Using the boundary condition

$$\hat{y} = 1 = b_0 + b_1(0)$$

yields $b_0 = 1$. Thus the linear model is

$$\hat{y} = 1 + b_1 x \qquad (8\text{-}46)$$

and the derivative is

$$\frac{d\hat{y}}{dx} = b_1$$

The error function of Eq. 8-41 is

$$e = b_1 - xy = b_1 - x(1 + b_1 x) \qquad (8\text{-}47)$$

Since there is only one unknown, b_1, only one derivative is needed for Eq. 8-36; the derivative is computed from Eq. 8-47 as

$$\frac{de}{db_1} = 1 - x^2 \qquad (8\text{-}48)$$

With Eq. 8-48, the derivative of Eq. 8-42 is

$$\int_0^x 2e \frac{de}{db_1} dx = \int_0^x 2[b_1 - x(1 + b_1 x)](1 - x^2)\, dx = 0 \qquad (8\text{-}49)$$

where the lower limit is established from the boundary condition. Dividing through by 2 yields

$$\left(b_1 x - \frac{x^2}{2} - \frac{2b_1 x^3}{3} + \frac{x^4}{4} + \frac{b_1 x^5}{5} \right)\Bigg|_0^x = 0 \qquad (8\text{-}50)$$

If we are interested in the value of y for the range $0 \le x \le 1$, then the integral of Eq. 8-50 with an upper limit of integration $x = 1$ yields $b_1 = \frac{15}{32}$. Thus the approximating function of Eq. 8-46 is

$$\hat{y} = 1 + \frac{15}{32} x \qquad (8\text{-}51)$$

A comparison of the approximating function of Eq. 8-51 and the true value given by Eq. 8-44 is given in Table 8-3. The linear model produces very poor estimates.

To improve the accuracy of estimates, a quadratic model is used instead of the linear model of Eq. 8-45:

$$\hat{y} = b_0 + b_1 x + b_2 x^2 \qquad (8\text{-}52)$$

The boundary condition again yields $b_0 = 1$. The derivative is

$$\frac{d\hat{y}}{dx} = b_1 + 2b_2 x$$

TABLE 8-3 A Solution Using a Linear Model for the Least-squares Method

x	True y Value (Eq. 8-44)	Numerical y Value (Eq. 8-51)	Error (%)
0	1.	1.	—
0.2	1.0202	1.0938	7.2
0.4	1.0833	1.1875	9.6
0.6	1.1972	1.2812	7.0
0.8	1.3771	1.3750	0.0
1.0	1.6487	1.4688	−10.9

and the error function is

$$e = b_1 + 2b_2x - xy = b_1 + 2b_2x - x(1 + b_1x + b_2x^2)$$
$$= b_1(1 - x^2) + b_2(2x - x^3) - x \tag{8-53}$$

Thus derivatives are computed for both b_1 and b_2 as follows:

$$\frac{\partial e}{\partial b_1} = 1 - x^2 \tag{8-54a}$$

$$\frac{\partial e}{\partial b_2} = 2x - x^3 \tag{8-54b}$$

The integral of Eq. 8-42 using the derivative of Eq. 8-54a is

$$\int_0^x [b_1(1 - x^2) + b_2 + (2x - x^3) - x](1 - x^2)\, dx = 0$$

$$\left[b_1x - \frac{2b_1x^3}{3} + b_2x^2 - \frac{3b_2x^4}{4} - \frac{x^2}{2} + \frac{b_1x^5}{5} + \frac{b_2x^6}{6} + \frac{x^4}{4} \right]\Bigg|_0^x = 0$$

Using $x = 1$ as the upper limit yields the following first of two normal equations:

$$\frac{8b_1}{15} + \frac{5b_2}{12} = \frac{1}{4} \tag{8-55a}$$

Using the derivative of Eq. 8-54b with the integral of Eq. 8-42 gives

$$2\int_0^x [b_1(1 - x^2) + b_2)(2x - x^3) - x](2x - x^3)\, dx = 0$$

$$\left[b_1x^2 - \frac{2b_1x^4}{4} + \frac{4b_2x^3}{3} - \frac{4b_2x^5}{5} - \frac{2x^2}{3} + \frac{b_1x^5}{5} + \frac{b_2x^7}{7} + \frac{x^5}{5} \right]\Bigg|_0^x = 0$$

For the upper limit $x = 1$, the second normal equation is

$$\frac{9b_1}{20} + \frac{71b_2}{105} = \frac{7}{15} \tag{8-55b}$$

Solving the two normal equations, Eqs. 8-55, yields the following quadratic approximating equation:

$$\hat{y} = 1 - 0.14669x + 0.78776x^2 \tag{8-56}$$

The values of y estimated from Eq. 8-56 are given in Table 8-4. This quadratic equation (that is, Eq. 8-56) gives more accurate estimates than the linear equation (Eq. 8-51). Unfortunately, it is not possible for any specific problem to know in advance the order of the polynomial that is necessary to provide sufficient accuracy.

TABLE 8-4   A Solution Using a Quadratic Model for the Least-squares Method

x	True y Value (Eq. 8-44)	Numerical y Value (Eq. 8-56)	Error (%)
0	1.	1.	—
0.2	1.0202	1.0022	−1.8
0.4	1.0833	1.0674	0.0
0.6	1.1972	1.1956	0.0
0.8	1.3771	1.3868	0.0
1.0	1.6487	1.6411	0.0

8.8 GALERKIN METHOD

The Galerkin method is similar to the least-squares method in that an approximating equation is necessary. The method uses the error function of Eq. 8-41 and an approximating equation such as Eq. 8-39. The normal equations are computed from the following:

$$\int_x w_i e \, dx = 0 \qquad i = 1, 2, \ldots, n \tag{8-57}$$

where n is the number of unknowns to be determined in Eq. 8-39, and w_i is a weighting factor. For the polynomial of Eq. 8-39, the weight w_i is $d\hat{y}/db_i$.

Example 8-9 Galerkin Method

The ordinary differential equation of Eq. 8-43 can be used to illustrate the Galerkin method. Using the quadratic approximating equation of Eq. 8-52 with $b_0 = 1$ and the error function of Eq. 8-53 in Eq. 8-57 yields the following system of integral equations:

$$\int_0^1 [b_1(1 - x^2) + b_2(2x - x^3) - x]x\, dx = 0 \tag{8-58a}$$

$$\int_0^1 [b_1(1 - x^2) + b_2(2x - x^3) - x]x^2\, dx = 0 \tag{8-58b}$$

which yields the following set of normal equations:

$$\frac{1}{12} b_1 + \frac{7}{15} b_2 = \frac{1}{3} \tag{8-59a}$$

$$\frac{2}{15} b_1 + \frac{1}{3} b_2 = \frac{1}{4} \tag{8-59b}$$

The solution of the normal equations yields the following approximating equation:

$$\hat{y} = 1 - 0.26316x + 0.85526x^2 \tag{8-60}$$

For the range $0 \le x \le 1$, the predicted values from Eq. 8-60 are given in Table 8-5.

TABLE 8-5 Solution Using a Quadratic Model for the Galerkin Method

x	True y Value (Eq. 8-44)	Numerical y Value (Eq. 8-60)	Error (%)
0	1.	1.	—
0.2	1.0202	0.9816	0.0
0.4	1.0833	1.0316	0.0
0.6	1.1972	1.1500	0.0
0.8	1.3771	1.3368	0.0
1.0	1.6487	1.5921	0.0

8.9 HIGHER-ORDER DIFFERENTIAL EQUATIONS

Many engineering problems require the solution of higher-order differential equations. For example, Eq. 8-2 is a second-order differential equation and is given by

$$\frac{d^2y}{dx^2} = f\left(x, y, \frac{dy}{dx}\right)$$

The function $f(\cdot)$ does not need to include all the parameters x, y, and dy/dx. The methods previously discussed can be used to solve higher-order differential equations after transforming them into systems of first-order differential equations. For example, Eq. 8-2 can be transformed into the following system of first-order differential equations:

$$\frac{dy_2}{dx} = f(x, y_1, y_2) \tag{8-61a}$$

$$\frac{dy_1}{dx} = y_2 \tag{8-61b}$$

where

$$y_1 = y \tag{8-62a}$$

and

$$y_2 = \frac{dy_1}{dx} = \frac{dy}{dx} \tag{8-62b}$$

The solution of this system of equations requires the boundary conditions of both y_1 and y_2 at the starting x value.

In general, any system of n equations of the following type can be solved using any of the previously discussed methods:

$$\frac{dy_1}{dx} = f_1(x, y_1, y_2, \dots, y_n) \tag{8-63a}$$

$$\frac{dy_2}{dx} = f_2(x, y_1, y_2, \dots, y_n) \tag{8-63b}$$

$$\frac{dy_3}{dx} = f_3(x, y_1, y_2, \dots, y_n) \tag{8-63c}$$

$$\text{M}$$

$$\frac{dy_n}{dx} = f_n(x, y_1, y_2, \dots, y_n) \tag{8-63d}$$

The solution procedure requires for the single-step methods the knowledge of n boundary conditions for $y^1, y^2, \ldots, y_n$ at the starting x value. Multistep methods require additional boundary conditions, as was discussed in previous sections. A system of equations is solved by applying any of the previously discussed methods for each equation in the system at each step. All the equations in the system should be solved at each step before proceeding to the next step. The discussion in the section is limited to solving higher-order differential equations and systems of equations based on step-based methods. Multistep methods can be similarly used.

Example 8-10 System of Equations for a Beam

A simply supported beam AB supports a uniformly distributed load of intensity 2 kips/ft as shown in Fig. 8-2. The bending moment at a distance X along the span of the beam is given by

$$M(X) = 10X - X^2 \tag{8-64}$$

The deflection and rotation at any point along the span of the beam can be determined by solving the following differential equation:

$$\frac{d^2Y}{dX^2} = \frac{M}{EI} = \frac{10X - X^2}{EI} \tag{8-65}$$

This equation can be expressed as

$$\frac{dZ}{dX} = \frac{M}{EI} = \frac{10X - X^2}{EI} \tag{8-66a}$$

$$\frac{dY}{dX} = Z \tag{8-66b}$$

where Y = deflection in feet, Z = rotation (or slope), E = modulus of elasticity of the material, and I = moment of inertia of the cross section of the beam. Assume that $EI = 3600$ kip/ft^2, and at $X = 0$, $Y = 0$ and $Z = -0.02314$. Using Euler's method, the following equations result:

$$Z_{i+1} = Z_i + f_2(X_i, Y_i, Z_i)h \tag{8-67a}$$

$$Y_{i+1} = Y_i + f_1(X_i, Y_i, Z_i)h \tag{8-67b}$$

Figure 8-2   Simply Supported Beam

TABLE 8-6   Deflection and Rotation along a Beam Using a Step Size of 0.1 Ft

X (ft)	$\dfrac{dZ}{dX}$	$Z = \dfrac{dY}{dX}$	Y (ft)	Exact Z	Exact Y (ft)
0	0	−0.0231481	0	−0.0231481	0
0.1	0.000275	−0.0231481	−0.0023148	−0.0231344	−0.0023144
0.2	0.0005444	−0.0231206	−0.0046296	−0.0230933	−0.004626
0.3	0.0008083	−0.0230662	−0.0069417	−0.0230256	−0.0069321
0.4	0.0010667	−0.0229854	−0.0092483	−0.0229319	−0.0092302
0.5	0.0013194	−0.0228787	−0.0115469	−0.0228125	−0.0115177
0.6	0.0015667	−0.0227468	−0.0138347	−0.0226681	−0.0137919
0.7	0.0018083	−0.0225901	−0.0161094	−0.0224994	−0.0160505
0.8	0.0020444	−0.0224093	−0.0183684	−0.0223067	−0.018291
0.9	0.002275	−0.0222048	−0.0206093	−0.0220906	−0.020511
1	0.0025	−0.0219773	−0.0228298	−0.0218519	−0.0227083
2	0.0044444	−0.0185565	−0.0434305	−0.0183333	−0.042963
3	0.0058333	−0.0134412	−0.0598019	−0.0131481	−0.0588194
4	0.0066667	−0.007187	−0.0704998	−0.0068519	−0.0688889
5	0.0069444	−0.0003495	−0.0746352	0.00000000	−0.072338
6	0.0066667	0.0065157	−0.0718747	0.0068519	−0.0688889
7	0.0058333	0.0128532	−0.0624406	0.0131481	−0.0588194
8	0.0044444	0.0181074	−0.0471107	0.0183333	−0.042963
9	0.0025	0.0217227	−0.0272183	0.0218519	−0.0227083
10	0.000000	0.0231435	−0.0046523	0.0231481	0.000000

where $f_1(X_i, Y_i, Z_i) = Z$ evaluated at Z_i, and $f_2(X_i, Y_i, Z_i) = M/EI$ evaluated at $M(X_i)$. Applying the above equations sequentially for a step size of 0.1 ft results in Table 8-6. The results of the example are also shown in Figs. 8-3 and 8-4. In this case, Euler's method results in accurate prediction of the deflection and rotation along the beam. The relatively high accuracy of the method can be attributed to the small step size.

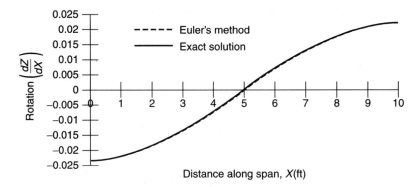

Figure 8-3   Rotation Along Span

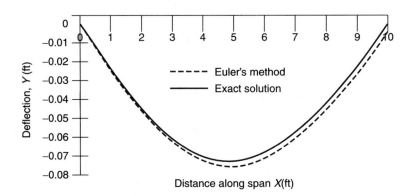

Figure 8-4   Deflection Along Span

8.10 BOUNDARY-VALUE PROBLEMS

The previously discussed methods for solving ordinary differential equations can be used to solve high-order differential equations as discussed in Section 8.9. These methods require the initial conditions to be at the same x value. For example, the bending moment at a distance x along the span of the beam in Example 8-10 is given by Eq. 8-64. The initial conditions were assumed to be at $x = 0$, $y = 0$, and $Z = -0.02314$. The specified initial condition for Z in Example 8-10 is not commonly known. However, in this problem it is known that at $x = 0$, $y = 0$ and at $x = 10$, $y = 0$. These two conditions have different x values. Therefore, they cannot be used in solving the differential equation according to the previously discussed methods. In this case, we are dealing with a boundary-value problem.

Boundary-value problems can be solved using other methods, such as the shooting method and finite-difference methods. In this section these methods are described.

8.10.1 Shooting Method

The shooting method is a trial-and-error method that utilizes any of the previously discussed methods for solving differential equations. According to this method, the boundary conditions are revised to make all of them at the same x value. Then the differential equation is solved, and the resulting solution at the original (known) boundary conditions is compared with the original boundary conditions. The difference between the solution at the boundary conditions and the original boundary conditions can be computed and used to re-revise the boundary conditions for another trial. Linear interpolation between two successive differences can be used to revise the conditions. This method becomes inefficient, especially for higher-order differential equations where several boundary conditions are needed.

Example 8-11 System of Equations for a Beam

The simply supported beam AB with a uniformly distributed load of intensity 2 kips/ft as shown in Fig. 8-2 was used in Example 8-10. The deflection and rotation at any point along the span of the beam can be determined by solving Eqs. 8-66. Assume that at $x = 0$, $y = 0$, and at $x = 10$, $y = 0$. This boundary-value problem can be solved using the shooting method in conjunction with Euler's method. The boundary conditions are revised to at $x = 0$, $y = 0$ and at $x = 0$, $Z = -0.02314$. Now the problem can be solved as was described in Example 8-10. The results are shown in Table 8-6 and Figs. 8-3 and 8-4. The solution produced $y = -0.0046523$ at $x = 10$, which is in error by 0.0046523 as compared to the specified boundary condition at $x = 10$, $y = 0$. The assumed boundary conditions can be revised to (second trial) $x = 0$, $y = 0$, and at $x = 0$, $Z = -0.0225$. The calculations are not shown herein, but the new y value at $x = 10$ is 0.0018292, which is better than -0.0046523. Using linear interpolation, the next trial is at $x = 0$, $Z = -0.0226806$. The resulting y value at $x = 10$ is 0.00002317, which is zero to four significant digits.

8.10.2 Finite-difference Methods

In these methods, the derivatives in the differential equation are replaced by finite-difference equations, which are described in Section 7.1. Then the resulting differential equation, which is now in a finite-difference form, is used at some interior points using a selected step size and at the specified boundary conditions. Each use of the finite-difference equation results in a linear equation in terms of the unknown solutions at the selected interior points. The resulting system of equations needs to be solved simultaneously to obtain the solution at the interior points. The system of linear equations can be solved using the numerical methods discussed in Chapter 5.

8.11 INTEGRAL EQUATIONS

The previously discussed methods for solving differential equations can be used to solve integration problems of the following type:

$$y = \int_{x_l}^{x_u} f(x, y)\, dx \tag{8-68}$$

where x_l = the lower limit of x, x_u = the upper limit of x, and $f(x, y)$ = any function of x and y. Equation 8-68 can be rewritten as

$$\frac{dy}{dx} = f(x, y) \tag{8-69}$$

The solution methods of Eq. 8-69 require the knowledge of the boundary condition, y at $x = x_l$. Any of the previously discussed methods can be used to solve the resulting differential equation. Several steps should be used between x_l and x_u to obtain an accurate result for the integral.

8.12 APPLICATIONS

8.12.1 Motion of a Falling Body

The mass system defined by Eq. 8-6 can be solved using a Taylor series expansion. For $g_r = 32.2$ ft/sec^2, a resistance R equal to 1 slug/sec, and a mass of 750 slugs, Eq. 8-6c can be written as

$$\frac{dV}{dt} = 32.2 - \frac{1}{750}V = 32.2 - 0.00133V \qquad \text{such that } V = 0 \text{ at } t = 0 \tag{8-70}$$

The solution to Eq. 8-70 would be a relationship between V and t. To provide an accurate estimate of the relationship between V and t, higher-order derivatives must be computed. The second and third derivatives are

$$\frac{d^2V}{dt^2} = -\frac{R}{m}\left(\frac{dV}{dt}\right)$$

$$= -\frac{R}{m}\left(g_r - \frac{R}{m}V\right) \tag{8-71}$$

$$\frac{d^3V}{dt^3} = \left(\frac{R}{m}\right)^2\left(g_r - \frac{R}{m}V\right) \tag{8-72}$$

Using only the terms including the first three derivatives of Eq. 8-9 yields

$$V(t) = V(t_0) + (t - t_0)\frac{dV}{dt} + \frac{(t - t_0)^2}{2!}\frac{d^2V}{dt^2} + \frac{(t - t_0)^3}{3!}\frac{d^3V}{dt^3} + \epsilon \tag{8-73}$$

in which ϵ is the error that results from not including the higher-order terms, $V(t)$ is the velocity at time t, and t_0 is the value of the independent variable for the boundary condition. For the boundary condition $V = 0$ at $t = 0$ and the derivatives of Eqs. 8-6c, 8-71, and 8-72, Eq. 8-73 can be rewritten as

$$V(t) = 0 + (t - 0)\left[g_r - \frac{R}{m}V\right] + \frac{(t-0)^2}{2!}\left[-\frac{R}{m}\left(g_r - \frac{R}{m}V\right)\right]$$
$$+ \left(\frac{R}{m}\right)^2\left(g_r - \frac{R}{m}V\right)\frac{(t-0)^3}{6} + \epsilon \tag{8-74}$$

Therefore, the velocity V is given by

$$V(t) = g_r t - \frac{R}{m}Vt - \frac{R}{2m}t^2\left(g_r - \frac{R}{m}V\right) + \frac{R^2}{m^2}\frac{t^3}{6}\left(g_r - \frac{R}{m}V\right) + \epsilon \tag{8-75}$$

Since V equals zero at the boundary condition, we get

$$V(t) = g_r t - \frac{R}{2m}t^2 g_r + \frac{R^2}{m^2}\frac{t^3}{6}g_r + \epsilon \tag{8-76}$$

For t equal to 3 sec, the velocity would be 96.407 ft/sec, with an error ϵ. Equation 8-76 could be used to plot a graph of V versus t. The true solution to Eq. 8-70 is

$$V(t) = \frac{g_r m}{R}[1 - e^{-(R/m)t}] \tag{8-77}$$

According to Eq. 8-77, the true velocity at $t = 3$ sec is 96.407 ft/sec, which is the same as the solution according to Eq. 8-76.

8.12.2 Electrical Circuit

For the electrical circuit shown in Fig. 8-5, the relationship between the dependent variable i (the current) and the independent variable t (time) is described by the following differential equation:

$$L\frac{di}{dt} + Ri = E, \qquad \text{such that } i = 0 \text{ at } t = 0 \tag{8-78a}$$

The differential equation can be rearranged as

$$\frac{di}{dt} = \frac{E - Ri}{L} \tag{8-78b}$$

Using Euler's equation, the current at any time t, denoted as $i(t)$, is given by

$$i(t) = i(t = t_0) + (t - t_0)\frac{di}{dt} \tag{8-79a}$$

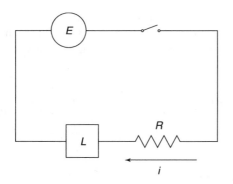

Figure 8-5 Circuit with Constant Voltage Source (E), a Resistance (R), and Inductance (L) in Series

$$i(t) = i(t = t_0) + \left[\frac{E - Ri}{L}\right](t - t_0) \qquad (8\text{-}79\text{b})$$

The bound on the error can be approximated by

$$\epsilon \cong \frac{(\Delta t)}{2}\frac{d^2 i}{dt^2} = \frac{(\Delta t)}{2}\left[\left(\frac{R}{L}\right)^2 i - \frac{RE}{L^2}\right] \qquad (8\text{-}80)$$

To illustrate the solution, assume that $L = 10$ henries, $E = 6$ volts, $R = 3$ ohms, and $\Delta t = 1$ second. Thus the current for times of 5 seconds and less are computed by Eqs. 8-79 as follows:

$$i(1) = 0 + (1 - 0)\frac{6 - 3(0)}{10} = 0.6 \text{ A}$$

$$i(2) = 0.6 + (1 - 0)\frac{6 - 3(0.6)}{10} = 1.02 \text{ A}$$

$$i(3) = 1.02 + (1 - 0)\frac{6 - 3(1.02)}{10} = 1.314 \text{ A}$$

$$i(4) = 1.314 + (1 - 0)\frac{6 - 3(1.314)}{10} = 1.5198 \text{ A}$$

$$i(5) = 1.5198 + (1 - 0)\frac{6 - 3(1.5198)}{10} = 1.66386 \text{ A}$$

The error bound at time 5 sec can be computed using Eq. 8-80

$$|\epsilon| \cong \left|\frac{5(1)}{2}\left[\left(\frac{3}{10}\right)^2(1.66386) - \frac{3(6)}{(10)^2}\right]\right| = 0.0756315 \text{ A}$$

An analytical solution for $i(5)$ yields 1.55374 A, which differs from the computed value by 0.11012 A. The true error in this case is larger than the error bound because the derivative evaluation in Eq. 8-80 was not performed at a point with an interval that maximizes the derivative (see Eq. 8-18).

8.12.3 One-dimensional Heat Flow

One-dimensional heat flow is governed by the equation

$$H = KA \frac{dT}{dr} \qquad (8\text{-}81)$$

in which K is the coefficient of thermal conductivity (calories per second-cm-°C), H is the quantity of heat (calories per second) flowing through a material, A is the area (cm²) perpendicular to the flow of heat, T is the temperature (°C), and r is a distance measurement (cm). For a pipe with a circular cross section carrying steam, the flow of heat would be perpendicular to the flow of steam (see Fig. 8-6). In such a case, the surface area A at a radius r from the center of the pipe would be $2\pi r$. Equation 8-81 can be used to calculate the temperature distribution across the cross section of the pipe. For example, if steam at 250°C is flowing through a pipe having internal (r_0) and external radii of 5 and 6 cm, respectively, and a coefficient of thermal conductivity of 0.0005, then Eq. 8-81 can be used to estimate the temperature distribution within the walls of the pipe. Placing Eq. 8-81 in standard form yields

$$\frac{dT}{dr} = \frac{H}{2\pi rK} \qquad \text{such that } T = T_0 \text{ at } r = r_0 \qquad (8\text{-}82)$$

in which T_0 is the temperature of the steam on the internal wall of the pipe. Thus, assuming a flow rate of 0.8 cal/sec yields

$$\frac{dT}{dr} = \frac{0.8}{2\pi(0.0005)r} \qquad \text{such that } T = 250°C \text{ at } r = 5 \text{ cm} \qquad (8\text{-}83)$$

Figure 8-6 Calculation of the Temperature Distribution within the Wall of a Steam Pipe

Using a step size of $\Delta r = 0.25$ cm, we can use the modified Euler's method to determine the temperature distribution within the wall of the pipe. For the wall thickness of 1 cm, four steps will be required. The calculations are as follows:

$$T_1 = T_0 + \frac{\Delta r}{2}\left[\left.\frac{dT}{dr}\right|_0 + \left.\frac{dT}{dr}\right|_1\right]$$

$$= 250 + \frac{0.25}{2}\left[\frac{-0.8}{2\pi(5)(0.0005)} + \frac{-0.8}{2\pi(5.25)(0.0005)}\right]$$

$$= 237.571°C$$

$$T_2 = T_1 + \frac{\Delta r}{2}\left[\left.\frac{dT}{dr}\right|_1 + \left.\frac{dT}{dr}\right|_2\right]$$

$$= 237.571 + \frac{0.25}{2}\left[\frac{-0.8}{2\pi(5.25)(0.0005)} + \frac{-0.8}{2\pi(5.5)(0.0005)}\right]$$

$$= 225.7205°C$$

$$T_3 = T_2 + \frac{\Delta r}{2}\left[\left.\frac{dT}{dr}\right|_2 + \left.\frac{dT}{dr}\right|_3\right]$$

$$= 225.7205 + \frac{0.25}{2}\left[\frac{-0.8}{2\pi(5.5)(0.0005)} + \frac{-0.8}{2\pi(5.75)(0.0005)}\right]$$

$$= 214.397°C$$

$$T_4 = T_3 + \frac{\Delta r}{2}\left[\left.\frac{dT}{dr}\right|_3 + \left.\frac{dT}{dr}\right|_4\right]$$

$$= 214.397 + \frac{0.25}{2}\left[\frac{-0.8}{2\pi(5.75)(0.0005)} + \frac{-0.8}{2\pi(6)(0.0005)}\right]$$

$$= 203.556°C$$

Thus the temperature at the outer wall of the pipe would be 203.556°C. The use of Euler's method would provide values of 237.268°, 225.1415°, 213.5666°, and 202.495°C for values of r of 5.25, 5.50, 5.75, and 6.00 cm, respectively.

8.12.4 Estimating Biochemical Oxygen Demand

Rivers and streams contain microorganisms that utilize organic materials, such as sewage, for growth. As part of the process, oxygen is utilized in the biological decomposition of the organic matter. The demand for oxygen is referred to as the biochemical oxygen demand (BOD). Environmental engineers use estimates of the oxygen demanded by the microorganisms as a measure of the health of the river or stream. For one or two days after organic pollution is discharged into the water body, the temporal relationship between the BOD (y) and time (t) is defined by the differential equation

$$\frac{dy}{dt} = -ky \tag{8-84}$$

where k is called the deoxygenation constant. At the time the organic matter is discharged into the water body, the initial BOD is y_0; letting this time be $t = 0$, the boundary condition is $y = y_0$ at $t = 0$.

The solution is easily computed analytically as

$$y_t = y_0 e^{-kt} \tag{8-85}$$

where y_t is the value of y at time t.

The least-squares method can be used to approximate the time variation of BOD. A quadratic model will be used:

$$y = b_0 + b_1 t + b_2 t^2 \tag{8-86}$$

If y_0 is 20 mg/L, then the boundary condition yields $b_0 = 20$. The time derivative of Eq. 8-86 is

$$\frac{dy}{dt} = b_1 + 2b_2 t \tag{8-87}$$

The error function is

$$\begin{aligned}
e = \frac{d\hat{y}}{dt} - \frac{dy}{dt} &= b_1 + 2b_2 t - (-ky) \\
&= b_1(1 + kt) + b_2(2t + kt^2) + 20k
\end{aligned} \tag{8-88}$$

Thus the derivatives of e with respect to b_1 and b_2 are, respectively,

$$\frac{\partial e}{\partial b_1} = 1 + kt \tag{8-89a}$$

and

$$\frac{\partial e}{\partial b_2} = 2t + kt^2 \tag{8-89b}$$

Thus the least-squares integrals are

$$\int_0^T [b_1(1 + kt) + b_2(2t + kt^2) + 20k](1 + kt)\, dt = 0 \tag{8-90a}$$

and

$$\int_0^T [b_1(1 + kt) + b_2(2t + kt^2) + 20k](2t + kt^2)\, dt = 0 \tag{8-90b}$$

where T is the longest time for which the solution is needed.

If the problem assumes $K = 0.25$ day^{-1} and $T = 2$ days, then the integrals of Eqs. 8-90a and 8-90b yield the following normal equations:

$$\frac{19}{6}b_1 + \frac{25}{4}b_2 = -12.5 \tag{8-91a}$$

and

$$\frac{25}{4}b_1 + \frac{226}{15}b_2 = -\frac{70}{3} \tag{8-91b}$$

Solution of the normal equations yields the following quadratic approximating equation:

$$\hat{y} = 20 - 4.9141t + 0.4898t^2 \tag{8-92}$$

Estimates of the BOD can be made with Eq. 8-92 for times in the range from 0 to 2 days and compared with the true values from Eq. 8-85. The comparison is given in Table 8-7. The small errors indicate that the numerical approximation provides accurate estimates.

TABLE 8-7   Comparison of BOD Estimates Made with the Least-squares Method and the True Solution

t (days)	True BOD (mg/L)	Estimated BOD (mg/L)	Error (mg/L)
0	20	20	0
0.25	18.788	18.802	0.014
0.50	17.650	17.665	0.015
0.75	16.581	16.590	0.009
1.00	15.576	15.576	0.000
1.25	14.632	14.623	-0.009
1.50	13.746	13.731	-0.015
1.75	12.913	12.900	-0.013
2.00	12.131	12.131	0.000

8.12.5 Accumulation of Eroded Soil in a Sediment Trap

Flood runoff from a construction site enters a sediment trap at a rate (q) of 25 ft^3/sec. The runoff contains a sediment load of 3.3 lb/ft^3. The tank initially contains 4000 ft^3 of water that is essentially sediment free. The turbulence from the runoff into the trap keeps the concentration of sediment fairly consistent with location and depth. The outlet permits stored water to drain at a rate of 5 ft³/sec.

Let S be the weight of the sediment in the trap at the end of time t. In the interval from t to $t + \Delta t$, $25\Delta t$ ft³ of runoff enters the trap and $5\Delta t$ ft³ of runoff flows out the trap. Thus the volume of water in the trap at time t is $4000 + (25 - 5)t$, and the concentration of sediment in the tank at time t is $S/(4000 + 20t)$. The change in sediment, inflow minus outflow, is ΔS:

$$\Delta S = \left(25\frac{\text{ft}^3}{\text{sec}}\right)\left(3.3\frac{\text{lb}}{\text{ft}^3}\right)(\Delta t \text{ sec}) - \frac{(5 \text{ ft}^3/\text{sec})(S \text{ lb})}{(4000 + 20t) \text{ ft}^3}(\Delta t \text{ sec})$$

or

$$\frac{\Delta S}{\Delta t} = 82.5 - \frac{5S}{4000 + 20t} \qquad (8\text{-}93)$$

As $\Delta t \to 0$, Eq. 8-93 becomes the following differential equation:

$$\frac{dS}{dt} = 82.5 - \frac{5S}{4000 + 20t} \qquad (8\text{-}94)$$

which can be solved to find the sediment in the sediment trap as a function of time.

The modified Euler's method is used to compute the accumulation of sediment in the trap for a 30-min storm; a time increment of 5 min ($\Delta t = 300$ sec) is used. The results are given in Table 8-8. The total accumulated sediment in the 30-min storm is 124,903 lb, which would have a volume of 757 ft³ ($\gamma = 165$ lb/ft³).

TABLE 8-8   Accumulation of Soil in a Trap by Modified Euler's Estimation

t (sec)	S_i (lb)	$\left.\dfrac{dS}{dt}\right|_i$	S_{i+1} (lb)	$\left.\dfrac{dS}{dt}\right|_{i+1}$	$\left.\dfrac{dS}{dt}\right|_{\text{average}}$
0	0	82.50	24,750	70.13	76.31
300	22,894	71.05	44,210	68.68	69.87
600	43,854	68.80	64,493	67.84	68.32
900	64,350	67.87	84,713	67.37	67.62
1200	84,637	67.39	104,853	67.08	67.23
1500	104,807	67.09	124,933	66.88	66.985
1800	124,903				

8.12.6 Growth Rate of Bacteria

A culture of bacteria increases at a rate that is proportional to the number of bacteria present at that instance. Assuming that the number doubles every 5 hr, a biomedical engineer can estimate the number of bacteria present at a future time using the differential equation

$$\frac{dy}{dt} = ky \tag{8-95}$$

where y = the number of bacteria present at time t. Since the number will double in 5 hr, it can be shown that $k = (\ln 2)/5$. Assuming that the culture consists of 1 unit at time $t = 0$, Eq. 8-95 becomes

$$\frac{dy}{dt} = \frac{(\ln 2)y}{5} \qquad \text{such that } y = 1 \text{ at } t = 0 \tag{8-96}$$

To estimate the number at the end of one hour, the fourth-order Runge–Kutta method was used with a step size $h = 1.0$. The following are the computational steps:

1. At $t = 0$, the slope is given by Eq. 8-33 as

$$S_1 = \frac{(\ln 2)(1)}{5} = 1.138629$$

2. The value of y at 0.5 hr according to Eq. 8-31a is

$$y(\text{at } t = 0.5 \text{ hr}) = y(0) + \frac{h}{2}S_1 = 1 + \frac{1.0}{2}(0.138629) = 1.0693147$$

3. The estimated slope at 0.5 hr (Eq. 8-31b) is

$$S_2 = \frac{(\ln 2)(1.0693147)}{5} = 0.1482385$$

4. The revised estimate of y at $t = 0.5$ hr is

$$y(0.5) = y(0) + \frac{h}{2}S_2 = 1 + \frac{1.0}{2}(0.1482385) = 1.074119$$

5. The revised estimate of the derivative is

$$S_3 = \frac{(\ln 2)(1.074119)}{5} = 0.1489045$$

6. The initial estimate of y at $t = 0.1$ is

$$y(1.0) = y(0) + hS_3 = 1 + 1.0(0.1489045) = 1.1489045$$

7. The slope estimated at $t = 1$ is

$$S_4 = \frac{(\ln 2)(1.1489045)}{5} = 0.159272$$

8. The final estimate of y at $t = 1$ is (Eq. 8-32)

$$y(1.0) = y(0) + \frac{h}{6}(S_1 + 2S_2 + 2S_3 + S_4)$$

$$= 1 + \frac{1}{6}[0.138629 + 2(0.148238) + 2(0.1489045) + 0.159272]$$

$$= 1.1486977$$

8.12.7 Bending Moment and Shear Force for a Beam

A simply supported beam AB supports a uniformly distributed load of intensity 2 kips/ft as shown in Fig. 8-7. The shear force V at a distance along the span of the beam is given by

$$V = 10 - 2X \tag{8-97a}$$

The bending moment at any point along the span of the beam can be determined by solving the following differential equation:

$$\frac{dM}{dX} = V = 10 - 2X \tag{8-97b}$$

where M = bending moment in kip-ft and V = shear force in kips. The true solution is

$$M = 10X - X^2 \tag{8-98}$$

Figure 8-7 Simply Supported Beam and

Assume that at $X = 0$, $M = 0$ and $V = -10$ kips. Using Euler's method, the following equation results:

$$M_{i+1} = M_i + f(X_i, V_i)h \tag{8-99}$$

where $f(X_i, V_i) = V(X_i)$. Applying the above equation for step sizes of 0.1 and 0.5 ft results in Table 8-9. The results of the example are also shown in Fig. 8-8.

Using the second-order (Eq. 8-22a), third-order (Eq. 8-32), and fourth-order (Eq. 8-28) Runge–Kutta methods, the following equations result:

Second order: $$M_{i+1} = M_i + \frac{1}{2}[V(X_i) + V(X_i + h)]h \tag{8-100}$$

Third order: $$M_{i+1} = M_i + \frac{1}{6}[V(X_i) + 4V(X_i + 0.5h) + V(X_i + h)]h \tag{8-101}$$

TABLE 8-9   Bending Moment for a Beam Using Euler's Method with Step Sizes of 0.1 and 0.5 Ft

x	$V = \dfrac{dM}{dX}$	M ($h = 0.1$)	M ($h = 0.5$)	M Exact
0	10	0	0	0
0.1	9.8	1		−0.99
0.2	9.6	1.98		−1.96
0.3	9.4	2.94		−2.91
0.4	9.2	3.88		−3.84
0.5	9	4.8	5	4.75
1	8	9.1	9.5	9
2	6	16.2	17	16
3	4	21.3	22.5	21
4	2	24.4	26	24
5	0	25.5	27.5	25
6	−2	24.6	27	24
7	−4	21.7	24.5	21
8	−6	16.8	20	16
9	−8	9.9	13.5	9
10	−10	1	5	0

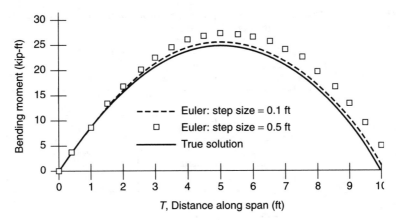

Figure 8-8   Bending Moment for a Beam

Fourth order: $M_{i+1} = M_i + \dfrac{1}{6}[V(X_i) + 4V(X_i + 0.5h) + V(X_i + h)]h$ (8-102)

Applying the above equations for step sizes of 0.1 and 0.5 ft results in Table 8-10. The results of the example are also shown in Fig. 8-9. Since the true solution is to the second order, the second-, third-, and fourth-order Runge–Kutta methods result in solutions without error.

8.12.8 Dynamic Response of a Structure

The one-story frame shown in Fig. 8-10 is displaced a distance of 0.7 in. and then released from rest (velocity = 0) to vibrate freely. The frame has the following properties: mass $m = 2$ kip·sec²/in., and stiffness $k = 40$ kips/in. This system can be modeled as a spring–mass system, in which the stiffness of the columns represents the stiffness of the spring, and the mass of the girders and floor loads represents the vibrating mass in the spring–mass system. Assuming the system to be undamped, that is, without internal friction that opposes the motion and without any external forces, the following equation of motion can be written:

$$m\frac{d^2Y}{dT^2} + kY = 0$$ (8-103)

In this application, the response of the system in the form of the horizontal displacement Y as a function of time T is of interest. Also, the velocity V as a function of time is of interest. The velocity is given by

$$V = \frac{dY}{dT}$$ (8-104)

TABLE 8-10   Bending Moment for a Beam Using the Runge–Kutta
Method with Step Size of 0.5 Ft

x	Second-order Runge–Kutta Method	Third-order Runge–Kutta Method	Fourth-order Runge–Kutta Method	True Solution
0	0	0	0	0
0.1	0.99	0.99	0.99	0.99
0.2	1.96	1.96	1.96	1.96
0.3	2.91	2.91	2.91	2.91
0.4	3.84	3.84	3.84	3.84
0.5	4.75	4.75	4.75	4.75
1	9	9	9	9
2	16	16	16	16
3	21	21	21	21
4	24	24	24	24
5	25	25	25	25
6	24	24	24	24
7	21	21	21	21
8	16	16	16	16
9	9	9	9	9
10	0	0	0	0

The second-order differential equation (Eq. 8-103) can be written as the following system
of two equations:

$$\frac{dY}{dT} = V \qquad (8\text{-}105a)$$

$$\frac{dV}{dT} = -\frac{kY}{m} \qquad (8\text{-}105b)$$

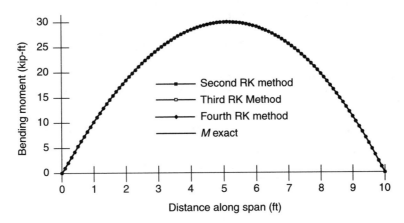

Figure 8-9   Bending Moment for a Beam Using the Runge–Kutta Methods

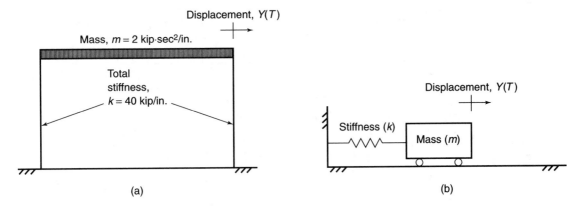

Figure 8-10   (a) One-story Frame and (b) Its Spring–Mass Model

Using Euler's method with three step sizes, 0.1, 0.05, and 0.005 sec, the response of the structure can be determined as shown in Table 8-11 and Fig. 8-11. The true solution is given by

$$Y = Y_0 \cos\left(\sqrt{\frac{k}{m}}\,t\right) + \frac{V_0}{k/m} \sin\left(\sqrt{\frac{k}{m}}\,t\right) \tag{8-106a}$$

$$V = -Y_0 \sqrt{\frac{k}{m}} \sin\left(\sqrt{\frac{k}{m}}\,t\right) + V_0 \cos\left(\sqrt{\frac{k}{m}}\,t\right) \tag{8-106b}$$

It can be observed from Table 8-11 and Fig. 8-11 that to produce accurate numerical results a small step size is required.

TABLE 8-11a   Velocity and Displacement for a One-story Frame Using Euler's Method with Step Size of 0.1 Sec

T (sec)	$\dfrac{dV}{dT}$	$V = \dfrac{dY}{dT}$ (in./sec)	V Exact (in./sec)	Y (in.)	Y Exact (in.)
0	−14	0	0	0.7	0.7
0.1	−14	−1.4	−1.3537978	0.7	0.6311589
0.2	−11.2	−2.8	−2.4413187	0.56	0.4381759
0.3	−5.6	−3.92	−3.0486595	0.28	0.1590087
0.4	2.24	−4.48	−3.0563632	−0.112	−0.1514339
0.5	11.2	−4.256	−2.4629144	−0.56	−0.432091
1	16.91648	6.8096	3.0405805	−0.845824	−0.1665639
1.5	−54.935552	−0.3050701	−1.2908213	2.7467776	0.6377217
2	45.803268	−16.456352	−1.4470025	−2.2901634	−0.6207328
2.5	63.412004	27.089275	3.0772121	−3.1706002	0.1286013

TABLE 8-11b   Velocity and Displacement for a One-story Frame Using Euler's Method with Step Size of 0.05 Sec

T (sec)	$\dfrac{dV}{dT}$	$V = \dfrac{dY}{dT}$ (in./sec)	V Exact (in./sec)	Y (in.)	Y Exact (in.)
0	−14	0	0	0.7	0.7
0.05	−14	−0.7	−0.6941812	0.7	0.6825728
0.1	−13.3	−1.4	−1.3537978	0.665	0.6311589
0.15	−11.9	−2.065	−1.946006	0.595	0.5483184
0.2	−9.835	−2.66	−2.4413187	0.49175	0.4381759
0.25	−7.175	−3.15175	−2.8150732	0.35875	0.3062158
0.5	10.513567	−3.2305438	−2.4629144	−0.5256783	−0.432091
1	7.0137979	4.8520768	3.0405805	−0.3506899	−0.1665639
2	30.118545	−4.8616409	−1.4470025	−1.5059272	−0.6207328
2.5	−0.1813191	10.60089	3.0772121	0.009066	0.1286013

Figure 8-11a   Velocity for a One-story Frame Using Euler's
Method with Step Size of 0.1 Sec

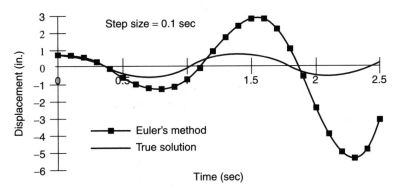

Figure 8-11b   Displacement for a One-story Frame Using Euler's
Method with Step Size of 0.1 Sec

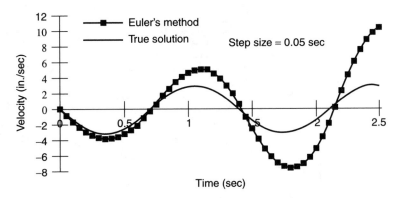

Figure 8-11c   Velocity for a One-story Frame Using Euler's
Method with Step Size of 0.05 Sec

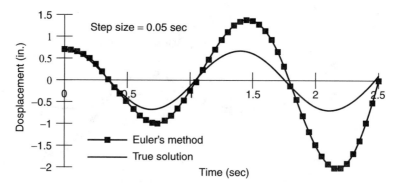

Figure 8-11d   Displacement for a One-story Frame Using Euler's Method with Step Size of 0.05 Sec

TABLE 8-11c   Velocity and Displacement for a One-story Frame Using Euler's Method with Step Size of 0.005 Sec

T (sec)	$\dfrac{dV}{dT}$	$V = \dfrac{dY}{dT}$ (in./sec)	V Exact (in./sec)	Y (in.)	Y Exact (in.)
0	-14	0	0	0.7	0.7
0.005	-14	-0.07	-0.0699942	0.7	0.699825
0.01	-13.993	-0.14	-0.1399533	0.69965	0.6993001
0.015	-13.979	-0.209965	-0.2098425	0.69895	0.6984256
0.02	-13.958004	-0.27986	-0.2796268	0.6979002	0.6972019
0.025	-13.930018	-0.34965	-0.3492713	0.6965009	0.6956296
0.5	8.8563255	-2.5259855	-2.4629144	-0.4428163	-0.432091
1	3.5126826	3.19585	3.0405805	-0.1756341	-0.1665639
2	13.709301	-1.6037152	-1.4470025	-0.685465	-0.6207328
2.5	-2.8853428	3.4880371	3.0772121	0.1442671	0.1286013

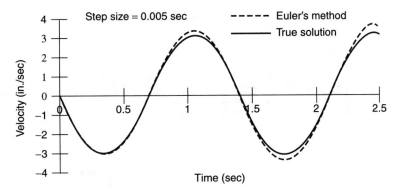

Figure 8-11e   Velocity for a One-story Frame Using Euler's
Method with Step Size of 0.005 Sec

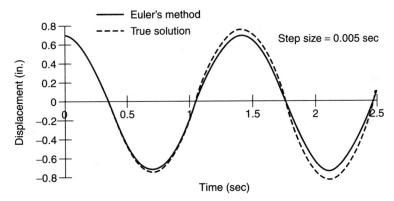

Figure 8-11f   Displacement for a One-story Frame Using Euler's
Method with Step Size of 0.005 Sec

8.12.9 Deflection of a Beam

A cantilevered beam *AB* supports a uniformly distributed load of intensity 1.5 kips/ft as
shown in Fig. 8-12. The deflection of any point along the span of the beam can be determined by solving the following differential equation:

$$\frac{d^4Y}{dX^4} = \frac{-w}{EI} \tag{8-107}$$

Figure 8-12   Cantilever Beam

where Y = deflection in feet, E = modules of elasticity of the material, and I = moment of inertia of the cross section of the beam. Equation 8-107 is a fourth-order differential equation. The following system of four differential equations can be developed for the purpose of solving for the deflection:

$$\frac{dY}{dX} = S \qquad\qquad (8\text{-}108a)$$

$$\frac{dS}{dX} = \frac{M}{EI} \qquad\qquad (8\text{-}108b)$$

$$\frac{dM}{dX} = V \qquad\qquad (8\text{-}108c)$$

$$\frac{dV}{dX} = -w \qquad\qquad (8\text{-}108d)$$

where S = slope of the tangent to the deflected shape of the beam, M = bending moment, V = shear force, and w = intensity of the uniform load. Assume that EI = 3,600 kip·ft², and at $X = 0$, $Y = 0$, $S = 0$, $V = 18$ kips, and $M = -108$ kip·ft. The boundary conditions were determined based on mechanics of materials and statics. The above four first-order equations can then be solved using any of the methods discussed in this chapter. Using Euler's equation, the solution can be expressed in the following form:

$$y_{i+1} = y_i + f(x_i, y_i)h \qquad\qquad (8\text{-}109)$$

Applying Eq. 8-109 sequentially for the four differential equations with a step size of 0.5 ft results in Table 8-12. The deflections for the example are shown in Fig. 8-13. The true solutions for shear, moment, rotation, and deflection are given, respectively, by

$$V = -wx + 18 \qquad\qquad (8\text{-}110a)$$

TABLE 8-12 Shear, Moment, Rotation, and Deflection for a Cantilever Beam with a Step Size of 0.1 Sec

X (ft)	$\dfrac{dV}{dX} = -w$ (kips/ft)	$V = \dfrac{dM}{dX}$ (kips)	V Exact (kips)	M (kip-ft)	M Exact (kip-ft)	$\dfrac{dS}{dX} = \dfrac{M}{EI}$	$S = \dfrac{dY}{dX}$	S Exact	Y (ft)	Y, Exact (ft)
0	-1.5	18	18	-108	-108	-0.03	0	0	0	0
0.5	-1.5	17.25	17.25	-99	-99.1875	-0.0275	-0.015	-0.0143837	0	-0.0035469
1	-1.5	16.5	16.5	-90.375	-90.75	-0.0251042	-0.02875	-0.0275694	-0.0075	-0.014184
1.5	-1.5	15.75	15.75	-82.125	-82.6875	-0.0228125	-0.0413021	-0.0396094	-0.021875	-0.0310254
2	-1.5	15	15	-74.25	-75	-0.020625	-0.0527083	-0.0505556	-0.042526	-0.0536111
2.5	-1.5	14.25	14.25	-66.75	-67.6875	-0.0185417	-0.0630208	-0.0604601	-0.0688802	-0.0814073
3	-1.5	13.5	13.5	-59.625	-60.75	-0.0165625	-0.0722917	-0.069375	-0.1003906	-0.1139063
4	-1.5	12	12	-46.5	-48	-0.0129167	-0.0879167	-0.0844444	-0.1768229	-0.1911111
5	-1.5	10.5	10.5	-34.875	-36.75	-0.0096875	-0.1	-0.0961806	-0.2679688	-0.281684
6	-1.5	9	9	-24.75	-27	-0.006875	-0.1089583	-0.105	-0.3703906	-0.3825
7	-1.5	7.5	7.5	-16.125	-18.75	-0.0044792	-0.1152083	-0.1113194	-0.4810677	-0.4908507
8	-1.5	6	6	-9	-12	-0.0025	-0.1191667	-0.1155556	-0.5973958	-0.6044444
9	-1.5	4.5	4.5	-3.375	-6.75	-0.0009375	-0.12125	-0.118125	-0.7171875	-0.7214063
10	-1.5	3	3	0.75	-3	0.0002083	-0.121875	-0.1194444	-0.8386719	-0.8402778
11	-1.5	1.5	1.5	3.375	-0.75	0.0009375	-0.1214583	-0.1199306	-0.9604948	-0.9600174
12	-1.5	0	0	4.5	0	0.00125	-0.1204167	-0.12	-1.0817188	-1.08

Figure 8-13   Deflection for a Cantilever Beam with a Step Size of 0.1 Sec

$$M = -wx^2/2 + 18x - 108 \qquad (8\text{-}110b)$$

$$S = \frac{1}{EI}\left(\frac{-w}{6}x^3 + 9x^2 - 108x\right) \qquad (8\text{-}110c)$$

and

$$Y = \frac{1}{EI}\left(\frac{-w}{24}x^4 + 3x^3 - 54x^2\right) \qquad (8\text{-}110d)$$

It can be observed from Fig. 8-12 that there is a good agreement between the numerical and analytical solutions.

8.13 PROBLEMS

8-1. Formulate a differential equation for each of the following geometric problems:

    a. The slope of the relationship between x and y equals 2.5 times the sum of x and y, with $y = 2$ at $x = 1$.

    b. The slope of the relationship between x and y equals a second-order polynomial $2x^2 + 3x - 7$, with $y = 3$ at $x = 0$.

    c. The slope of the slope of the relationship between y and x equals $2x^2 + 3y^3$, with $y = 1$ and $dy/dx = 0$ at $x = 1$.

For Problems 8-2 through 8-6, estimate values of the function $y(x)$ using a Taylor series expansion for the range and increment (Δx) shown. Also, compare the accuracy of the solution by comparing the computed values with the true values computed from the corresponding true solutions.

Problem	Differential Equation	Boundary Condition	Δx	Range	True Solution
8-2.	$\dfrac{dy}{dx} = 5x^4 - 12x^2 + 6$	$y = 3$ at $x = 1$	0.05	$1.0 \le x \le 1.3$	$y = x^5 - 4x^3 + 6x$
8-3.	$\dfrac{dy}{dx} = 1 - y$	$y = 2$ at $x = 0$	0.1	$0 \le x \le 0.5$	$y = 1 + e^{-x}$
8-4.	$\dfrac{dy}{dx} = 3x^2 y$	$y = 1$ at $x = 0$	0.1	$0 \le x \le 0.6$	$y = e^{x^3}$
8-5.	$\dfrac{d^2 y}{dx^2} = 2x$	$y = 0$ and $dy/dx = 0$ at $x = 0$	0.1	$0 \le x \le 0.5$	$y = x^3/3$
8-6.	$\dfrac{dy}{dx} = e^x + xe^x + 1$	$y = 0$ at $x = 0$	0.2	$0 \le x \le 1$	$y = xe^x + x$

8-7. Formulate a differential equation for the set of curves that intersects orthogonally the set of parabolas $y^2 = 9bx$, where b is a constant. Then evaluate the curve of the set that passes through the point $y = 3$ at $x = 0$ using the Taylor series expansion for the interval from 0 to 0.6 using an increment of 0.1.

For Problems 8-8 through 8-15, determine the function $y(x)$ using Euler's one-step method for the range and increment (Δx) shown. Also, compare the accuracy of the solution by comparing the computed values with the true values computed from the corresponding true solutions.

Problem	Differential Equation	Boundary Condition	Δx	Range	True Solution
8-8.	$\dfrac{dy}{dx} + y - 1 = 0$	$y = 2$ at $x = 0$	0.2	$0 \le x \le 1$	$y = 1 + e^{-x}$
8-9.	$\dfrac{dy}{dx} + y - 1 = 0$	$y = 2$ at $x = 0$	0.1	$0 \le x \le 1$	$y = 1 + e^{-x}$
8-10.	$\dfrac{dy}{dx} - 5x + 2 = 0$	$y = 1$ at $x = 1$	0.1	$1 \le x \le 1.5$	$y = 2.5x^2 - 2x + 0.5$
8-11.	$(1 + x^2)\dfrac{dy}{dx} - xy = 0$	$y = 5$ at $x = 2$	0.1	$2 \le x \le 3$	$y = \sqrt{5(1 + x^2)}$
8-12.	$(x^2 - x)\dfrac{dy}{dx} + y^2 - y = 0$	$y = 2$ at $x = 2$	0.05	$2 \le x \le 2.25$	$(x - 1)(y - 1) = 0.25xy$

Problem	Differential Equation	Boundary Condition	Δx	Range	True Solution
8-13.	$x + y\dfrac{dy}{dx} - 2x^3 - 2xy^2 = 0$	$y = 1$ at $x = 0$	0.1	$0 \le x \le 0.5$	$\ln(x^2 + y^2) - 2x^2 = 0$
8-14.	$2xy\dfrac{dy}{dx} + x^2 - 3y^2 = 0$	$y = 2$ at $x = 1$	0.2	$1 \le x \le 2$	$y = \sqrt{3x^3 + x^2}$
8-15.	$(x^2 + y^2)\dfrac{dy}{dx} - xy = 0$	$y = 0.584$ at $x = 1$	0.05	$1 \le x \le 1.5$	$\ln y - \dfrac{x^2}{2y^2} = -2$

8-16. If interest is added continuously to the principal P at an annual rate r, estimate the total account at the end of each year for 10 years. Use an interest rate of 12% on a principal of $100,000. Use modified Euler's method.

8-17. The rate at which a radioactive isotope decomposes is at any instant proportional to the remaining mass of the isotope. If a piece initially has a mass of 10 milligrams and the mass decreases to half its original mass in 1000 years, estimate the decay of the piece for a 100-year period using a time increment of 10 years. Use modified Euler's method.

For Problems 8-18 through 8-25, estimate values of the function $y(x)$ using the fourth-order Runge–Kutta method for the range and increment (Δx) shown. Also, compare the accuracy of the solution by comparing the computed values with the values computed from the corresponding true solutions.

Problem	Differential Equation	Boundary Condition	Δx	Range	True Solution
8-18.	$\dfrac{dy}{dx} + y - 1 = 0$	$y = 2$ at $x = 0$	0.2	$0 \le x \le 1$	$y = 1 + e^{-x}$
8-19.	$\dfrac{dy}{dx} = x^2 - 2x + 7$	$y = 6$ at $x = 1$	0.2	$1 \le x \le 2$	$y = \dfrac{x^3}{3} - x^2 + 7x - \dfrac{1}{3}$
8-20.	$\dfrac{dy}{dx} = x^2 - 2x + 7$	$y = 6$ at $x = 1$	0.1	$1 \le x \le 2$	$y = \dfrac{x^3}{3} - x^2 + 7x - \dfrac{1}{3}$
8-21.	$(1 + x^2)\dfrac{dy}{dx} - xy = 0$	$y = 2$ at $x = 1$	0.25	$1 \le x \le 2$	$y = \sqrt{2(1 + x^2)}$
8-22.	$x + y\dfrac{dy}{dx} - 2x^3 - 2xy^2 = 0$	$y = 1$ at $x = 0$	0.1	$0 \le x \le 1$	$\ln(x^2 + y^2) - 2x^2 = 0$
8-23.	$2xy\dfrac{dy}{dx} + x^2 - 3y^2 = 0$	$y = 2$ at $x = 1$	0.1	$1 \le x \le 2$	$3x^3 + x^2 - y^2 = 0$

Problem	Differential Equation	Boundary Condition	Δx	Range	True Solution
8-24.	$(x^2 - x)\dfrac{dy}{dx} + y^2 - y = 0$	$y = 3$ at $x = 3$	0.05	$3 \le x \le 3.3$	$(x-1)(y-1) = \dfrac{4xy}{9}$
8-25.	$(x^2 + y^2)\dfrac{dy}{dx} - xy = 0$	$y = 0.584$ at $x = 1$	0.05	$1 \le x \le 1.5$	$\ln y - \dfrac{x^2}{2y^2} = -2$

8-26. If the angular velocity w of a cylinder rotating in a viscous fluid decreases at a rate proportional to w and if w is initially 100 revolutions per minute, estimate w versus time (t) for 10 min, at an increment of 1 min if the angular velocity decreases to 90 rpm after 1 min; that is, the proportionality constant can be shown to be $\ln(0.9)$. Use the fourth-order Runge–Kutta method.

8-27. Find the total accumulated soil in the sediment trap discussed in Section 8.12.5 using (a) Euler's method and (b) the fourth-order Runge–Kutta method.

8-28. Solve Problem 8-8 using the Euler–trapezoidal (predictor–corrector) method.

8-29. Solve Problem 8-10 using the Euler–trapezoidal (predictor–corrector) method.

8-30. Solve Problem 8-12 using the Euler–trapezoidal (predictor–corrector) method.

8-31. Solve Problem 8-14 using the Euler–trapezoidal (predictor–corrector) method.

8-32. Solve Problem 8-15 using the Milne–Simpson (predictor–corrector) method. Use the Euler–trapezoidal (predictor–corrector) method to obtain the needed boundary conditions to start the Milne–Simpson method.

8-33. Solve Problem 8-8 using the least-squares method with a quadratic model (that is, $\hat{y} = b_0 + b_1 x + b_2 x^2$). Assess the accuracy of the model as compared to the true solution.

8-34. Solve Problem 8-10 using the least-squares method with a quadratic model (i.e, $\hat{y} = b_0 + b_1 x + b_2 x^2$). Assess the accuracy of the model as compared to the true solution.

8-35. Solve Problem 8-8 using the Galerkin method with a quadratic model (i.e., $\hat{y} = b_0 + b_1 x + b_2 x^2$). Assess the accuracy of the model as compared to the true solution.

8-36. Solve Problem 8-10 using the Galerkin method with a quadratic model (i.e., $\hat{y} = b_0 + b_1 x + b_2 x^2$). Assess the accuracy of the model as compared to the true solution.

8-37. Using the Euler method with $h = 0.5$, find the values of y and y' at $x = 1.0$ for the following differential equation:

$$\frac{d^2 y}{dx^2} - 0.05\frac{dy}{dx} + 0.15y = 0$$

The boundary conditions are at $x = 0$, $y = 1$, and $y' = 0$.

8-38. Solve the following differential equation:

$$\left(\frac{dy}{dx}\right)^2 - xy^2 - 4 = 0$$

Use Euler's method with a step size (h) = 0.1, $y = 5$ at $x = 1$, and from $x = 1$ to $x = 1.5$.

8-39. The beam example of Section 8.12.7 involves a boundary-value problem. Define the boundary-value problem for the example and then solve it using Euler's method in conjunction with the shooting method.

8-40. The beam example of Section 8.12.7 involves a boundary-value problem. Define the boundary-value problem for the example and then solve it using the second Runge–Kutta method in conjunction with the shooting method.

8-41. The beam example of Section 8.12.7 involves a boundary-value problem. Define the boundary-value problem for the example and then solve it using the finite-difference method.

8-42. Evaluate the following integral using a step size of 0.1 for $x_0 = 1$:

$$y = 0.5 + \int_0^{x_0} \frac{1}{\sqrt{2\pi}} \exp(-0.5x^2) \, dx$$

where $\pi = 3.1416$ and $y = 0.5$ at $x = 0$.

9

Data Description and Treatment

9.1 INTRODUCTION

It is common in engineering to deal with certain types of dispersion and uncertainty by collecting data and information about some variables that are needed to solve a problem of interest. The data can then be used to establish some understanding about the relations among the different variables. After collecting the data, it is necessary to utilize techniques for describing, treating, and analyzing them. The objective of this chapter is to introduce some of the techniques that are commonly used in engineering to describe and summarize data.

9.2 CLASSIFICATION OF DATA

Graphical analyses are widely used in engineering for both making initial assessments of data characteristics and presenting results to be used by engineers. There are a number of ways of graphically presenting data, each having advantages for specific types of data. Therefore, before discussing the graphical methods, it is necessary to review methods for classifying data.

Data can be measured on one of four scales: nominal, ordinal, interval, and ratio. The four scales are given in order of numerical value. Variables defined on one scale can be reduced to a lower scale of measurement, but cannot be described on a higher scale of measurement. However, when a variable is transformed to a lower scale of measurement, there is an accompanying loss of information.

In addition to the scale of measurement, data can be classified based on their dimensionality. The dimensionality is a measure of the number of axes needed to graphically present the data. Graphical analyses are usually limited to use with one-, two-, and three-dimensional data.

9.2.1 Nominal Scale

The nominal scale of measurement is at the lowest level, because there is no order to the data. Measurements consist of simply identifying the sample as belonging to one of several categories. Nominal measurement scales are both discrete and qualitative. However, numbers may be assigned to the categories for the purpose of coding.

Frequently used examples of variables measured on a nominal scale include (1) gender: female or male; (2) political affiliation: Republican, Democrat, Independent, or other; and (3) college major: engineering, sciences, physical education, or other. Engineering data are sometimes provided using a nominal scale, for example, (1) project failed or did not fail; (2) fatal and nonfatal accidents; and (3) land use: urban, rural, forest, institutional, commercial, other.

9.2.2 Ordinal Scale

The ordinal scale of measurement is considered to be a higher scale than the nominal scale because it has the added property that there is order among the groups. However, the magnitude of the differences between groups is not meaningful. For example, military ranks are measured on an ordinal scale. The major is above the sergeant and the sergeant is above the private, but we cannot say that a major is two or three times higher than a sergeant.

Variables of interest in engineering that are measured on an ordinal scale include the infiltration potential of soil texture classes and hazard classifications for dam design (high hazard, moderate hazard, low hazard). Soils are classified into one of several categories, such as sand, sandy loam, clay loam, and clay. In this respect, soil texture is measured on a nominal scale. However, if we consider the infiltration potential of the soil, then we can put the soil textures in order according to the infiltration potential, high to low.

9.2.3 Interval Scale

The interval scale of measurement has the characteristics of the ordinal scale, in addition to having a meaningfulness in the separation between any two numbers on the scale. Temperature is defined on the interval scale. We recognize that a difference in temperature of 5°C is less than a difference of 10°C. Values on an interval scale may be treated with arithmetic operators. For example, the mean value of a set of test grades requires addition and division.

Engineering data are frequently recorded on an interval scale. The yield strength of steel, the compression strength of concrete, and the shear stress of soil are variables measured on an interval scale. The annual number of traffic fatalities and the number of lost worker-hours on construction sites due to accidents are also engineering variables recorded on an interval scale.

9.2.4 Ratio Scale

The ratio scale represents the highest level of measurement. In addition to the characteristics of the interval scale, the ratio scale has a true zero point as its origin, not like the interval scale, for which the zero point is set by some standard. For example, in the interval scale the zero point for temperature (°C) is set at the point where water freezes. However, it could have been either set at the point where water boils or based on some other substance.

The standard deviation, which is discussed in Section 9.5.2, is measured on a ratio scale. The zero point is that for which there is no variation. The coefficient of a variation and many dimensionless ratios such as the Mach or Reynolds numbers are measured on a ratio scale.

9.2.5 Dimensionality of Data

The dimensionality of data was defined as the number of axes needed to represent the data. Tabular data with one value per classification is an example of one-dimensional data. For example, if a transportation engineer gives the number of fatal traffic accidents for each state in 1991, the variable is described using a one-dimensional, interval scale data. It can be represented in tabular form as a function of the state. It could also be represented pictorially with the number of fatalities as the ordinate (vertical axis) and the state as the abscissa (horizontal axis). In this case, the ordinate is on an interval scale, while the abscissa is on a nominal scale; when presented this way, it appears as a two-dimensional graph.

Two-dimensional plots are very common in engineering. In Chapter 8, the solution of a dependent variable y could be plotted as a function of the independent variable x. In this case, both variables are expressed on an interval scale. As another example, the corrosion rate of structural steel as a function of the length of time that the steel has been exposed to the corrosive environment is a two-dimensional plot. If we have data for different types of steel (carbon steel, copper steel, and weathered steel), the three relationships can be presented on the same graph, with the steel types identified. In this case, steel type is a nominal variable; so the two-dimensional plot includes two variables on interval scales and one variable on a nominal scale.

9.3 GRAPHICAL DESCRIPTION OF DATA

The first step in data analysis is often a graphical study of the characteristics of the data sample. Depending on the objectives of the analysis and the nature of the problem under consideration, one or more of the following graphical descriptors are commonly used:

area charts, pie charts, bar charts, column charts, scatter diagrams, line charts, combinations, and three-dimensional charts. The selection of a graphical descriptor type should be based on (1) its intended readers, (2) the type of data, (3) the dimensionality of the problem, and (4) the ability of the graphical descriptor to emphasize certain characteristics or relations of the parameters of the problem. In this section, examples are used to illustrate the different types.

9.3.1 Area Charts

Area charts are useful for three-dimensional data that include both nominal and interval independent variables, with the value of the dependent variable measured on an interval scale and cumulated over all values of the nominal variable. The independent variable measured on an interval scale is shown on the abscissa. At any value along the abscissa, the values of the dependent variable are cumulated over the independent variable measured on the nominal scale.

Example 9-1 Area Chart for Traffic Analysis

A traffic engineer is interested in analyzing the traffic at an intersection. An incoming vehicle to the intersection can proceed in one of the following directions: straight, left turn, or right turn (U-turns are assumed to be illegal). The vehicles were counted at the intersection and classified according to direction. The counts were established for 24 hours in a typical business day. The results are shown in Fig. 9-1 in the form of an area chart. The area chart shows that at 0800 approximately 50 vehicles made a right turn, approximately 30 vehicles made either a right or left turn, and approximately 135 vehicles went through the intersection. Therefore, the total is about 215 vehicles.

Figure 9-1   Traffic at an Intersection

9.3.2 Pie Charts

Pie charts are commonly used to graphically present data recorded as fractions, percentages, or proportions. The 100% of the circular pie chart is separated into pie slices based on fractions or percentages. The variable presented in a pie chart is measured on an interval scale.

Example 9-2 Pie Chart for Shipment Breakdown

A shipping company facilitates for customers the transfer of any items from any location to another within the United States. The company transfers 25%, 30%, and 45% of the items by air, ground, and sea transportation, respectively. In this case, the variable is the form of transportation, with values recorded as a percentage of all transportation. The breakdown of these items by shipping method is shown in Fig. 9-2 in the form of a pie chart.

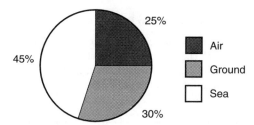

Figure 9-2   Shipping Methods

9.3.3 Bar Charts

Bar charts are also useful for data recorded on an interval scale with one or more independent variables recorded on nominal or ordinal scales. The dependent variable can be a magnitude or a fraction. Both one- and two-dimensional bar charts can be used, with a dimension that is based on the number of independent variables.

Example 9-3 Bar Chart for Reinforcing Steel Production

A reinforcing-steel manufacturer provides steel of three different yield strengths: 40, 50, and 60 ksi. The production manager would like to keep track of production for both reinforcing-steel type and the four quarters of a production year. Thus the amount of steel produced is the dependent variable. There are two independent variables, the steel type and the quarters. The dependent variable is on an interval scale, while the independent variables are on ordinal and nominal scales, respectively. A bar chart, in this case, can meet the requirements of the manager as shown in Fig. 9-3. In this case, the bars are shown vertically because the descriptors can be easily included on the abscissa and as a side note.

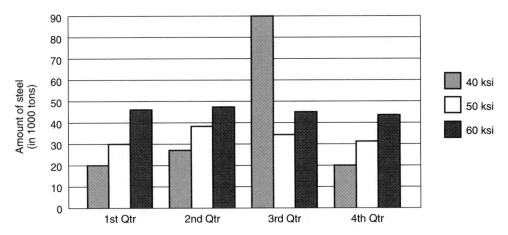

Figure 9-3   Reinforcing Steel Production by Yield Strength and Quarter

Example 9-4   Bar Chart for Capacity of Desalination Plants

In this case, the dependent variable is the capacity (in million gallons per day, mgd) of water from desalination plants worldwide. The independent variable is the process used in the desalination. Figure 9-4 shows the distribution. The bar chart, in this example, is shown sideways to facilitate presenting the descriptors of the six processes.

Figure 9-4   Worldwide Desalination Capacity (mgd) for Available Processes

9.3.4 Column Charts

Column charts are very similar to bar charts, but with the added constraint that the dependent variable is expressed as a percentage (or fraction) of a total. In this case, one of the independent variables is used for the abscissa, while the dependent variable is shown as percentages (or fractions) of the second independent variable. The dependent variable is shown as the ordinate.

Example 9-5 Column Chart for Reinforcing Steel Production

An alternative method for displaying the reinforcing steel production of Example 9.3 is by column charts. Figures 9.5a and b show example charts. Note that, in Fig. 9.5a, steel production is displayed as a percentage of the total steel produced for each quarter.

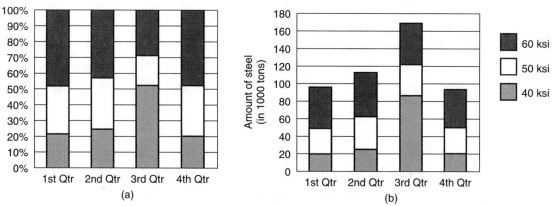

Figure 9-5   Reinforcing Steel Production by Yield Strength and Quarter

9.3.5 Scatter Diagrams

When both the independent and dependent variables are measured on interval or ratio scales, the data are best presented with scatter plots. Usually, the variable to be predicted (dependent variable) is shown on the ordinate and the independent variable on the abscissa. The ranges for the variables on axes are based on either the minimum and maximum values of the measured data, possible values for the variables, or values that may be expected to occur in the future.

Example 9-6 Scatter Diagram for Yield Strength and Carbon Content

The yield strength of steel depends on several factors, such as the carbon content. An experimental program was designed to understand this relationship. Figure 9-6 shows the measured values. Since we are interested in the yield strength as a function of the carbon content, the yield strength is placed on the ordinate. The limits of the plot were set by the values of the data. An alternative plot with a range of 0 to 200 ksi and 0% to 0.1% carbon content could also be used to display the data. It is evident from this figure that under similar test conditions, the yield strength increases as the carbon content increases.

9.3.6 Line Graphs

Line graphs are commonly used to illustrate mathematical equations. With both variables measured on interval or ratio scales, the variable to be predicted is usually shown as the ordinate. Line graphs are frequently used for design work when the design method is relatively simple.

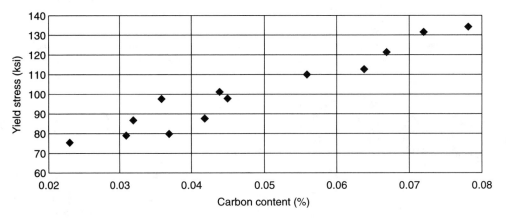

Figure 9-6   Yield Strength and Carbon Content

Example 9-7 Line Chart for Peak Discharge Rates

For one section of the state of Maryland, peak discharge rates (Q in ft³/sec, cfs) can be estimated as a function of drainage area (A in mile²) by the following equations:

$$Q_2 = 55.1A^{0.672} \qquad (9\text{-}1a)$$

$$Q_{10} = 172A^{0.667} \qquad (9\text{-}1b)$$

$$Q_{100} = 548A^{0.662} \qquad (9\text{-}1c)$$

in which the subscripts on Q reflect the return frequency (T in years) of the storm. Thus Q_{100} is the 100-year peak discharge. Designs with these equations are usually limited to drainage areas of less than 100 square miles. Figure 9-7 shows the peak discharge for drainage areas up to 100 square miles. The two-dimensional line chart shows the dependent variable on the ordinate, one independent variable as the abscissa, and the second independent variable in the legend taking three values. All three variables are measured on interval scales.

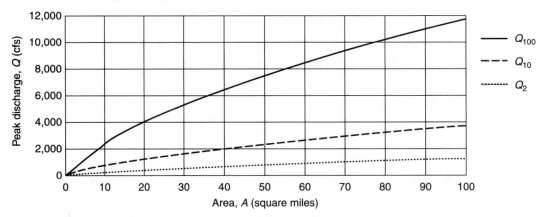

Figure 9-7   Peak Discharge Rates versus Drainage Area and Return Frequency

Example 9-8 Line Chart for Yield Strength and Carbon Content

To establish the relationship between yield strength and carbon content, two independent laboratories were requested to perform similar tests on similar sets of specimens. The results from the two laboratories are shown in Fig. 9-8. The length of the lines connecting the points reflects the magnitude of the variations that are reported based on measurements made at different laboratories.

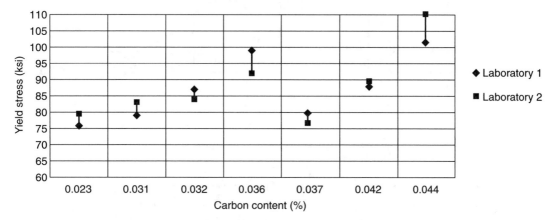

Figure 9-8   Line Chart for Yield Strength and Carbon Content

9.3.7 Combination Charts

In combination charts, two or more of the previously discussed graphical methods are used to present data. For example, a line graph and bar chart can be combined in the same plot. A combination chart that includes both a scatter plot and a line graph is also commonly used to present experimental data and theoretical (or fitted) prediction equations.

Example 9-9 Operation of a Marine Vessel

The annual number of operation-hours of a high-speed Coast Guard patrol boat is of interest. The annual number of hours was recorded for 10 years. The measured data are shown in Fig. 9-9 as vertical bars; they represent sample information. An analytical model is used to fit the data; this is shown in the figure as a solid line representing an assumed frequency model for the population. The combination chart in this case provides an effective presentation tool.

Example 9-10 Sample Distribution and Probability Function

A combination graph is useful in comparing sample measurements and an assumed probability function. Measurements of daily evaporation (in./day) were collected for each day in a month. A histogram of 30 values is shown in Fig. 9-10. An increment of 0.02 in./day is used to develop this histogram. The measured evaporation ranges from 0.046 to 0.157 in./day; thus lower and upper bounds of 0.04 and 0.16 in./day are used. Based on the figure, it is reasonable to assume that each increment on the abscissa of the graph is equally likely. Therefore, a uniform frequency of 5, which results from 30 measurements divided by 6 intervals, can be used as the model.

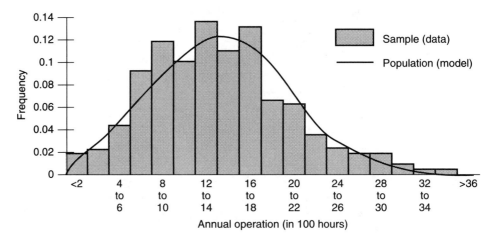

Figure 9-9   Combination Chart for Operational Profile

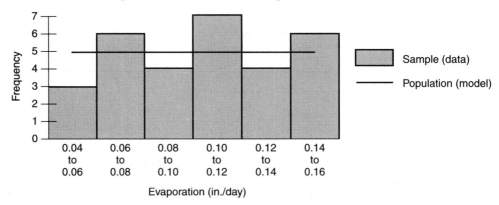

Figure 9-10   Combination Chart for Evaporation Data

Example 9-11 Corrosion Penetration versus Exposure Time

Measurements of corrosion were made yearly on a set of 10 steel specimens. The measured data are plotted in the combination graph of Fig. 9-11. A power model was fitted to the data (see Section 10.9) and is also shown in Fig. 9-11. The power model is

$$\text{penetration} = 28.075t^{0.39943} \tag{9-2}$$

where t = time exposure. The combination chart of Fig. 9-11 is useful because it shows both the measured data and a graph that can be used to make predictions. The scatter of the measured points about the fitted line indicates the accuracy of the model.

9.3.8 Three-dimensional Charts

Any of the charts described in the previous examples can be displayed in three dimensions. However, three-dimensional charts are commonly used to describe the relationships among three variables. For example, Fig. 9-12a shows a three-dimensional pie chart

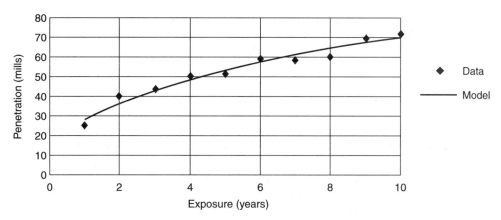

Figure 9-11   Combination Chart for Corrosion Prediction

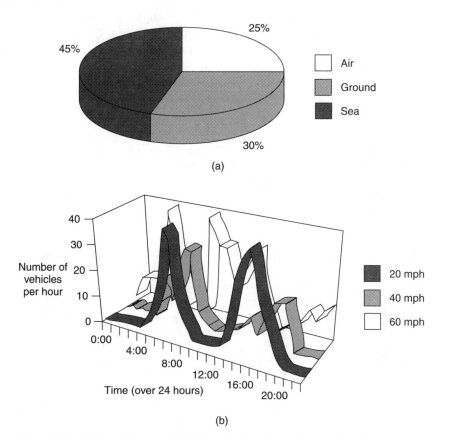

Figure 9-12   (a) Three-dimensional Pie Chart; (b) Three-dimensional Surface Chart

for the data of Example 9-2. Another example is shown in Fig. 9-12b, in which the speed (in mph) and number of vehicles per hour passing through an intersection are shown at different times over a 24-hour period.

9.4 HISTOGRAMS AND FREQUENCY DIAGRAMS

Histograms and frequency diagrams are special types of charts that are commonly used to display and describe data. They are developed from the data for some variables of interest. A histogram is a plot (or tabulation) of the number of data points versus selected intervals or values for a parameter. A frequency diagram (or frequency histogram) is a plot (or tabulation) of the frequency of occurrence versus selected intervals or values of the parameter. The number of intervals (k) can be subjectively selected depending on the sample size (n). The number of intervals can be approximately determined as

$$k = 1 + 3.3 \log_{10}(n) \tag{9-3}$$

Also, the number of intervals can depend on the level of dispersion in the data. The frequency diagrams can be derived from the histogram by dividing the number of data points that correspond to each interval by the sample size. The meaning of these diagrams and their usefulness and development are illustrated in the following example.

Example 9-12 Grades of Students

Students are always interested in the frequency histogram of test scores. This can be developed by determining the number of test scores in various groups, such as the letter grades or intervals of 10 points. If a test is given to 50 students and the numbers of scores are tabulated for each grade level (i.e., A, B, C, D, and F), a histogram can be plotted. If the number of students receiving grades of A, B, C, D, and F were 5, 11, 18, 10, and 6, respectively, a graph of the number of students versus the grade level indicates that the grades have a bell-shaped plot. The resulting histogram and frequency diagrams are shown in Figs. 9-13a and b, respectively. Alternatively, the frequency of grades could be computed for intervals, such as 0–10, 11–20. 21–30, . . . , 91–100. The frequency could be plotted against these intervals.

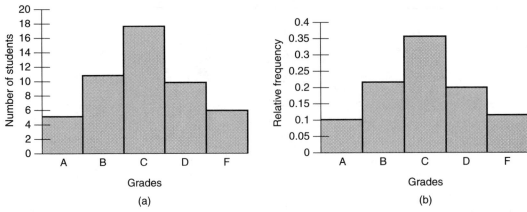

Figure 9-13   (a) Histogram of Grades; (b) Frequency Histogram of Grades

The effectiveness of a graphical analysis in identifying the true shape of the frequency diagram depends on the sample size and the intervals selected to plot the abscissa. For small samples, it is difficult to separate the data into groups that provide a reliable indication of the frequency of occurrence. With small samples, the configuration of the histogram may be very different for a small change in the intervals selected for the abscissa; for example, the grade distribution might appear very different if the intervals 0–15, 16–25, 26–35, . . . , 76–85, 86–100 were used instead of the 10-point intervals described previously.

9.5 DESCRIPTIVE MEASURES

In engineering it is sometimes desirable to characterize some data by certain descriptive measures (or statistics). These measures, which take numerical values, can be easily communicated to others and quantify the main characteristics of the data.

Most data analyses include the following three descriptive measures at the fundamental level:

1. Central tendency measures
2. Dispersion measures
3. Percentile measures

In this section an introductory description of these measures is provided. A formal discussion of these measures is provided in probability and statistics textbooks. Additionally, box-and-whisker plots are introduced as a graphical means of presenting these measures.

9.5.1 Central Tendency Measures

A very important descriptor of data is the central tendency measure. The following three types can be used:

1. Average value
2. Median value
3. Mode value

The average value is the most commonly used central tendency descriptor. For n observations, if all observations are given equal weights, the average value is given by

$$\overline{X} = \frac{1}{n} \sum_{i=1}^{n} x_i \tag{9-4}$$

where x_i = a sample point, and $i = 1, 2, \ldots, n$.

The median value x_m is defined as the point that divides the data into two equal parts; that is, 50% of the data are above x_m and 50% are below x_m. The median value can be determined by ranking the n values in the sample in decreasing order, 1 to n. If n is an odd number, the median is the value with a rank of $(n + 1)/2$. If n is an even number, the median equals the average of the two middle values, that is, those with ranks $n/2$ and $(n/2) + 1$.

The mode value x_d is defined as the point of the highest percent of frequency of occurrence. This point can be determined with the aid of the frequency histogram.

Although these measures convey certain information about the underlying sample, they do not completely characterize the underlying variables. Two variables can have the same mean, but different histograms. Thus measures of central tendency cannot fully characterize the data. Other characteristics are also important and necessary.

Example 9-13 Mean Value of Grades

Consider the grades of the students discussed in Example 9-9. Assume the following grade points (levels) that correspond to the letter grades:

$$A = 4$$
$$B = 3$$
$$C = 2$$
$$D = 1$$
$$F = 0$$

Therefore, the average grade of the class can be computed as follows:

$$\overline{X} = \frac{4 + 4 + 4 + 4 + 4 + 3 + 3 + L\ + 3 + 2 + 2 + L\ + 2 + 1 + 1 + L\ + 1 + 0 + 0 + L\ + 0}{50} = 1.98$$

(9-5)

In this equation, there are 5 of the 4 value, 11 of the 3 value, 18 of the 2 value, 10 of the 1 value, and 6 of the 0 value. Therefore, this equation can be written as

$$\overline{X} = \frac{5(4) + 11(3) + 18(2) + 10(1) + 6(0)}{50} = 1.98 \qquad (9-6)$$

The average value indicates that on the average the class is at the C level. By inspecting the frequency histogram in Fig. 9-13b, the median value is also C, since 32% of the grades are in the A and B levels and 32% of the grades in the D and F levels. Therefore, the C grade divides the grades into two equal percentages. This method of finding the median is proper for values on an ordinal scale, but not on a continuous scale. Again, the C grade is the mode, since it has the highest frequency of 36%. In this example, the average, median, and mode values have about the same value. However, in general, these values can be different.

9.5.2 Dispersion Measures

The dispersion measures describe the level of scatter in the data about the central tendency location. The most commonly used measures are the variance and other quantities that are derived from it. For n observations in a sample that are given equal weight, the variance (S^2) is given by

$$S^2 = \frac{1}{n-1} \sum_{i=1}^{n} (x_i - \overline{X})^2 \qquad (9-7)$$

The units of the variance are the square of the units of the variable x; for example, if the variable is measured in pounds per square inch (psi), the variance has units of (psi)^2. Computationally, the variance of a sample can be determined using the following alternative equation:

$$S^2 = \frac{1}{n-1}\left[\sum_{i=1}^{n} x_i^2 - \frac{1}{n}\left(\sum_{i=1}^{n} x_i\right)^2\right] \tag{9-8}$$

Equation 9-8 provides the same answer as Eq. 9-7 when computations are made using an appropriate number of significant digits. Equations 9-7 and 9-8 provide an estimate of the variance that is an average of the squared difference between the x values and their average value. In these equations, $(n - 1)$ is used to compute the average, instead of (n), in order to obtain an unbiased estimate of the variance.

There are two commonly used derived measures based on the variance: the standard deviation and the coefficient of variation. By definition, the standard deviation (S) is the square root of the variance as follows:

$$S = \sqrt{\frac{1}{n-1}\left[\sum_{i=1}^{n} x_i^2 - \frac{1}{n}\left(\sum_{i=1}^{n} x_i\right)^2\right]} \tag{9-9}$$

It has the same units as both the underlying variable and the central tendency measures. Therefore, it is a useful descriptor of the dispersion or spread of a sample of data.

The coefficient of variation (cov or δ) is a normalized quantity based on the standard deviation and the mean. Therefore, it is dimensionless. The cov is defined as

$$\text{cov} = \frac{S}{\overline{X}} \tag{9-10}$$

It is also used as an expression of the standard deviation in the form of a percent of the average value. For example, consider $\overline{X}$ and S to be 50 and 20, respectively; therefore, $\text{cov}(X) = 0.4$ or 40%. In this case, the standard deviation is 40% of the average value.

Example 9-14 Dispersion Measures of Concrete Strength

A sample of five tests was taken to determine the compression strength (in ksi) of concrete. Test results are 2.5, 3.5, 2.2, 3.2, and 2.9 ksi. Compute the variance, standard deviation, and coefficient of variation of concrete strength. The mean value of concrete strength $\overline{X}$ is given by

$$\overline{X} = \frac{2.5 + 3.5 + 2.2 + 3.2 + 2.9}{5} = 2.86 \text{ ksi} \tag{9-11}$$

The variance of concrete strength is computed using Eq. 9.8 as follows:

$$S^2 = \frac{2.5^2 + 3.5^2 + 2.2^2 + 3.2^2 + 2.9^2 - \dfrac{(2.5 + 3.5 + 2.2 + 3.2 + 2.9)^2}{5}}{5-1} = 0.273 \text{ ksi}^2 \tag{9-12}$$

Therefore, the standard deviation is given by

$$S = \sqrt{0.273} = 0.52249 \text{ ksi} \qquad (9\text{-}13)$$

The coefficient of variation can be computed as follows:

$$\delta \text{ or } \text{cov}(X) = \frac{S}{\overline{X}} = \frac{0.52249}{2.86} = 0.183 \qquad (9\text{-}14)$$

The relatively large coefficient of variation, 18.3%, suggests that the average value is not re-liable and that additional measurements might be needed. If, with additional measurements, the coefficient of variation remains large, relatively large factors of safety should be used for projects that use the concrete.

9.5.3 Percentiles

A p percentile value (x_p) for a parameter or variable based on a sample is the value of the parameter such that $p\%$ of the data is less or equal to x_p. On the basis of this definition, the median value is considered to be the 50th percentile value. It is common in engineering to have interest in the 10th, 25th, 50th, 75th, or 90th percentile values for a variable.

Example 9-15 Operation of a Marine Vessel

The annual number of operation-hours of a high-speed Coast Guard patrol boat was dis-cussed in Example 9-9. The annual number of hours was recorded for 10 years. An analyti-cal model is used to fit the data; this is shown in Fig. 9-9 as a solid line representing an assumed frequency model for the population. The model is also shown in Fig. 9-14. Also shown in this figure is the 25 percentile value, which is 702 hours. The 75, 90, and 10 per-centile values are 1850 hours, 2244 hours, and 578 hours, respectively. The calculation of the percentile values, in this example, requires the knowledge of the equation of the model and probabilistic analysis. The model used in Fig. 9-14 is called the normal probability density

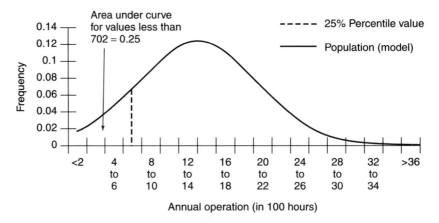

Figure 9-14   Percentile Values for the Operational Profile

function. The area under this model is a measure of likelihood or probability. The total area under the model is 1. Therefore, the 25 percentile value is the x value such that the area under the model up to this value is 0.25. This value is 702 hours, as shown in Fig. 9-14.

9.5.4 Box-and-whisker Plots

A box-and-whisker plot is a graphical method for showing the distribution of sampled data, including the central tendency (mean and median), dispersion, percentiles (10, 25, 75, and 90 percentiles), and the extremes (minimum and maximum). Additionally, it can be used to show the bias about the standard value and, if the figure includes multiple plots for comparison, the relative sample size.

To construct a box-and-whisker plot, the following characteristics of a data set need to be computed:

1. Mean and median of the sample
2. Minimum and maximum value in the sample
3. The 90, 75, 25, and 10 percentile values

The plot consists of a box, the upper and lower boundaries of which define the 75 and 25 percentiles, and upper and lower whiskers, which extend from the ends of the box to the extremes, as shown in Fig. 9-15. At the 90 and 10 percentiles, bars, which are one-half of the width of the box, are placed perpendicular to the whiskers. The mean and median are indicated by solid and dashed lines that are the full width of the box, respectively.

Figure 9-15 shows the box-and-whisker plot for the maximum daily ozone concentration. The mean and median values are 59 and 52 ppb (parts per billion), respectively. The 10, 25, 75, and 90 percentile points are 24, 36, 79, and 97 ppb, respectively.

If a figure includes more than one box-and-whisker plot and the samples from which each plot is derived are of different sizes, then the width of the box can be used to indicate the sample size, with the width of the box increasing as the sample sizes in-

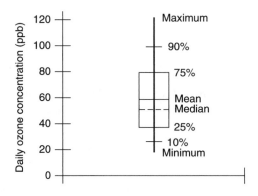

Figure 9-15 Box-and-whisker Plot for Maximum Daily Ozone Concentration (ppb)

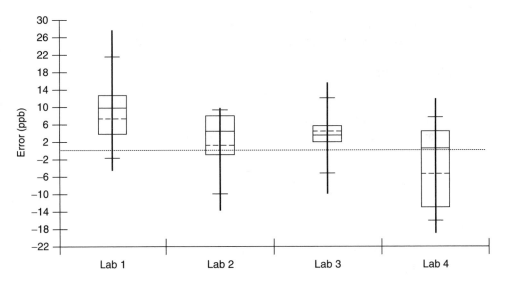

Figure 9-16 Display of Multiple Box-and-whisker Plots

crease. Figure 9-16 shows box-and-whisker plots of samples of a toxic chemical analyzed at four different laboratories. Since the samples were of different concentrations, the distributions are presented as the differences between the true concentration and the concentration reported by the laboratory (i.e., error). The difference between the line for zero error and the solid line for the mean in each plot represents the corresponding lab's bias. Lab 1 tends to overpredict by almost 10 ppb. While lab 2 shows a slight positive bias, it shows a skewed distribution of results, with many values underestimating the true value. Laboratories 1, 2, 3, and 4 processed the same number of samples each. Therefore, the widths of the boxes are the same, reflecting the equal sample sizes. However, it should be noted that greater accuracy of data descriptors can be expected for the descriptors based on larger sample sizes.

9.6 APPLICATIONS

9.6.1 Two Random Samples

The two random samples shown in Table 9-1 are used to illustrate the meaning of dispersion in data. Both samples have the same mean value of 10. The mean value for sample 1 can be computed as

$$\text{mean of sample 1} = \frac{15(9) + 15(11)}{30} = 10$$

TABLE 9-1   Two Random Samples

Sampled Value	Number of Occurrences	
	Sample 1	Sample 2
5	0	3
6	0	2
7	0	4
8	0	0
9	15	3
10	0	4
11	15	2
12	0	6
13	0	2
14	0	2
15	0	2

Similarly, the mean for sample 2 is

$$\text{mean of sample 2} \; = \; \frac{3(5) + 2(6) + 4(7) + \cdots + 2(13) + 2(14) + 2(15)}{30} = 10$$

Although both samples have the same mean and the same sample size, they are different in their levels of scatter or dispersion. Figure 9-17 shows the histograms for the two sam-

Figure 9-17   Histograms for Two Samples

ples, which show clearly the different levels of scatter. The variances, standard deviations, or coefficients of variation can be computed and used to measure the dispersion. The variance for sample 1 can be computed as

$$\text{variance of sample 1} = \frac{15(9 - 10)^2 + 15(11 - 10)^2}{30 - 1} = 1.034$$

Similarly, the variance for sample 2 is

variance of sample 2

$$= \frac{3(5 - 10)^2 + 2(6 - 10)^2 + 4(7 - 10)^2 + \cdots + 2(13 - 10)^2 + 2(14 - 10)^2 + 2(15 - 10)^2}{30 - 1} = 9.379$$

Therefore, the variance of sample 2 is about nine times the variance of sample 1. The standard deviations for samples 1 and 2 are 1.017 and 3.063, respectively, and the coefficient of variations are 0.10 and 0.31, respectively. The larger scatter in sample 2 could reflect natural variation in the value of the variable or a difficulty in making precise measurements. In some cases, variation is not desirable in a sample of data if it reflects a lack of precision. In other cases, sample variation is necessary to ensure that the sample is representative of the full range of the underlying population.

9.6.2 Stage and Discharge of a River

The stage of a river is defined as its flow depth, and the discharge is defined as the volume of flow. For the Little Patuxent River near Guilford, Maryland, the stage in meters (m) and the discharge in cubic meters per second (cms) were obtained as shown in Table 9-2. The table shows the stage and discharge for the years 1933 to 1989. Descriptive statistics of stage and discharge are shown in Table 9-3. The frequency histogram for the stage is shown in Fig. 9-18. The average, which is indicated on Table 9-3, is 2.84 m. The median stage is 2.71 m. The standard deviation and coefficient of variation are 0.71 m and 0.25, respectively. The maximum and minimum values (extreme values) are 5.60 and 1.69 m, respectively.

The frequency histogram for the discharge is shown in Fig. 9-19. The average, which is indicated on Table 9-3, is 54.82 cms. The median discharge is 34.8 cms. The standard deviation and coefficient of variation are 53.78 cms and 0.98, respectively. The maximum and minimum values (extreme values) are 351.10 and 15.90 cms, respectively.

By comparing the two histograms in Figs. 9-18 and 9-19, it can be observed that the relative dispersion in discharge is larger than that in stage. The coefficient of variation of discharge is 0.98, which is four times larger than the coefficient of variation of stage (0.25). Also, it can be observed that the frequency histogram for stage is almost bell shaped, whereas the histogram for discharge is highly skewed and appears as an exponential decay function. The closeness of the average stage (2.84 m) to the median stage (2.71 m) is an indication of a symmetric histogram. On the other hand, the average discharge (54.82 cms) is considerably different from the median discharge (34.8 cms), indicating a lack of symmetry in the discharge measurements.

TABLE 9-2 Stage and Discharge for the Little Patuxent River, Guilford, Maryland

Year	Stage (m)	Discharge (cms)	Year	Stage (m)	Discharge (cms)
1933	3.81	119.2	1962	2.78	36.2
1934	2.86	41.9	1963	2.25	23.2
1935	2.35	25.9	1964	2.7	32.8
1936	2.74	37.4	1965	2.35	24.8
1937	3.14	56.6	1966	2.72	34
1938	3.08	51.5	1967	2.71	33.4
1939	2.41	27.4	1968	2.86	39.9
1940	3.51	77.6	1969	2.02	19.4
1941	1.87	15.9	1970	2.02	19.4
1942	3.25	61.7	1971	3.49	86.9
1943	2.79	39.6	1972	5.6	351.1
1944	3.19	58.3	1973	3.07	53.2
1945	3.72	107.9	1974	2.65	31.4
1946	2.52	30.6	1975	4.08	152
1947	2.44	23.1	1976	3.08	54.1
1948	3.02	43.9	1977	2.14	21.3
1949	2.22	19.9	1978	3.63	103.9
1950	1.95	15.9	1979	3.91	132.5
1951	3.19	55.8	1980	2.65	31.4
1952	4.04	150.1	1981	2.5	28
1953	3.11	56.6	1982	2.18	22
1954	1.91	18.3	1983	3.33	78.7
1955	3.69	107.3	1984	2.65	42.2
1956	2.55	28.6	1985	2.6	40.5
1957	1.9	17.3	1986	1.69	19.3
1958	2.75	34.8	1987	2.42	34.8
1959	2.14	21.4	1988	2.42	34.8
1960	2.54	28.3	1989	4.01	143.5
1961	2.49	27.4			

TABLE 9-3   Descriptive Statistics of Stage and Discharge for the Little Patuxent River, Guilford, Maryland

Parameter	Stage (m)	Discharge (cms)
Average	2.84	54.82
Median	2.71	34.8
Mode	2.65	34.8
Standard deviation	0.71	53.78
Sample variance	0.505	2892.521
Coefficient of variation	0.25	0.98
Range	3.91	335.2
Minimum	1.69	15.9
Maximum	5.6	351.1
Count	57	57

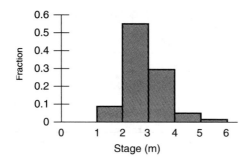

Figure 9-18   Frequency Histogram of the Stage of a River

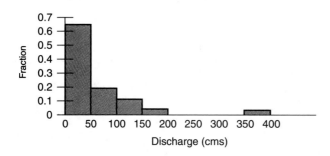

Figure 9-19   Frequency Histogram of the

9.7 PROBLEMS

9-1. For each of the measurement scales, identify five variables that are measured with the scale.

9-2. Using age as a variable of interest, identify a function that would be measured on each of the scales: ordinal, interval, and ratio.

9-3. Using the copper content of steel as a variable of interest, identify one function that would be measured for each of the four measurement scales.

9-4. For the eight methods of graphical analysis given in Section 9.3, develop a classification system for distinguishing among them. The classification system should center on basic, yet important, characteristics.

9-5. Using the data from Problem 9-12 and the following population totals, construct an area chart that shows the total and the breakdown of the total population in rural, suburban, and central city.

Year	1900	1910	1920	1930	1940	1950	1960	1970
Total ($\times 10^6$)	76	92	106	123	132	151	179	203

9-6. The following data are the solid waste (millions of tons) produced annually in the United States:

Municipal trash and garbage	150
Industrial	350
Mining	1700
Agriculture	2300

Construct a pie chart to present the data. Discuss the merits of presenting the data in a pie chart versus the tabular summary given.

9-7. The following data give the age distribution of United States citizens as a function of age group. Select a method for graphing the results. Interpret the results. Then combine the values into three meaningful groups and provide a graphical analysis of the distribution. How do the two analyses differ in their information content and emphasis?

Age group	< 5	5–9	10–14	15–19	20–24	25–29	30–34
Percentage	7.2	7.4	8.1	9.3	9.4	8.6	7.8

Age group	35–39	40–44	45–49	50–54	55–59	60–64	≥ 65
Percentage	6.2	5.2	4.9	5.2	5.1	4.5	11.3

9-8. A local highway department compiled the following percentages from accident records according to traffic control method (flashing red light, two-way stop signs, or four-way stop signs) and accident severity (loss of life, major damage, minor damage):

Traffic Control Method	Loss of Life	Major Damage	Minor Damage	Total
Flashing red light	23	41	36	100
Two-way stop signs	18	39	43	100
Four-way stop signs	12	21	67	100

    a. Present the data using three pie charts, one for each of the traffic control methods.

    b. Present all the data in a bar chart in a way that emphasizes differences between accident severity.

    c. Present all the data in a bar chart in a way that emphasizes differences between the traffic control methods.

    d. Present the data as a column chart.

    e. Discuss the advantages and disadvantages of each of the above graphical methods with respect to these data.

9-9. Compare the graphical analyses of Figs. 9-3, 9-5a, and 9-5b, and identify the circumstances under which each would be the most appropriate for making decisions.

9-10. The following data provide the estimated remaining strippable resources and reserves of bituminous coal in the United States (billions of short tons). Create a bar chart to display the data.

Alaska	0.9
Rocky Mountains and Northern Great Plains	1.1
Interior and Gulf provinces	32
Eastern province	27

9-11. Create column charts to present the following estimates of United States production of bituminous and lignite coal from surface and underground mines:

Year	Surface Mines	Underground Mines
1940	43	418
1950	123	393
1960	131	285
1970	264	339

Use one column chart to emphasize the temporal variation of total production. Use a second column chart to emphasize the proportion produced from surface mines. Use a third column chart to emphasize the total production from surface versus underground mines.

9-12. The following percentages indicate the change in rural, suburban, and central city United States populations from 1900 to 1970. Present the data graphically to emphasize the decline in the proportion of the population living in rural areas. Present the data graphically to illustrate any association between the change in central city population and the increase in the proportion living in suburban areas.

Year	Rural	Suburban	Central City
1900	58	16	26
1910	55	17	28
1920	51	18	31
1930	46	20	34
1940	46	22	32
1950	45	24	31
1960	40	30	30
1970	34	38	28

9-13. Plot the data shown in the following table using a column chart:

	Number of Constructed Bridges by Year		
Superstructure Type	1989	1990	1991
Steel	5	10	12
Concrete	10	6	7
Pre-stressed Concrete	4	6	5
Total	19	22	24

9-14. Using the bridge data in Problem 9-13, show the number of bridges as a function of year and by bridge type using a line chart, a pie chart, and three-dimensional surface chart.

9-15. The following concrete strength data (in ksi) were collected using an ultrasonic nondestructive testing method at different locations of an existing structure: 3.5, 3.2, 3.1, 3.5, 3.6, 3.2, 3.4, 2.9, 4.1, 2.6, 3.3, 3.5, 3.9, 3.8, 3.7, 3.4, 3.6, 3.5, 3.5, 3.7, 3.6, 3.8, 3.2, 3.4, 4.2, 3.6, 3.1, 2.9, 2.5, 3.5, 3.4, 3.2, 3.7, 3.8, 3.4, 3.6, 3.5, 3.2, 3.6, and 3.8. Plot a histogram and a frequency diagram for concrete strength.

9-16. For the data of Problem 9-15, determine the central tendency measures, that is, the average value, median, and mode.

9-17. For the data of Problem 9-15, determine the dispersion measures, that is, the variance, standard deviation and coefficient of variation.

9-18. Create a box-and-whisker plot of the data in Problem 9-19. Also create a frequency histogram of the data. Discuss and compare the information content of the two graphical analyses.

9-19. The following data are the maximum daily ozone concentrations for the months indicated. Graph the data in a way that will emphasize the monthly variation in the concentration. Also graph the data in a way that emphasizes the annual variation. Does the effect of monthly variation make it more difficult to assess the importance of the annual variation? Explain.

Year	Feb.	Apr.	June	Aug.	Oct.	Dec.
1980	61	72	77	83	64	55
1981	63	71	78	87	66	58
1982	64	72	78	86	66	59
1983	68	72	80	89	72	63
1984	74	73	85	94	68	66
1985	73	74	86	92	70	64
1986	67	73	85	90	67	64
1987	66	72	83	86	63	59
1988	62	68	81	87	65	57
1989	59	66	82	86	62	51
1990	56	65	80	86	60	50

10

Curve Fitting and Regression Analysis

10.1 INTRODUCTION

In many engineering problems, variation in the value of a variable is associated with variation in one or more additional variables. By establishing the relationship between two or more variables, it may be possible to reduce the uncertainty in an estimate of the variable of interest. Thus, before making a decision based on the analysis of observations on a random variable alone, it is wise to consider the possibility of systematic associations between the variable of interest and other variables that have a causal relationship with it. These associations among variables can be understood using correlation and regression analyses.

To summarize these concepts, the variation in the value of a variable may be either random or systematic in its relationship with another variable. Random variation represents uncertainty. If the variation is systematically associated with one or more other variables, the uncertainty in estimating the value of a variable can be reduced by identifying the underlying relationship. Correlation and regression analyses are important methods to achieve these objectives. They should be preceded by a graphical analysis to determine (1) if the relationship is linear or nonlinear, (2) if the relationship is direct or indirect, and (3) if there are any extreme events that might control the relationship.

10.2 CORRELATION ANALYSIS

Correlation analysis provides a means of drawing inferences about the strength of the relationship between two or more variables. That is, it is a measure of the degree to which the values of these variables vary in a systematic manner. Thus it provides a quantitative index of the degree to which one or more variables can be used to predict the values of another variable.

It is just as important to understand what correlation analysis cannot do as it is to know what it can do. Correlation analysis does not provide an equation for predicting the value of a variable. Also, correlation analysis does not indicate whether a relationship is causal; that is, it is necessary for the investigator to determine whether there is a cause-and-effect relationship between the variables. Correlation analysis only indicates whether the degree of common variation is significant.

Correlation analyses are most often performed after an equation relating one variable to another has been developed. This happens because the correlation coefficient, which is a quantitative index of the degree of common linear variation, is used as a measure of the goodness of fit between the prediction equation and the data sample used to derive the prediction equation. But correlation analyses can be very useful in model formulation, and every model development should include a correlation analysis.

10.2.1 Graphical Analysis

The first step in examining the relationship between variables is to perform a graphical analysis. Visual inspection of the data can identify the following information:

1. Degree of common variation, which is an indication of the degree to which the two variables are related
2. Range and distribution of the sample data points
3. Presence of extreme events
4. Form of the relationship between the two variables
5. Type of relationship

Each of these factors is of importance in the analysis of sample data and in decision making.

When variables show a high degree of association, we assume that there is a causal relationship. If there is a physical reason to suspect that a causal relationship exists, the association demonstrated by the sample data provides empirical support for the assumed relationship. Common variation implies that, when the value of one of the variables is changed, the value of the other variable will change in a systematic manner. For example, an increase in the value of one variable occurs when the value of the other variable increases. If the change in the one variable is highly predictable from a given change in the other variable, there is a high degree of common variation. Figure 10-1 shows graphs of different samples of data for two variables having different degrees of common variation. The common variation is measured in the figure using the correlation coefficient (R). In Fig. 10-1a and e the degree of common variation is very high; thus the variables are said to be correlated. In Fig. 10-1c there is no correlation between the two variables because, as the value of X is increased, it is not certain whether Y will increase or decrease. In Fig. 10-1b and d, the degree of correlation is moderate; in Fig. 10-1b, evidently Y will increase as X is increased, but the exact change in Y for a change in X is difficult to estimate

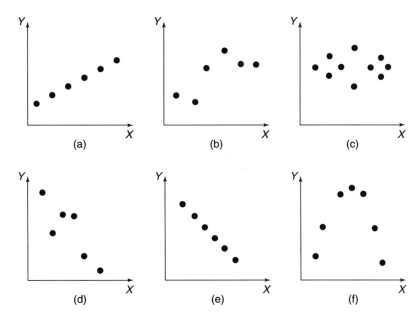

Figure 10-1   Different Degrees of Correlation between Variables (X and Y): (a) $R = 1.0$;
(b) $R = 0.5$; (c) $R = 0.0$; (d) $R = -0.5$; (e) $R = -1.0$; (f) $R = 0.3$

from just looking at the graph. A more quantitative description of the concept of common variation is needed.

It is important to use a graphical analysis to identify the range and distribution of the sample data points so that the stability of the relationship can be assessed and so that we can assess the ability of the data sample to represent the distribution of the population. If the range of the data is limited, a computed relationship may not be stable; that is, it may not apply to the distribution of the population. The data set of Fig. 10-2 shows a case where the range of the sample is much smaller than the expected range of the population. If an attempt is made to use the sample to project the relationship between the two random variables, a small change in the slope of the relationship causes a large change in the predicted estimate of Y for values of X at the extremes of the range of the population. A graph of two variables might alert the investigator to a sample in which the range of the sample data may cause stability problems in a derived relationship between two random variables, especially when the relationship is projected beyond the range of the sample data.

It is important to identify extreme events in a sample of data for several reasons. First, extreme events can dominate a computed relationship between two variables. For example, in Fig. 10-3a, the extreme point suggests a high correlation between X and Y; in this case, the cluster of points acts like a single observation. In Fig. 10-3b, the extreme point causes a poor correlation between the two random variables; since the mean of the cluster of points is nearly the same as the value of Y of the extreme point, the data of Fig. 10-3b suggest that a change in X is not associated with a change in Y. A correlation coefficient is more sensitive to an extreme point when the sample size is small. An extreme

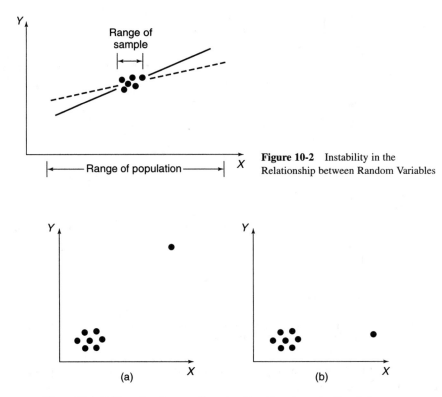

Figure 10-2   Instability in the
Relationship between Random Variables

Figure 10-3   Effect of an Extreme Event in a Data Sample on the Correlation:
(a) High Correlation; (b) Low Correlation

event may be due to (1) errors in recording or plotting the data or (2) a legitimate observation in the tail of the distribution. Therefore, an extreme event must be identified and
the reason for the event determined. Otherwise, it is not possible to properly interpret the
results of the correlation analysis.

Relationships can be linear or nonlinear. Since the statistical methods to be used for
the two forms of a relationship differ, it is important to identify the form. Additionally,
the most frequently used correlation coefficient depends on a linear relationship existing
between the two random variables; thus low correlation may result for a nonlinear relationship even when the relationship is obvious. For example, the bivariate relationship of
Fig. 10-1f suggests a very predictable trend in the relationship between Y and X; but the
correlation is poor and is certainly not as good as that in Fig. 10-1a, even though the two
plots suggest equal levels of predictability.

Graphs relating pairs of variables can be used to identify the type of the relationship. Linear trends can be either direct or indirect, with an indirect relationship indicating
a decrease in Y as X increases. This information is useful for checking the rationality of

the relationship, especially when dealing with data sets that include more than two variables. A variable that is not dominant in the physical relationship may demonstrate a physically irrational relationship with another variable because of the values of the other variable affecting the physical relationship.

10.2.2 Bivariate Correlation

Correlation is the degree of association between the elements of two samples of data, that is, between observations on two variables. Correlation coefficients provide a quantitative index of the degree of linear association. Examples of variables that are assumed to have a causal relationship and significant correlation are (1) the cost of living and wages and (2) the volumes of rainfall and flood runoff. However, examples that have no cause-and-effect relationship but may have significant correlation are (1) the crime rate and the sale of chewing gum over the past two decades and (2) annual population growth rates in nineteenth-century France and annual cancer death rates in the United States in the twentieth century.

Many indexes of correlation exist. The method that is used most frequently is the *Pearson product–moment* correlation coefficient. Nonparametric correlation indexes include the contingency coefficient, the Spearman rank correlation coefficient, and the Kendall rank correlation coefficient. Only the Pearson correlation coefficient is considered in this chapter.

10.2.3 Separation of Variation

A set of observations on a variable Y has a certain amount of variation, which may be characterized by the variance of the sample of size n. The variance equals the sum of squares of the deviations of the observations from the mean of the observations divided by the degrees of freedom $(n - 1)$. Ignoring the degrees of freedom, the variation in the numerator can be separated into two parts: (1) variation associated with a second variable X and (2) variation not associated with X. That is, the total variation (TV), which equals the sum of the squares of the sample data points about the mean of the data points, is separated into the variation that is explained by variation in the second variable (EV) and the variation that is not explained, that is, the unexplained variation, UV. Thus TV can be expressed as

$$TV = EV + UV \tag{10-1}$$

Using the general form of the variation of a random variable, each of the three terms in this equation can be represented by a sum of squares as follows:

$$\sum_{i=1}^{n} (Y_i - \overline{Y})^2 = \sum_{i=1}^{n} (\hat{Y}_i - \overline{Y})^2 + \sum_{i=1}^{n} (\hat{Y}_i - Y_i)^2 \tag{10-2}$$

where Y = an observation on the random variable, $\hat{Y}$ = the value of Y estimated from the best linear relationship with the second variable X, and $\bar{Y}$ = the mean of the observations on Y. These variation components are shown in Fig. 10-4; the dashed lines illustrate the deviations of Eq. 10-2. Figure 10-4a shows the variation of the observations about the mean value; this represents the variation that exists in the sample and that may potentially be explained by the variable X; it reflects the left side of Eq. 10-2. The variation in Fig. 10-4b represents the variation of the line about the mean and thus corresponds to the first term of the right side of Eq. 10-2. If all the sample points fall on the line, the explained variation will equal the total variation. The variation in Fig. 10-4c represents the variation of the points about the line; therefore, it corresponds to the second term on the right side of Eq. 10-2. This is the variation that is not explained by the relationship between Y and X. If this unexplained variation equals the total variation, the line does not explain any of the variation in the sample, and the relationship between X and Y is not significant.

10.2.4 Correlation: Fraction of Explained Variation

The separation of variation concept is useful for quantifying the correlation coefficient by Pearson. Specifically, dividing both sides of Eq. 10-1 by the total variation TV gives

$$1 = \frac{EV}{TV} + \frac{UV}{TV} \tag{10-3}$$

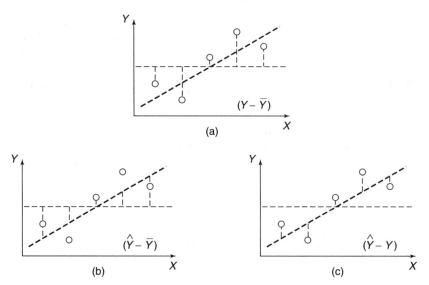

Figure 10-4   Separation of Variation: (a) Total Variation; (b) Explained Variation; (c) Unexplained Variation

The ratio EV/TV represents the fraction of the total variation that is explained by the linear relationship between Y and X; this is called the *coefficient of determination* and is given by

$$R^2 = \frac{EV}{TV} = \frac{\sum_{i=1}^{n}(\hat{Y}_i - \bar{Y})^2}{\sum_{i=1}^{n}(Y_i - \bar{Y})^2} \tag{10-4}$$

The square root of the ratio is the correlation coefficient, R. If the explained variation equals the total variation, the correlation coefficient will equal 1. If the relationship between X and Y is inverse, as in Fig. 10-1e, and the explained variation equals the total variation in magnitude, R will equal -1. These represent the extremes, but both values indicate a perfect association, with the sign only indicating the direction of the relationship. If the explained variation equals zero, R equals zero. Thus a correlation coefficient of zero, which is sometimes called the null correlation, indicates no linear association between the two variables X and Y.

10.2.5 Computational Form for Correlation Coefficient

While Eq. 10-4 provides the means to compute a value of the correlation coefficient, it can be shown that Eq. 10-4 can be rearranged to the following form using a linear model for $\hat{Y}$ that minimizes the unexplained variation (UV):

$$R = \frac{\sum_{i=1}^{n} X_i Y_i - \frac{1}{n}\left(\sum_{i=1}^{n} X_i\right)\left(\sum_{i=1}^{n} Y_i\right)}{\sqrt{\sum_{i=1}^{n} X_i^2 - \frac{1}{n}\left(\sum_{i=1}^{n} X_i\right)^2}\sqrt{\sum_{i=1}^{n} Y_i^2 - \frac{1}{n}\left(\sum_{i=1}^{n} Y_i\right)^2}} \tag{10-5}$$

This equation is used most often because it does not require prior computation of the means, and thus the computational algorithm is easily programmed for a computer. The linear model $\hat{Y}$ that minimizes UV is based on the principle of least squares, as discussed in Section 10.4.

Example 10-1 Computational Illustration for Correlation Coefficient

A sample of five observations, which is given as the values of X and Y in Table 10-1, is used to illustrate the computational methods outlined herein. The last row of Table 10-1 contains column summations. Using Eq. 10-4, the correlation is

$$R = \sqrt{\frac{8.10}{10}} = 0.9 \tag{10-6}$$

TABLE 10-1   Calculation of the Correlation Coefficient

X	Y	$X - \bar{X}$	$Y - \bar{Y}$	$(X - \bar{X})$ $(Y - \bar{Y})$	$\hat{Y}$	$\hat{Y} - \bar{Y}$	$(Y - \bar{Y})^2$	$(Y - \bar{Y})^2$	XY	X^2	Y^2
1	2	−2	−1	2	1.2	−1.8	3.24	1	2	1	4
2	1	−1	−2	2	2.1	−0.9	0.81	4	2	4	1
3	3	0	0	0	3.0	0.0	0.00	0	9	9	9
4	4	1	1	1	3.9	0.9	0.81	1	16	16	16
5	5	2	2	4	4.8	1.8	3.24	4	25	25	25
15	15			9			8.10	10	54	55	55

The predicted values, $\hat{Y}$, were obtained by fitting a regression equation to the data points; the regression method is discussed in the remaining sections of this chapter. Therefore, Eq. 10-5 yields

$$R = \frac{54 - \dfrac{15(15)}{5}}{\sqrt{55 - \dfrac{15(15)}{5}}\sqrt{55 - \dfrac{15(15)}{5}}} = 0.9 \qquad (10\text{-}7)$$

Thus 81% (based on $0.9^2 = 0.81$) of the total variation in Y is explained by variation in X.

10.3 INTRODUCTION TO REGRESSION

In many engineering projects, estimates of variables are required. For example, in the design of irrigation projects, it is necessary to provide estimates of evaporation. If we are fortunate to have a past record of daily evaporation rates from a pan, the mean of these observations multiplied by some proportionality constant (a pan coefficient) may be our best estimate. However, the error in such an estimate can be significant. Because the standard deviation of the observations is a measure of the variation in the past, it is our best estimate of the future variation and thus the error in a single measurement. The error in a mean value is also a function of the standard deviation.

If a variable can be related to other variables, it may be possible to reduce the error in a future estimate. For example, evaporation is a function of the temperature and humidity of the air mass that is over the water body. Thus, if measurements of the air temperature and the relative humidity are also available, a relationship, or model, can be developed. The relationship may provide a more accurate estimate of evaporation for the conditions that may exist in the future.

The process of deriving a relationship between a random variable and measured values of other variables is called *optimization*, or *model calibration*. It makes use of the fundamental concepts of calculus and numerical analysis. The objective of optimization is to find the values of a vector of unknowns that provide the minimum or maximum value of some function. Before examining regression analysis, it is necessary to discuss the fundamentals of statistical optimization.

10.3.1 Elements of Statistical Optimization

In statistical optimization, the function to be optimized is called the *objective function*, which is an explicit mathematical function that describes what is considered the optimal solution. There is a mathematical model that relates a random variable, called the *criterion* or *dependent variable*, to a vector of unknowns, called regression coefficients, and a vector of *predictor variables*, which are sometimes called *independent variables*. The predictor variables are usually variables that have a causal relationship with the criterion variable. For example, if we are interested in using measurements of air temperature and humidity to predict evaporation rates, the evaporation would be the criterion variable and the temperature and relative humidity would be the predictor variables. While the objective function of statistical optimization corresponds to the explicit function, the vector of unknowns in statistical optimization corresponds to the unknowns in the explicit function. The unknowns are the values that are necessary to transform values of the predictor variable(s) into a predicted value of the criterion variable. It is important to note that the objective function and the mathematical model are two separate explicit functions. The third element of statistical optimization is a data set. The *data set* consists of measured values of the criterion variable and the predictor variable(s). In summary, statistical optimization requires (1) an objective function, which defines what is meant by best fit; (2) a mathematical model, which is an explicit function relating a criterion variable to vectors of unknowns and predictor variable(s); and (3) a matrix of measured data.

As an example, we may attempt to relate evaporation (E), the criterion variable, to temperature (T), which is the predictor variable, using the equation

$$\hat{E} = b_0 + b_1 T \tag{10-8}$$

in which b_0 and b_1 = the unknown coefficients, and $\hat{E}$ = the predicted value of E. To evaluate the unknowns b_0 and b_1, a set of simultaneous measurements on E and T would be made. For example, if we were interested in daily evaporation rates, we may measure both the total evaporation for each day in a year and the corresponding mean daily temperature; this would give us 365 observations from which we could estimate the two regression coefficients. An objective function would have to be selected to evaluate the unknowns; for example, regression that minimizes the sum of the squares of the differences between the predicted and measured values of the criterion variable. Other criterion functions, however, may be used.

10.3.2 Zero-intercept Model

Assume that the criterion variable Y is related to a predictor variable X using the linear model

$$\hat{Y} = bX \tag{10-9}$$

in which b = the unknown and $\hat{Y}$ = the predicted value of Y. The objective function, F, is defined as minimizing the sum of the squares of the differences (e_i) between the predicted values $(\hat{Y}_i)$ and the measured values (Y_i) of Y as follows:

$$F = \min \sum_{i=1}^{n} e_i^2 = \min \sum_{i=1}^{n} (\hat{Y}_i - Y_i)^2 \tag{10-10}$$

where n = the sample size.

To derive a value for b, differentiate $\sum_{i=1}^{n} e_i^2$ with respect to the unknown, set the derivative equal to zero, and solve for the unknown. Substituting Eq. 10-9 into Eq. 10-10 yields

$$F = \min \sum_{i=1}^{n} (\hat{Y}_i - Y_i)^2 = \min \sum_{i=1}^{n} (bX_i - Y_i)^2 \tag{10-11}$$

The derivative of $\sum_{i=1}^{n} e_i^2$ with respect to the unknown b is

$$\frac{\partial \sum_{i=1}^{n} e_i^2}{\partial b} = 2 \sum_{i=1}^{n} (bX_i - Y_i)(X_i) \tag{10-12}$$

Setting Eq. 10-12 to 0 and solving for b yields

$$b = \frac{\sum_{i=1}^{n} X_i Y_i}{\sum_{i=1}^{n} X_i^2} \tag{10-13}$$

The value of b can be computed using the two summations, $\sum_{i=1}^{n} X_i Y_i$ and $\sum_{i=1}^{n} X_i^2$.

Example 10-2 Fitting a Zero-intercept Model

The following table gives the data that can be used for fitting the value of b:

i	X_i	Y_i
1	3	2
2	5	3
3	6	4

The following table shows the necessary computations:

i	X_i	X_i^2	Y_i	X_iY_i
1	3	9	2	6
2	5	25	3	15
3	6	36	4	24
Column summation		70		45

Therefore, the value of b equals $\frac{45}{70}$ or 0.64286. The following equation could then be used to derive a future estimate of Y for a future value of X:

$$\hat{Y} = 0.64286X \tag{10-14}$$

It is important to note that Eq. 10-13 is valid only for the linear model of Eq. 10-9 that does not include an intercept; Eq. 10-13 would not be valid for Eq. 10-8, since Eq. 10-8 includes an intercept coefficient (b_0).

10.3.3 Regression Definitions

The objective of regression is to evaluate the coefficients of an equation relating the criterion variable to one or more other variables, which are called the *predictor variables*. The predictor variables are variables in which their variation is believed to cause or agree with variation in the criterion variable. A predictor variable is often called an *independent variable*; this is a misnomer in that independent variables are usually neither independent of the criterion variable nor independent of other predictor variables.

The most frequently used linear model relates a criterion variable Y to a single predictor variable X by the equation

$$\hat{Y} = b_0 + b_1X \tag{10-15}$$

in which b_0 = the intercept coefficient and b_1 = the slope coefficient; b_0 and b_1 are called *regression coefficients* because they are obtained from a regression analysis. Because Eq. 10-15 involves two variables, Y and X, it is sometimes referred to as the *bivariate model*. The intercept coefficient represents the value of Y when X equals zero. The slope coefficient represents the rate of change in Y with respect to change in X. Whereas b_0 has the same dimensions as Y, the dimensions of b_1 equal the ratio of the dimensions of Y to X.

The linear multivariate model relates a criterion variable to two or more predictor variables:

$$\hat{Y} = b_0 + b_1X_1 + b_2X_2 + \cdots + b_pX_p \tag{10-16}$$

in which p = the number of predictor variables, X_i = the ith predictor variable, b_i = the ith slope coefficient, and b_0 is the intercept coefficient, where $i = 1, 2, \ldots, p$. The coefficients b_i are often called *partial regression coefficients* and have dimensions equal to the ratio of the dimensions of Y to X_i. Equation 10-15 is a special case of Eq. 10-16.

10.4 PRINCIPLE OF LEAST SQUARES

10.4.1 Definitions

The values of the slope and intercept coefficients of Eqs. 10-15 and 10-16 can be computed using the principle of least squares. The principle of least squares is a process of obtaining *best* estimates of the coefficients and is referred to as a regression method. Regression is the tendency for the expected value of one of two jointly correlated random variables to approach more closely the mean value of its set than any other. The principle of least squares is used to regress Y on either X or the X_i values of Eqs. 10-15 and 10-16, respectively. To express the principle of least squares, it is important to define the error, e, or residual, as the difference between the predicted and measured values of the criterion variable:

$$e_i = \hat{Y}_i - Y_i \tag{10-17}$$

in which $\hat{Y}_i$ = the ith predicted value of the criterion variable, Y_i = the ith measured value of Y, and e_i = the ith error. It is important to note that the error is defined as the measured value of Y subtracted from the predicted value. Some computer programs use the measured value minus the predicted value. This definition is avoided because it indicates that a positive residual implies underprediction. With Eq. 10-17 a positive residual indicates overprediction, while a negative residual indicates underprediction. The objective function for the principle of least squares is to minimize the sum of the squares of the errors:

$$F = \min \sum_{i=1}^{n} (\hat{Y}_i - Y_i)^2 \tag{10-18}$$

in which n = the number of observations on the criterion variable, that is, the sample size.

10.4.2 Solution Procedure

After inserting the model of $\hat{Y}$, the objective function of Eq. 10-18 can be minimized by taking the derivatives with respect to each unknown, setting the derivatives equal to zero, and then solving for the unknowns. The solution requires the model for predicting Y_i to be substituted into the objective function. It is important to note that the derivatives are taken with respect to the unknowns b_i and not the predictor variables X_i.

To illustrate the solution procedure, the model of Eq. 10-15 is substituted into the objective function of Eq. 10-18, which yields

$$F = \min \sum_{i=1}^{n} (b_0 + b_1 X_i - Y_i)^2 \tag{10-19}$$

The derivatives of the sum of squares of the errors with respect to the unknowns b_0 and b_1 are, respectively,

$$\frac{\partial \Sigma_{i=1}^{n} (\hat{y}_i - y_i)^2}{\partial b_0} = 2 \sum_{i=1}^{n} (b_0 + b_1 X_i - Y_i) = 0 \qquad (10\text{-}20a)$$

$$\frac{\partial \Sigma_{i=1}^{n} (\hat{y}_i - y_i)^2}{\partial b_1} = 2 \sum_{i=1}^{n} (b_0 + b_1 X_i - Y_i) = 0 \qquad (10\text{-}20b)$$

Dividing each equation by 2, separating the terms in the summations, and rearranging yield the set of normal equations:

$$nb_0 + b_1 \sum_{i=1}^{n} X_i = \sum_{i=1}^{n} Y_i \qquad (10\text{-}21a)$$

$$b_0 \sum_{i=1}^{n} X_i + b_1 \sum_{i=1}^{n} X_i^2 = \sum_{i=1}^{n} X_i Y_i \qquad (10\text{-}21b)$$

All the summations in Eq. 10-21 are calculated over all values of the sample. Sometimes the index values of the summation are omitted when they refer to summation over all elements in a sample of size n. Also, the subscripts for X and Y can be omitted but are inferred. The two unknowns b_0 and b_1 can be evaluated by solving the two simultaneous equations as follows:

$$b_1 = \frac{\sum_{i=1}^{n} X_i Y_i - \frac{1}{n} \sum_{i=1}^{n} X_i \sum_{i=1}^{n} Y_i}{\sum_{i=1}^{n} X_i^2 - \frac{1}{n} \left(\sum_{i=1}^{n} X_i \right)^2} \qquad (10\text{-}22a)$$

$$b_0 = \bar{Y} - b_1 \bar{X} = \frac{\sum_{i=1}^{n} Y_i}{n} - \frac{b_1 \sum_{i=1}^{n} X_i}{n} \qquad (10\text{-}22b)$$

Example 10-3 Least-squares Analysis

The data of Table 10-2 can be used to illustrate the solution of the least-squares principle for a data set of five observations on two variables. Substituting the sums of Table 10-2 into the normal equations of Eq. 10-21 gives

$$5b_0 + 50b_1 = 10 \qquad (10\text{-}23a)$$

$$50b_0 + 510b_1 = 109 \qquad (10\text{-}23b)$$

Solving Eqs. 10-23 for the unknowns (b_0 and b_1) yields the model

$$\hat{Y} = -7 + 0.9X \qquad (10\text{-}24)$$

The errors, which are shown in Table 10-2, can be computed by substituting the measured values of X into Eq. 10-24. In solving a linear model using the least-squares principle, the sum of the errors always equals zero, which is shown in Table 10-2. The sum of the squares of the errors equals 1.9, which is a minimum.

TABLE 10-2 Least-squares Computations

i	X_i	Y_i	X_i^2	Y_i^2	X_iY_i	e_i	e_i^2
1	12	4	144	16	48	-0.2	0.04
2	11	3	121	9	33	-0.1	0.01
3	10	2	100	4	20	0.0	0.00
4	9	0	81	0	0	1.1	1.21
5	8	1	64	1	8	-0.8	0.64
Column summation	50	10	510	30	109	0.0	1.90

Note that if a line is passed through the first three observations of Table 10-2 the error for each of these three observations is zero. Such a model would be

$$\hat{Y} = -8 + 1.0X \tag{10-25}$$

While three of the residuals are zero, the sum of the squares of the errors for Eq. 10-25 equals 2.0, which is greater than the corresponding sum for Eq. 10-24. Thus the least-squares solution always provides the straight line with the smallest sum of squares of the errors.

10.5 RELIABILITY OF THE REGRESSION EQUATION

Having evaluated the coefficients of the regression equation, it is of interest to evaluate the reliability of the regression equation. The following selected criteria can be assessed in evaluating the model: (1) the correlation coefficient, (2) the standard error of estimate, and (3) the rationality of the coefficients and the relative importance of the predictor variables, both of which can be assessed using the standardized partial regression coefficients.

10.5.1 Correlation Coefficient

As suggested in Section 10.2, the correlation coefficient (R) is an index of the degree of linear association between two random variables. The magnitude of R indicates whether the regression provides accurate predictions of the criterion variable. Thus R is often computed before the regression analysis is performed in order to determine whether it is worth the effort to perform the regression. However, R is always computed after the regression analysis because it is an index of the goodness of fit. The correlation coefficient measures the degree to which the measured and predicted values agree and is used as a measure of the accuracy of future predictions. It must be recognized that if the measured data are not representative of the population, that is, data that will be observed in the future, the correlation coefficient cannot be indicative of the accuracy of future predictions.

The square of the correlation coefficient (R^2) equals the percentage of the variance in the criterion variable that is explained by the predictor variable. Because of this physical interpretation, R^2 is a meaningful indicator of the accuracy of predictions.

10.5.2 Standard Error of Estimate

In the absence of additional information, the mean is the best estimate of the criterion variable; the standard deviation S_Y of Y is an indication of the accuracy of prediction. If Y is related to one or more predictor variables, the error of prediction is reduced from S_Y to the standard error of estimate, S_e. Mathematically, the standard error of estimate equals the standard deviation of the errors, has the same units as Y, and is given by

$$S_e = \sqrt{\frac{1}{v} \sum_{i=1}^{n} (\hat{Y}_i - Y_i)^2}$$ (10-26)

in which v = the degrees of freedom, which equals the sample size minus the number of unknowns estimated by the regression procedure. For the bivariate model of Eq. 10-15, $p = 1$ and $v = n - 2$. For the general linear model with an intercept, Eq. 10-16, there are $(p + 1)$ unknowns; thus $v = n - p - 1$. In terms of the separation of variation concept discussed previously, the standard error of estimate equals the square root of the ratio of the unexplained variation to the degrees of freedom. It is important to note that S_e is based on $(n - p - 1)$ degrees of freedom, while the error S_Y is based on $(n - 1)$ degrees of freedom. Thus in some cases S_e may be greater than S_Y. To assess the reliability of the regression equation, S_e should be compared with the bounds of zero and S_Y. If S_e is near S_Y, the regression has not been successful. If S_e is much smaller than S_Y and is near zero, the regression analysis has improved the reliability of prediction.

The standard error of estimate is sometimes computed using the following relationship:

$$S_e = S_Y \sqrt{1 - R^2}$$ (10-27)

Equation 10-27 must be considered as only an approximation to Eq. 10-26 because R is based on n degrees of freedom and S_Y is based on $(n - 1)$ degrees of freedom.

Using the separation-of-variation concept, the exact relationship between S_e, S_Y, and R can be computed as follows:

$$TV = EV + UV$$ (10-28)

The total variation is related to the variance of Y by

$$S_y^2 = \frac{TV}{n - 1}$$ (10-29)

The square of the correlation coefficient is the ratio of the explained variation (EV) to the total variation (TV):

$$R^2 = \frac{EV}{TV} \tag{10-30}$$

The standard error of estimate is related to the unexplained variation (UV) by

$$S_e^2 = \frac{UV}{n - p - 1} \tag{10-31}$$

Equation 10-29 can be solved for TV, which can then be substituted into both Eqs. 10-28 and 10-30. Equation 10-30 can then be solved for EV, which can be substituted into Eq. 10-28. Equation 10-31 can be solved for UV, which is also substituted into Eq. 10-28. Solving for S_e^2 from Eq. 10–28 yields

$$S_e^2 = \frac{n - 1}{n - p - 1} S_y^2 (1 - R^2) \tag{10-32}$$

Thus Eq. 10-32 is a more exact relationship than Eq. 10-27. However, for large sample sizes the difference between the two estimates is small.

Although S_e may actually be greater than S_Y, in general, S_e is within the range from zero to S_Y. When a regression equation fits the data points exactly, S_e equals zero; this corresponds to a correlation coefficient of 1. When the correlation coefficient equals zero, S_e equals S_Y. As was indicated previously, S_e may actually exceed S_Y when the degrees of freedom have a significant effect. The standard error of estimate is often preferred to the correlation coefficient because S_e has the same units as the criterion variable and its magnitude is a physical indicator of the error; the correlation coefficient is only a standardized index and does not properly account for degrees of freedom lost because of the regression coefficients.

10.5.3 Standardized Partial Regression Coefficients

Because the partial regression coefficient, b_1 of Eq. 10-15, is dependent on the units of both Y and X, it is often difficult to use its magnitude to measure the rationality of the model. The partial regression coefficient can be standardized by

$$t = \frac{b_1 S_X}{S_Y} \tag{10-33}$$

in which t is called the standardized partial regression coefficient, and S_X and S_Y are the standard deviations of the predictor and criterion variables, respectively. Because the units of b_1 are equal to the ratio of the units of Y to the units of X, t is dimensionless. For the bivariate regression model, t equals the correlation coefficient of the regression of the standardized value of Y on the standardized value of X. This suggests that t is a measure of the relative importance of the corresponding predictor variable. The value of t has the same sign as b_1. Therefore, if the sign of t is irrational, we can conclude that the model is not rational. Although it is more difficult to assess the rationality of the magnitude of a regression coefficient than it is to assess the rationality of the sign, the magnitude of the standardized partial regression coefficient can be used to assess the rationality of the model. For rational models, t must vary in the range $-1 \leq t \leq 1$. Because t equals R for the bivariate model of Eq. 10-15, values of t outside this range should not occur; this is not true for multivariate models, such as Eq. 10-16.

Example 10-4 Errors in Regression Models

The data of Table 10-2 were previously used to compute the standard error of estimate and the correlation coefficient; they can also be used to illustrate the computation of the standardized partial regression coefficient, which is computed using Eq. 10-33 as

$$t = 0.9\sqrt{\frac{510 - \dfrac{(50)^2}{5}}{30 - \dfrac{(10)^2}{5}}} = 0.9 \tag{10-34}$$

Using the equations given previously for computing the correlation coefficient, it is easy to show that t equals R.

10.5.4 Assumptions Underlying the Regression Model

The principle of least squares assumes that the errors, that is, the differences between the predicted and measured values of the criterion variable, (1) are independent of each other, (2) have zero mean, (3) have a constant variance across all values of the predictor variables, and (4) are normally distributed. If any of these assumptions are violated, we must assume that the model structure is not correct. Violations of these assumptions are easily identified using statistical analyses of the residuals.

When the sum of the residuals does not equal zero, it reflects a bias in the model. The regression approach, when applied analytically, requires the sum of the residuals for a linear model to equal 0. For a nonlinear model, the sum of the errors may not equal 0, which suggests bias. When an inadequate number of significant digits is computed for the regression coefficients, the sum of the residuals may not equal 0, even for a linear model. However, a model may be biased even when the sum of the residuals equals zero. For example, Fig. 10-5 shows an X–Y plot in which the trend of the data is noticeably nonlinear, with the linear regression line also shown. If the errors e_i are computed and plotted versus the corresponding values of the predictor variable, a noticeable trend appears (Fig. 10-6). The errors are positive for both low and high values of X, while the errors show a negative bias of prediction for intermediate values of X. While the sum of the residuals equals 0, the trends in the residuals suggest a biased model that should be replaced by a model that has a different structure. Figure 10-7 shows data with a trend in the error variance; specifically, as X increases, the error variance increases.

The residuals (i.e., $\hat{Y}_i - Y_i$) are an important criterion in assessing the validity of a bivariate regression equation. Plots of the residuals may indicate an incorrect model structure, such as a nonlinear relationship between the criterion variable and one of the

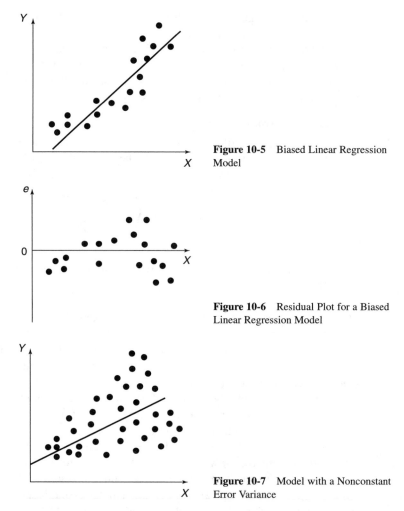

Figure 10-5   Biased Linear Regression Model

Figure 10-6   Residual Plot for a Biased Linear Regression Model

Figure 10-7   Model with a Nonconstant Error Variance

predictor variables. Additionally, the residuals may suggest a need to consider another predictor variable that was not included in the model, requiring the use of multiple regression (see Section 10.8). A plot of the residuals should be made to assess the assumptions of zero mean, constant error variation, independence, and normality. A plot of the residuals versus the predicted value of the criterion variable (i.e., e_i versus $\hat{Y}_i$) may identify either a violation of the assumption of constant variance or a lack of independence of the observations.

The fourth assumption deals with the independence of the observations. This is not usually a problem except when time or space is the predictor variable. In such cases, the measurement of Y for one time period may not be completely independent of the measurement for an adjacent time period.

10.6 CORRELATION VERSUS REGRESSION

Before discussing other topics in regression, it is important to emphasize the differences between correlation and regression. Regression is a means of calibrating the unknown coefficients of a prediction equation; whereas correlation provides a measure of goodness of fit. Thus regression is a method for model calibration, while correlation would have usefulness in model formulation and model verification. When using regression, it is necessary to specify which variable is the criterion and which is the predictor; when using correlation, it is not necessary to make such a distinction. The distinction is necessary with regression because a regression equation is not transformable, unless the correlation coefficient equals 1.0. That is, if Y is the criterion variable when the equation is calibrated, the equation cannot be rearranged algebraically to get an equation for predicting X. Specifically, Eq. 10-16 can be rearranged to predict X as

$$Y = b_0 + b_1 X \qquad (10\text{-}35a)$$

$$Y - b_0 = b_1 X \qquad (10\text{-}35b)$$

Therefore,

$$X = \frac{Y - b_0}{b_1} \qquad (10\text{-}35c)$$

or

$$X = -\frac{b_0}{b_1} + \frac{1}{b_1} Y \qquad (10\text{-}35d)$$

$$X = a_0 + a_1 Y \qquad (10\text{-}35e)$$

If X is regressed on Y, the resulting coefficients will not be the same as the coefficients obtained by regressing Y on X (Eq. 10-35a) and then setting $a_0 = -b_0/b_1$ and $a_1 = 1/b_1$. The correlation for the regression of Y on X is the same as the correlation for the regression of X on Y.

Computationally, the correlation coefficient is a function of the explained and total variation; the slope coefficient of a regression equation is related to the correlation coefficient by

$$b_1 = \frac{RS_Y}{S_X} \qquad (10\text{-}36)$$

in which S_X and S_Y are the standard deviations of X and Y, respectively.

After a regression equation is calibrated, it is very important to examine the rationality of the regression coefficients; specifically, is the predicted value of Y rational for all reasonable values of X? Since the slope coefficient of the regression equation represents the rate of change of Y with respect to X, is the effect of X on Y rational? A regression equation that is not considered rational should be used with caution, if at all. In addition to checking for rationality, the goodness-of-fit statistics, R and S_e, should be computed. If the expected accuracy is not acceptable, we may elect to collect more data or to develop a model that uses other predictor variables.

10.7 APPLICATIONS OF BIVARIATE REGRESSION ANALYSIS

10.7.1 Estimating Trip Rate

A traffic planner needs estimates of trip rates for residential areas. The planner conducts studies in 10 residential areas of different densities of development, with the following measured rates:

X	3	4	8	8	13	15	18	22	24	27
Y	4.5	3.3	3.5	2.3	3.8	2.6	2.7	1.6	1.9	1.7

where X = residential density (households per acre) and Y = trip rate (daily trips per household). The regression of trip rate on density yields the following linear equation:

$$Y = 4.100 - 0.09226X \qquad (10\text{-}37)$$

The negative slope coefficient indicates that the trip rate decreases by 0.092 daily trips per household for each unit increase in the number of households per acre. The standard error of estimate is 0.603, which is an improvement over the standard deviation of 0.965 ($S_e/S_y = 0.625$). The correlation coefficient is -0.808, which yields a normalized explained variance (R^2) of 0.653. Figure 10-8 shows the measured data and the regression line; the moderate variation of the measured points about the line confirms the moderate correlation between residential density and trip rate.

10.7.2 Breakwater Cost

A construction engineering firm has recently had a number of contracts for the construction of breakwaters in coastal environments. To reduce the effort in developing cost estimates for future clients, the company compiled a data record of 14 recent breakwater

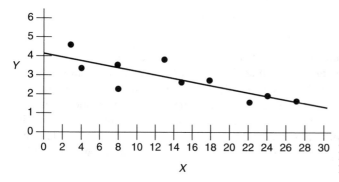

Figure 10-8 Bivariate Regression of Trip Rate (Y, Daily Trips per Household) on Residential Density (X, Households per Acre)

construction projects. The data consist of the length of the breakwater and the cost for rock, rock spurs, groins, marsh plants, and beach fill (see Table 10-3). The cost (Y) was regressed on the length of the breakwater (X):

$$\hat{Y} = 253.4 + 0.08295X \tag{10-38}$$

The standard error of estimate is $\$149.1 \times 10^3$, which is slightly greater than the standard deviation of the measured costs ($S_y = \$145.5 \times 10^3$); that is, $S_e/S_y = 1.024$. Since S_e is greater than S_y, it would be more accurate to use the mean cost ($\$335.7 \times 10^3$) than the value estimated with the regression equation. If we use Eq. 10-5 to compute the correlation coefficient, we get a value of 0.176; however, Eq. 10-32 shows that we have essentially no reliability in the estimated values. Thus, for practical purposes, we have a correlation of zero. The adjusted correlation computed with Eq. 10-32 is a more realistic measure of reliability than is the value of R computed with the unadjusted, but more commonly used, equation (Eq. 10-5), because Eq. 10-32 accounts for differences in the degrees of freedom.

Why does the computed equation give such poor results? Certainly, breakwater length is a determinant of cost. However, it appears that other factors are more important and necessary to get accurate estimates of the cost.

TABLE 10-3   Bivariate Regression
Analysis of Breakwater Cost

X	Y	$\hat{Y}$	e
450	230	291	61
625	490	305	−185
670	150	309	159
730	540	314	−226
810	280	321	41
880	120	326	206
1020	380	338	−42
1050	170	340	170
1100	530	345	−185
1175	290	351	61
1230	460	355	−105
1300	230	361	131
1350	470	365	−105
1510	360	379	19

10.7.3 Stress–Strain Analysis

Table 10-4 includes nine measurements of the observed stress and the resulting axial strain for an undisturbed sample of a Boston Blue Clay. A linear equation was computed by regressing the strain (Y) on the stress (X):

$$\hat{Y} = -8.29 + 0.6257X \tag{10-39}$$

The equation has a standard error of estimate of 1.04%, which is much smaller than the standard deviation of the measured strains; that is, $S_e/S_y = 0.35$. The correlation coefficient of 0.945 also suggests that the linear equation provides a reasonable fit to the measured data. An R^2 of 0.89 indicates that almost 90% of the total variation is explained by the linear regression; the standard error ratio of 0.35 suggests that the fit is not as good as suggested by the R^2. The predicted strains and residuals are given in Table 10-4 and shown in Fig. 10-9. While most of the errors are small, there is a noticeable bias in the residuals, with negative residuals for small and large stresses. While the mean bias ($\bar{e}$) is zero, the local biases indicate that a nonlinear structure may be more appropriate. Another problem with the linear model is the negative estimates of strain that occur for stresses of less than 13.244 psi. The negative strain is physically not meaningful, so the model is not rational for small strains.

10.7.4 Project Cost versus Time

For large projects that require years to complete, the construction engineer needs to schedule project finances just as it is necessary to schedule construction activities. While costs vary with time, a construction engineer believes that a linear model can be used to

TABLE 10-4 Bivariate Regression
Analysis of Stress–Strain Data

X	Y	$\hat{Y}$	e
10.8	0.1	−1.5	−1.6
14.1	0.4	0.5	0.1
16.3	1.1	1.9	0.8
17.5	1.4	2.7	1.3
18.6	2.7	3.4	0.7
19.8	3.6	4.1	0.5
22.1	5.7	5.5	−0.2
23.8	6.8	6.6	−0.2
24.3	8.3	6.9	−1.4

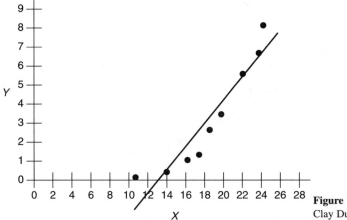

Figure 10-9   Axial Strain (Y) of a Boston Blue Clay Due to Stress (X)

represent the relationship between the value of work completed (Y) and the time since the start of the project (X). Since no work would be completed at the start of the project, a zero-intercept model would seem appropriate. The engineer compiled semiannual cost for a recent 6-year project; the data are given in Table 10-5. The linear, zero-intercept, bivariate model was fit to the data as follows:

$$\hat{Y} = 0.50421X \tag{10-40}$$

where X = time in months from start of project, and $\hat{Y}$ = predicted value of work in $\$ \times 10^6$. The model is shown in Fig. 10-10 and the residuals are given in Table 10-5. The model underpredicts for the middle part of the period, while it overpredicts for the early and latter parts of the time period. The standard error of estimate is $\$7.73 \times 10^6$, which gives $S_e/S_y = 0.757$; this indicates that the model does not provide highly accurate estimates of Y. Additionally, the sum of the errors is not zero, so the model is biased.

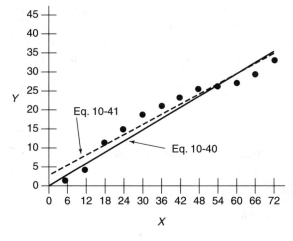

Figure 10.10   Regression of Project Cost (Y, $\$10^6$) on Time from Start of Project (X, Months) for Zero-intercept (Eq. 10-40) and Linear (Eq. 10-41) Models

TABLE 10-5 Bivariate Regression Analyses
of Project Cost (Y) versus Time (X)

		Eq. 10-40		Eq. 10-41	
X	Y	$\hat{Y}$	e	$\hat{Y}$	e
6	1.5	3.03	1.53	5.10	3.60
12	4.3	6.05	1.75	7.84	3.54
18	11.2	9.08	−2.12	10.59	−0.61
24	15.5	12.10	−3.40	13.33	−2.17
30	19.1	15.13	−3.97	16.07	−3.03
36	21.8	18.15	−3.65	18.81	−2.99
42	24.0	21.18	−2.82	21.55	−2.45
48	26.2	24.20	−2.00	24.30	−1.90
54	26.9	27.23	0.33	27.04	0.14
60	27.8	30.25	2.45	29.78	1.98
66	30.1	33.28	3.18	32.52	2.42
72	33.8	36.30	2.50	35.26	1.46

X = time (months) from start of project, Y = value
of work completed ($\$ \times 10^6$), $\hat{Y}$ = predicted value,
and e = error in predicted value = $\hat{Y} - Y$.

Whereas a linear model with an intercept will provide an unbiased model, a zero-intercept model can be biased, as it is for this example with a mean error of $-\$0.52 \times 10^6$. This means that, on the average, Eq. 10-40 tends to underpredict by $\$520,000$. A correlation coefficient should not be computed for a biased model. The correlation coefficient assumes a linear model with an intercept, so it would be incorrect to use either Eq. 10-5 or 10-32 to compute a correlation coefficient for Eq. 10-40.

Because of the poor accuracy of the model of Eq. 10-40, a linear model with an intercept was fit to the data of Table 10-5:

$$Y = 2.36 + 0.4570X \qquad (10\text{-}41)$$

This model is also shown in Fig. 10-10 and the residuals are given in Table 10-5. Equation 10-41 is unbiased and has a standard error of $\$2.651 \times 10^6$, which gives $S_e/S_y = 0.26$. The model explains (R^2) 94% of the variation in Y and has a correlation coefficient of 0.969. In spite of the reasonably accurate estimates suggested by the goodness-of-fit statistics, the model still has local biases in the residuals (see Table 10-5), with

TABLE 10-6   Effect of Extreme Event on Regression

		Eq. 10-42		Eq. 10-43	
X	Y	$\hat{Y}$	e	$\hat{Y}$	e
640	0.063	0.066	0.003	—	—
1930	0.061	0.050	−0.011	0.058	−0.003
1980	0.057	0.049	−0.008	0.055	−0.002
2010	0.049	0.049	0.000	0.053	0.004
2080	0.043	0.048	0.005	0.049	0.006
2120	0.048	0.048	0.000	0.046	−0.002
2200	0.039	0.047	0.008	0.041	0.002
2250	0.042	0.046	0.004	0.038	−0.004

X = distance from nuclear power facility, Y = concentration of ^{241}Am in soil samples, $\hat{Y}$ = predicted value of Y, e = residual = $\hat{Y} - Y$.

underpredictions for the range $18 \leq X \leq 48$ months. However, even with the local biases, the model of Eq. 10-41 may be sufficiently accurate for making project planning estimates. The nonzero intercept of Eq. 10-40 suggests a cost of 2.36×10^6 at the time the project is initiated. While a value of $0 is expected, the nonzero intercept may suggest that the model may not be realistic for the initial part of a project.

In comparing the two models, the standard error ratio would suggest that the zero-intercept model is considerably less accurate than the nonzero-intercept model. However, in comparing the residuals of the two models, there appears to be very little difference. The averages of the absolute values of the residuals are 2.5 and 2.2 for Eqs. 10-40 and 10-41, respectively. The nonzero-intercept model is slightly better, but greater accuracy may be achieved with another model structure.

10.7.5 Effect of Extreme Event

Small samples very often contain an extreme event, which can greatly influence the resulting equation. Table 10-6 contains eight measurements of the concentration of ^{241}Am in soil samples and the distance of the soil sample from the nuclear facility. The eight measurements were used to regress the ^{241}Am concentration (Y) on the distance (X):

$$\hat{Y} = 0.07366 - 0.0000123X \tag{10-42}$$

The negative slope coefficient indicates that the concentration decreases with distance. The standard error of estimate is 0.00694, which gives $S_e/S_y = 0.764$. This suggests poor prediction ability. The correlation coefficient of -0.706 means that 50% of the total variation in Y is explained by the regression equation. The predicted values of Y and the errors for Eq. 10-42 are shown in Table 10-6. Except for the value for a distance of 640, the predicted values range from 0.046 to 0.050 picocuries per gram (pCi/g); however, the measured values for these distances ranged from 0.039 to 0.061 pCi/g. These predicted values are not much different from the mean of 0.050 pCi/g. The batch of seven points acts as a single point.

A second model was developed using seven of the eight values. By eliminating the extreme event, the following model was fitted:

$$\hat{Y} = 0.1755 - 0.0000611X \tag{10-43}$$

This model has a standard error of 0.00412 pCi/g ($S_e/S_y = 0.51$). The correlation coefficient in this case is -0.885, which means that 78.3% of the total variation in the seven measurements was explained by the regression equation of Eq. 10-43. The residuals for Eq. 10-43 are small, and the predicted values range from 0.038 to 0.058 pCi/g, which is much greater than the range for the model of Eq. 10-42. But for a distance of 640, which is the measured value not included in fitting Eq. 10-43, the value predicted with Eq. 10-43 is 0.136 pCi/g; this is considerably higher than the measured value of 0.063 pCi/g.

In summary, in spite of the goodness-of-fit statistics for both models, neither model is an accurate representation of the system. This is a result of both the small sample size ($n = 8$) and the distribution of the data. It is undesirable to have a cluster of points and then one extreme event. Such a distribution of values makes it difficult to attain a model that is representative over a wide range of values of the predictor variable.

10.7.6 Variable Transformation

Studies have shown that fuel consumption (Y, gallons per mile) is related to the vehicle velocity (X, miles per hour) by

$$\hat{Y} = b_0 + \frac{b_1}{X} \tag{10-44}$$

By creating a new variable $W = 1/X$, a relationship between Y and X can be fitted. The data of Table 10-7 can be used to illustrate the fitting of a model that requires a transformation of the predictor variable.

A linear regression analysis of Y on W produces the following equation:

$$\hat{Y} = 0.03939 + \frac{0.9527}{X} \tag{10-45}$$

The standard error of estimate for Eq. 10-45 is 0.0141 gal/mi, which is 17.5% of the standard deviation of Y (i.e., $S_e/S_y = 0.175$). Accurate estimates are also suggested by the correlation coefficient of 0.986 ($R_2 = 0.972$). The residuals (see Table 10-7) are small and a trend is not apparent.

In this case, only the predictor variable was transformed, which does not create

TABLE 10-7   Bivariate Regression of Fuel
Consumption (Y) on Speed (X)

X	W	Y	$\hat{Y}$	e
3.4	0.294	0.32	0.319	−0.001
8.1	0.123	0.14	0.157	0.017
16.4	0.0610	0.13	0.0975	−0.0325
23.7	0.0422	0.094	0.0796	−0.0144
31.3	0.0319	0.062	0.0698	0.0078
38.8	0.0258	0.066	0.0640	−0.0020
42.3	0.0236	0.051	0.0619	0.0109
49.0	0.0204	0.056	0.0588	0.0028
54.6	0.0183	0.055	0.0568	0.0018
58.2	0.0172	0.049	0.0558	0.0068
66.1	0.0151	0.051	0.0538	0.0028

X = fuel consumption (gallons per mile),
$W = 1/X$, and Y = mean speed (miles per hour).

problems. Problems can arise when it is necessary to transform the criterion variable in order to put the relationship into a linear form.

10.8 MULTIPLE REGRESSION ANALYSIS

What options are available to improve the accuracy of predictions if the accuracy from a bivariate regression is still not sufficient for the design problem? Multivariate systems are a possible option to this problem. Where one predictor variable may not be adequate, several predictor variables may provide sufficient prediction accuracy. Thus one reason for using multivariate models is to reduce the standard error of estimate.

Multivariate data analyses are also of value when a goal of the analysis is to examine the relative importance of the predictor variables. If a predictor variable is not included in the model, a regression analysis cannot determine the sensitivity of the criterion variable to variation in that predictor variable. Also, if a predictor variable that has not been included in the model is correlated with another predictor variable that is included in the model, the regression coefficient for the former variable may reflect the effects of both the former and latter predictor variables. Thus predictor variables that have only slight correlation with the criterion variable are often included in the model so that a complete analysis may be made.

How does a multivariate analysis compare with a bivariate analysis? Actually, the two are very similar. First, the same least-squares objective function is used to calibrate the regression coefficients. Second, bivariate correlations are still computed. Third, the data should be properly screened using graphical analyses prior to selecting a model structure. The major difference between multivariate and bivariate analyses is the necessity to account for the interdependence (i.e., correlation) of the predictor variables.

10.8.1 Correlation Matrix

The first step in a multivariate analysis is to perform a graphical analysis. This includes developing plots of the criterion variable versus each of the predictor variables in the data set. Also, plots of each pair of predictor variables should be made; such plots will help to identify the characteristics of the interdependence between predictor variables. It will become evident that the interdependence is extremely important in the analysis of multivariate systems. All the plots should be examined for the characteristics identified previously for bivariate analyses, for example, extreme events, nonlinearity, and random scatter.

After a graphical analysis, the bivariate correlation coefficients should be computed for each pair of variables; this includes the correlation between the criterion variable and each predictor variable, as well as the correlation between each pair of predictor variables. The correlations are best presented in matrix form. The correlation matrix is a means of presenting in an organized manner the correlations between pairs of variables in a data set; it appears as

$$
\begin{array}{c|cccccc}
 & X_1 & X_2 & X_3 & \cdots & X_p & Y \\
\hline
X_1 & 1.0 & r_{12} & r_{13} & \cdots & r_{1p} & r_{1Y} \\
X_2 & & 1.0 & r_{23} & \cdots & r_{2p} & r_{2Y} \\
X_3 & & & 1.0 & \cdots & r_{3p} & r_{3Y} \\
\vdots & & & & \ddots & \vdots & \vdots \\
X_p & & & & & 1.0 & r_{pY} \\
Y & & & & & & 1.0
\end{array}
\qquad (10\text{-}46)
$$

Note that the correlation matrix, which includes p predictor variables (X_i, $i = 1, 2, \ldots, p$) and the criterion variable (Y), is shown in upper-triangular form because the matrix is symmetric (i.e., $r_{ij} = r_{ji}$); also, the elements on the principal diagonal equal 1.0 because the correlation between a variable with itself is unity. The matrix is $(p + 1) \times (p + 1)$.

Example 10-5 Sediment Yield Data

The sediment data base includes four predictors and a criterion variable, the sediment yield. The ultimate objective would be to develop a model for estimating the sediment yield from small watersheds. The predictor variables are (1) the precipitation/temperature ratio (X_1), which reflects the vegetation potential of the area; (2) the average watershed slope (X_2),

which reflects the erosion potential due to the momentum of surface runoff; (3) the soil particle size index (X_3), which reflects the coarse-particle composition of the soil; and (4) the soil
aggregation index (X_4), which reflects the dispersion characteristics of the small particles.
Since vegetation retards erosion, we would expect r_{1y} to be negative. Similarly, coarse particles are more difficult to transport, so r_{3Y} should also be negative. As the slope increases, the
water will move with a greater velocity and more soil can be transported, so r_{2Y} should be
positive. Erosion should increase as the dispersion characteristics of small particles increase,
so r_{4y} should be positive.

     The correlation matrix for 37 measurements of these five variables is given in Table
10-8. In general, the intercorrelations are low, with the largest in absolute value being for r_{14}
(-0.445); this suggests that more than one predictor variable will probably be used to obtain
accurate estimates of the criterion variable. Many of the intercorrelations are very low, less
than 0.2; this is usually a desirable characteristic. The predictor–criterion correlations are
low to moderate; the largest is 0.570. Using the square of the predictor–criterion correlations,
the fraction of variance explained ranges from 0.062 to 0.325. These low values would also
suggest that more than one predictor variable will be necessary to accurately estimate the
sediment yield.

Example 10-6 Evaporation Data

The evaporation data set includes four predictor variables and a criterion variable, which is
the pan evaporation from a small pond in southeastern Georgia. The predictor variables are
(1) the mean daily temperature in degrees Fahrenheit (°F), X_1; (2) the wind speed in miles
per day, X_2; (3) the radiation in equivalent inches, X_3; and (4) the vapor pressure deficit, X_4.
The wind speed attempts to define the rate at which the drier air is moved over the pond to
replace the more moist air. The temperature and the radiation measure the potential energy of
the water molecules and their potential for escaping from the water surface. The vapor pressure deficit represents the potential for the overlying air mass to accept water molecules.

     The correlation matrix, which was based on 354 daily measurements, is shown in
Table 10-9. The correlation matrix provides a sharp contrast to that from the sediment data
because the matrix of Table 10-9 is characterized by high intercorrelation and moderate
predictor–criterion correlations. The high intercorrelations suggest that only one or two predictor variables will be necessary to estimate evaporation rates. The moderate predictor–
criterion correlations suggest that the potential accuracy of predictions is only moderate. The
fraction of the variance explained by X_4 is 0.403. Because most of the intercorrelations are
high, it is unlikely that the variation explained by the four predictor variables will be much
greater than the 40% explained by just one of the predictor variables.

TABLE 10-8 Correlation Matrix of the Sediment Data Base

	X_1	X_2	X_3	X_4	Y
X_1: Precipitation/temperature ratio	1.000	0.340	-0.167	-0.445	-0.297
X_2: Average slope		1.000	-0.051	-0.185	0.443
X_3: Soil particle size			1.000	0.069	-0.253
X_4: Soil aggregation index				1.000	0.570
Y: Sediment yield					1.000

TABLE 10-9   Correlation Matrix for the Evaporation Data Base

	X_1	X_2	X_3	X_4	Y
X_1: Temperature (°F)	1.000	−0.219	0.578	0.821	0.581
X_2: Wind speed (miles/per day)		1.000	−0.261	−0.304	−0.140
X_3: Radiation (equivalent inches)			1.000	0.754	0.578
X_4: Vapor pressure deficit				1.000	0.635
Y: Pan evaporation (inches)					1.000

10.8.2 Calibration of the Multiple Linear Model

The three components of regression analysis are the model, the objective function, and the data set. The data set consists of a set of n observations on p predictor variables and one criterion variable, where n should be, if possible, at least four times greater than p. The data set can be viewed as a matrix having dimensions of n by $(p + 1)$. The principle of least squares is used as the objective function. The model, in raw score form, is

$$\hat{Y} = b_0 + b_1 X_1 + b_2 X_2 + \cdots + b_p X_p \tag{10-47}$$

in which X_j ($j = 1, 2, \ldots, p$) are the predictor variables, b_j ($j = 1, 2, \ldots, p$) are the partial regression coefficients, b_0 is the intercept coefficient, and Y is the criterion variable. Using the least-squares principle and the model of Eq. 10-47, the objective function becomes

$$F = \min \sum_{i=1}^{n} e_i^2 = \min \sum_{i=1}^{n} \left(b_0 + \sum_{j=1}^{p} b_j X_{ij} - Y_i \right)^2 \tag{10-48}$$

in which F is the value of the objective function. It should be noted that the predictor variables include two subscripts, with i indicating the observation and j the specific predictor variable.

The method of solution is to take the $(p + 1)$ derivatives of the objective function, Eq. 10-48, with respect to the unknowns, b_j ($j = 0, 1, \ldots, p$), setting the derivatives equal to zero and solving for the unknowns. A set of $(p + 1)$ normal equations is an intermediate result of this process.

As an example, consider the case where $p = 2$; thus Eq. 10-47 reduces to

$$\hat{Y} = b_0 + b_1 X_1 + b_2 X_2 \tag{10-49}$$

Also, the objective function, Eq. 10-48, is given by

$$F = \min \sum_{i=1}^{n} e_i^2 = \min \sum_{i=1}^{n} (b_0 + b_1 X_{i1} + b_2 X_{i2} - Y_i)^2 \tag{10-50}$$

The resulting derivatives are

$$\frac{\partial \sum_{i=1}^{n} e_i^2}{\partial b_0} = 2 \sum_{i=1}^{n} (b_0 + b_1 X_{i1} + b_2 X_{i2} - Y_i)(1) = 0 \qquad (10\text{-}51a)$$

$$\frac{\partial \sum_{i=1}^{n} e_i^2}{\partial b_1} = 2 \sum_{i=1}^{n} (b_0 + b_1 X_{i1} + b_2 X_{i2} - Y_i)(X_{i1}) = 0 \qquad (10\text{-}51b)$$

$$\frac{\partial \sum_{i=1}^{n} e_i^2}{\partial b_2} = 2 \sum_{i=1}^{n} (b_0 + b_1 X_{i1} + b_2 X_{i2} - Y_i)(X_{i2}) = 0 \qquad (10\text{-}51c)$$

Rearranging Eqs. 10-51 yields a set of normal equations:

$$n b_0 + b_1 \sum_{i=1}^{n} X_{i1} + b_2 \sum_{i=1}^{n} X_{i2} = \sum_{i=1}^{n} Y_i \qquad (10\text{-}52a)$$

$$b_0 \sum_{i=1}^{n} X_{i1} + b_1 \sum_{i=1}^{n} X_{i1}^2 + b_2 \sum_{i=1}^{n} X_{i1} X_{i2} = \sum_{i=1}^{n} X_{i1} Y_i \qquad (10\text{-}52b)$$

$$b_0 \sum_{i=1}^{n} X_{i2} + b_1 \sum_{i=1}^{n} X_{i1} X_{i2} + b_2 \sum_{i=1}^{n} X_{i2}^2 = \sum_{i=1}^{n} X_{i2} Y_i \qquad (10\text{-}52c)$$

The solution of the three simultaneous equations yields values of b_0, b_1, and b_2.

10.8.3 Standardized Model

When the approach of the previous section is used and the means and standard deviations of the predictor and criterion variables are significantly different, round-off error, which results from the inability to maintain a sufficient number of significant digits in the computations, may cause the partial regression coefficient to be erroneous. Thus most multiple regression analyses are made using a standardized model:

$$Z_Y = t_1 Z_1 + t_2 Z_2 + \cdots + t_p Z_p \qquad (10\text{-}53)$$

in which t_j ($j = 1, 2, \ldots, p$) are called standardized partial regression coefficients, and Z_y and the Z_j ($j = 1, 2, \ldots, p$) are the criterion variable and the predictor variables, respectively, expressed in standardized form; specifically, for $i = 1, 2, \ldots, n$, they are computed by

$$Z_Y = \frac{Y_i - \bar{Y}}{S_Y} \qquad (10\text{-}54)$$

and

$$Z_j = \frac{X_{ij} - \bar{X}_j}{S_j} \qquad (10\text{-}55)$$

in which S_y is the standard deviation of the criterion variable and S_j ($j = 1, 2, \ldots, p$) are the standard deviations of the predictor variables. It can be shown that the standardized partial regression coefficients (i.e., the t_j's) and the partial regression coefficients (i.e., the b_j's) are related by

$$b_j = \frac{t_j S_Y}{S_j} \qquad (10\text{-}56)$$

The intercept coefficient can be computed by

$$b_0 = \overline{Y} - \sum_{j=1}^{p} b_j \overline{X}_j \qquad (10\text{-}57)$$

Thus the raw score model of Eq. 10-47 can be computed directly from the standardized model, Eq. 10-53, and Eqs. 10-56 and 10-57.

   Although the differences between regression and correlation have been emphasized, the correlation matrix can be used in solving for the standardized partial regression coefficients. The solution is represented by

$$\mathbf{R}_{11}\,\mathbf{t} = \mathbf{R}_{12} \qquad (10\text{-}58)$$

in which $\mathbf{R}_{11}$ is the $p \times p$ matrix of intercorrelations between the predictor variables, $\mathbf{t}$ is a $p \times 1$ vector of standardized partial regression coefficients, and $\mathbf{R}_{12}$ is the $p \times 1$ vector of predictor–criterion correlation coefficients. Since $\mathbf{R}_{11}$ and $\mathbf{R}_{12}$ are known while $\mathbf{t}$ is unknown, it is necessary to solve the matrix equation, Eq. 10-58, for the $\mathbf{t}$ vector. This involves premultiplying both sides of Eq. 10-58 by $\mathbf{R}_{11}^{-1}$ (i.e., the inverse of $\mathbf{R}_{11}$) and using the matrix identities, $\mathbf{R}_{11}^{-1} \cdot \mathbf{R}_{11} = \mathbf{I}$ and $\mathbf{I} \cdot \mathbf{t} = \mathbf{t}$, where $\mathbf{I}$ is the unit matrix, as follows:

$$\mathbf{R}_{11}^{-1} \cdot \mathbf{R}_{11} \cdot \mathbf{t} = \mathbf{R}_{11}^{-1} \cdot \mathbf{R}_{12} \qquad (10\text{-}59a)$$

$$\mathbf{I} \cdot \mathbf{t} = \mathbf{R}_{11}^{-1} \cdot \mathbf{R}_{12} \qquad (10\text{-}59b)$$

$$\mathbf{t} = \mathbf{R}_{11}^{-1} \cdot \mathbf{R}_{12} \qquad (10\text{-}59c)$$

It is important to recognize from Eq. 10-59c that the elements of the $\mathbf{t}$ vector are a function of both the intercorrelations and the predictor–criterion correlation coefficients. If $\mathbf{R}_{11} = \mathbf{I}$, then $\mathbf{R}_{11}^{-1} = \mathbf{I}$, and $\mathbf{t} = \mathbf{R}_{21}$; this suggests that since the t_j values serve as weights on the standardized predictor variables, the predictor–criterion correlations also reflect the importance, or "weight," that should be given to a predictor variable. However, when $\mathbf{R}_{11}$ is very different from $\mathbf{I}$ (i.e., when the intercorrelations are significantly different from zero), the t_j values will provide considerably different estimates of the importance of the predictors than would be indicated by the elements of $\mathbf{R}_{12}$.

10.8.4 Intercorrelation

It is evident from Eq. 10-59c that intercorrelation can have a significant effect on the t_j values and thus on the b_j values. In fact, if the intercorrelations are significant, the t_j values can be irrational. It should be evident that irrational regression coefficients can lead to irrational predictions. Thus it is important to assess the rationality of the coefficients.

The irrationality results from the difficulty in taking the inverse of the $\mathbf{R}_{11}$ matrix that is necessary for Eq. 10-59c; this corresponds to the round-off-error problem associated with the solution of the normal equations of Eqs. 10-52. A matrix in which the inverse cannot be evaluated is called a singular matrix. A near-singular matrix is one in which one or more pairs of the standardized normal equations are nearly identical (i.e., linearly dependent).

The determinant of a square matrix, such as a correlation matrix, can be used as an indication of the degree of intercorrelation. The determinant is a unique scalar value that characterizes the intercorrelation of the $\mathbf{R}_{11}$ matrix; that is, the determinant is a good single-valued representation of the degree of linear association between the normal equations.

Consider the following four matrices:

$$\mathbf{A}_1 = \begin{bmatrix} 1 & 0 & 0 \\ 0 & 1 & 0 \\ 0 & 0 & 1 \end{bmatrix} \tag{10-60}$$

$$\mathbf{A}_2 = \begin{bmatrix} 1.0 & 0.5 & 0.5 \\ 0.5 & 1.0 & 0.5 \\ 0.5 & 0.5 & 1.0 \end{bmatrix} \tag{10-61}$$

$$\mathbf{A}_3 = \begin{bmatrix} 1.0 & 0.9 & 0.9 \\ 0.9 & 1.0 & 0.9 \\ 0.9 & 0.9 & 1.0 \end{bmatrix} \tag{10-62}$$

$$\mathbf{A}_4 = \begin{bmatrix} 1.0 & 0.9 & 0.0 \\ 0.9 & 1.0 & 0.0 \\ 0.0 & 0.0 & 1.0 \end{bmatrix} \tag{10-63}$$

The determinants of these matrices are

$$|\mathbf{A}_1| = 1 \tag{10-64a}$$

$$|\mathbf{A}_2| = 0.5 \tag{10-64b}$$

$$|\mathbf{A}_3| = 0.028 \tag{10-64c}$$

$$|\mathbf{A}_4| = 0.19 \tag{10-64d}$$

These values indicate that (1) the determinant of a correlation matrix will lie between 0.0 and 1.0; (2) if the intercorrelations are zero, the determinant of a correlation matrix equals 1.0; (3) as the intercorrelations become more significant, the determinant approaches zero; and (4) when any two rows of a correlation matrix are nearly identical, the determinant approaches zero.

10.8.5 Criteria for Evaluating a Multiple Regression Model

After a multiple regression model has been calibrated, we may ask how well the linear model represents the observed data. The following criteria should be used in answering this question: (1) the rationality of the coefficients; (2) the coefficient of multiple determination (R^2); (3) the standard error of estimate, which is usually compared with S_Y; (4) the relative importance of the predictor variables; and (5) the characteristics of the residuals. The relative importance of each of these five criteria may vary with the problem, as well as with the analyst.

The coefficient of multiple determination equals the fraction of the variation in the criterion variable that is explained by the regression equation. Thus it is the square of the correlation coefficient. Mathematically, it is defined as

$$R^2 = \frac{\sum_{i=1}^{n}(\hat{Y}_i - \overline{Y})^2}{\sum_{i=1}^{n}(Y_i - \overline{Y})^2} \tag{10-65}$$

It may be computed by

$$R^2 = \sum_{j=1}^{p} t_j r_{jY} \tag{10-66}$$

in which r_{jY} is the predictor–criterion correlation for predictor j. The value of R^2 is always in the range from zero to 1.0, with a value of zero indicating that Y is not related to any of the predictor variables. The coefficient of multiple determination must be at least as large as the square of the largest predictor–criterion correlation; if the intercorrelations are high, there will be little difference between the two, which would indicate that including more than one predictor variable in the regression equation does little to improve the accuracy of predictions. If the intercorrelations are low, including more than one predictor variable will probably provide greater accuracy than a bivariate model.

The standard error of estimate (S_e) is defined as the square root of the sum of the squares of the errors divided by the degrees of freedom (ν):

$$S_e = \left[\frac{1}{\nu}\sum_{i=1}^{n}(\hat{Y}_i - Y_i)^2\right]^{0.5} \tag{10-67}$$

Previously, the degrees of freedom used to compute S_e for a bivariate model was $(n-2)$; however, in general, the degrees of freedom are defined as

$$\nu = n - q \tag{10-68}$$

in which q is the number of unknowns. For the case where there are p partial regression coefficients and one intercept coefficient, $q = p + 1$ and $\nu = n - p - 1$. The S_e has the same units as the criterion variable and should be compared with S_Y in assessing the accuracy of prediction.

If we use Eq. 10-56 to compute t_i for the bivariate case (i.e., one predictor variable), it will be evident that t, the standardized partial regression coefficient, equals the correlation coefficient. Even in the case where $p > 1$, t_i is still a measure of the common variation. However, Eq. 10-59c indicates that the t_i values are a function of both the intercorrelation and the predictor–criterion correlations. When the intercorrelations are significant (i.e., $\mathbf{R}_{11}$ of Eq. 10-59c is significantly different from $\mathbf{I}$, the identity matrix), the t_i values may be irrational; thus they would not be valid measures of the relative importance of the corresponding predictor variable. Since the t value corresponds to a correlation coefficient in the bivariate case, it is not unreasonable to provide a similar interpretation for a multiple regression; thus t_i values should be considered irrational in magnitude if they are greater than 1.0 in absolute value. Even values of t near 1.0 may indicate irrationality. If a t value is irrational, the corresponding b value must be considered irrational, since b is a scaled value of t.

10.8.6 Analysis of Residuals

The residuals (i.e., $\hat{Y}_i - Y_i$) were shown to be an important criterion in assessing the validity of a bivariate regression equation; the same concepts apply in the analysis of the residuals for a multiple regression equation. Plots of the residuals may indicate an incorrect model structure, such as a nonlinear relationship between the criterion variable and one of the predictor variables. Additionally, the residuals may suggest a need to consider another predictor variable that was not included in the model. Plots of the residuals should be made to assess the assumptions of zero mean, constant error variance, independence, and normality. A plot of the residuals versus the predicted value of the criterion variable (i.e., e_i versus $\hat{Y}_i$) may identify either a violation of the assumption of constant variance or a lack of independence of the observations.

10.8.7 Computational Example

Although we rarely perform a multiple regression without the aid of a computer, it is instructive to see the manual computations at least one time. Table 10-10 provides values for the criterion variable (Y) and two predictor variables (X_1 and X_2); the sample consists of six observations (i.e., $n = 6$). Figure 10-11 shows the graphical relationship between each pair of variables; in general, linear associations are evident, although the scatter about the apparent linear relationships is significant. The correlation between the two predictor variables appears to be high relative to the degree of linearity between the criterion variable and the predictor variables.

After examining the plots, the correlation matrix ($\mathbf{R}$) should be computed:

$$\mathbf{R} = \begin{array}{c|ccc} & X_1 & X_2 & Y \\ \hline X_1 & 1.000 & 0.776 & 0.742 \\ X_2 & & 1.000 & 0.553 \\ Y & & & 1.000 \end{array} \qquad (10\text{-}69)$$

TABLE 10-10 Data Base for Example of a Two-predictor Model

Y	X_1	X_2	X_1^2	X_1X_2	X_1Y	X_2^2	X_2Y
2	1	2	1	2	2	4	4
2	2	3	4	6	4	9	6
3	2	1	4	2	6	1	3
3	5	5	25	25	15	25	15
5	4	6	16	24	20	36	30
6	5	4	25	20	30	16	24
Column summation 21	19	21	75	79	77	91	82

The correlation coefficient of 0.776 between X_1 and X_2 indicates that the intercorrelation is high. The determinant of the intercorrelation matrix equals 0.398, which should be considered as moderate; that is, it suggests that we should carefully examine the resulting regression equation for rationality.

The summations shown in Table 10-10 can be used to develop the normal equations using (Eqs. 10-52), as follows:

$$19b_0 + 75b_1 + 79b_2 = 77 \tag{10-70a}$$

$$21b_0 + 79b_1 + 91b_2 = 82 \tag{10-70b}$$

$$6b_0 + 19b_1 + 21b_2 = 21 \tag{10-70c}$$

The normal equations can be solved to obtain values for the coefficients of the following equation:

$$\hat{Y} = 1.30 + 0.75X_1 - 0.05X_2 \tag{10-71}$$

Similarly, the correlation matrix could be used to develop the normal equations for the standardized model:

$$\begin{bmatrix} 1.000 & 0.776 \\ 0.776 & 1.000 \end{bmatrix} \begin{bmatrix} t_1 \\ t_2 \end{bmatrix} = \begin{bmatrix} 0.742 \\ 0.553 \end{bmatrix} \tag{10-72}$$

which yields

$$t_1 + 0.776t_2 = 0.742 \tag{10-73a}$$

$$0.776t_1 + t_2 = 0.553 \tag{10-73b}$$

Solving for t_1 and t_2 yields the following standardized equation:

$$Z_y = 0.786Z_1 - 0.057Z_2 \tag{10-74}$$

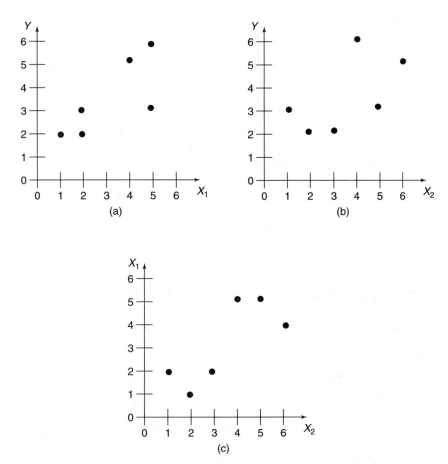

Figure 10-11   Graphs of Variables for Computational Example: (a) Y versus X_1; (b) Y versus X_2; (c) X_1 versus X_2

Both the raw score equation (Eq. 10-71) and the standardized equation (Eq. 10-74) indicate that as X_2 increases Y decreases; this is in conflict with both the correlation coefficient and the graphical relationship shown in Fig. 10-11. This irrationality is the result of the high intercorrelation between the predictor variables.

The correlation coefficient can be computed from either Eq. 10-65 or 10-66. Using Eq. 10-66 yields

$$R^2 = t_1 r_{1Y} + t_2 r_{2Y} \tag{10-75a}$$

$$= 0.786(0.742) + (-0.057)(0.553) = 0.552 \tag{10-75b}$$

Therefore, R equals 0.743. Predictor X_1 explained 55.1% (i.e., 0.742^2) of the variation in Y by itself. Thus the use of X_2 is not justified because it does not make a significant contribution to the explained variation and, even more important, it results in an irrational model. The standard error of estimate is

$$S_e = S_Y \sqrt{\frac{n-1}{n-p-1}(1-R)^2} = 1.643\sqrt{\frac{5}{3}(1-0.552)^2} \tag{10-76a}$$

$$= 1.643(0.864) = 1.420 \tag{10-76b}$$

Thus the two-predictor model reduced the error variance by 25.3%; the standard error of estimate is about 86% of the standard deviation of the criterion variable. For the one-predictor model with X_1 as the predictor variable, the standard error of estimate would be 1.231, which is less than the standard error of estimate for the two-predictor model; this occurs because the ratio $(n-1)/(n-p-1)$ is much less for the one-predictor model. This example illustrates the importance of computing the standard error of estimate when the sample size is small. The standard error provides a more realistic measure of the goodness of fit than the correlation coefficient does.

10.9 REGRESSION ANALYSIS OF NONLINEAR MODELS

In most empirical analyses, linear models are attempted first because of the relative simplicity of linear analysis. Also, linear models are easily applied, and the statistical reliability is easily assessed.

Linear models may be rejected because of either theoretical considerations or empirical evidence. Specifically, theory may suggest a nonlinear relationship between a criterion variable and one or more predictor variables; for example, biological growth curves used by environmental engineers are characterized by nonlinear forms. When a model structure cannot be identified by theoretical considerations, empirical evidence can be used in model formulation and may suggest a nonlinear form. For example, the hydrologic relationship between peak discharge and the drainage area of a watershed has been found to be best represented by a log–log equation, which is frequently referred to as a power model. A nonlinear form may also be suggested by the residuals that result from a linear analysis; that is, if a linear model produces nonrandomly distributed residuals (i.e., a bias is apparent), the underlying assumptions are violated. A nonlinear functional form may produce residuals that satisfy the assumptions that underlie the principle of least squares.

One very common use of nonlinear models is the fitting of existing design curves that have significant nonlinearity. The equations are then used as part of computer software that performs a design procedure that, in the past, was solved manually.

10.9.1 Common Nonlinear Alternatives

Linear models were separated into bivariate and multivariate; the same separation is applicable to nonlinear models. It is also necessary to separate nonlinear models on the basis of the functional form. Although polynomial and power models are the most frequently used nonlinear forms, it is important to recognize that other model structures are available and may actually provide the correct structure. In addition to the power and polynomial structures, forms such as a square root, exponential, and logarithmic may provide the best fit to a set of data. Since the polynomial and power forms are so widely used, it is of importance to identify their structure:

1. Bivariate
   a. Polynomial

$$\hat{Y} = b_0 + b_1 X + b_2 X^2 + \cdots + b_p X^p \tag{10-77}$$

   b. Power

$$\hat{Y} = b_0 X^{b_1} \tag{10-78}$$

2. Multivariate
   a. Polynomial

$$\hat{Y} = b_0 + b_1 X_1 + b_2 X_2 + b_3 X_1^2 + b_4 X_2^2 + b_5 X_1 X_2 \tag{10-79}$$

   b. Power

$$\hat{Y} = b_0 X_1^{b_1} X_2^{b_2} \ldots X_p^{b_p} \tag{10-80}$$

In these relationships, Y is the criterion variable, X is the predictor variable in the bivariate case, X_i is the ith predictor variable in the multivariate case, b_j ($j = 0, 1, \ldots, p$) is the jth regression coefficient, and p is either the number of predictor variables or the order of the polynomial. The bivariate forms are just special cases of the multivariate forms.

10.9.2 Calibration of Polynomial Models

The power and polynomial models are widely used nonlinear forms because they can be transformed in a way that makes it possible to use the principle of least squares. Although the transformation to a linear structure is desirable from the standpoint of calibration, it has important consequences in terms of assessing the goodness-of-fit statistics.

The bivariate polynomial of Eq. 10-77 can be calibrated by forming a new set of predictor variables:

$$W_i = X^i \qquad \text{for } i = 1, 2, \ldots, p \tag{10-81}$$

This results in the model

$$\hat{Y} = b_0 + b_1 W_1 + b_2 W_2 + \cdots + b_p W_p \tag{10-82}$$

The coefficients b_j can be estimated using a standard linear multiple regression analysis. The multivariate polynomial of Eq. 10-79 can also be solved using a multiple regression analysis for a set of transformed variables. For the model given by Eq. 10-79, the predictor variables are transformed as follows:

$$W_1 = X_1 \tag{10-83a}$$

$$W_2 = X_2 \tag{10-83b}$$

$$W_3 = X_1^2 \tag{10-83c}$$

$$W_4 = X_2^2 \tag{10-83d}$$

$$W_5 = X_1 X_2 \tag{10-83e}$$

The revised model has the form

$$\hat{Y} = b_0 + \sum_{i=1}^{5} b_i W_i \tag{10-84}$$

It is important to note that the polynomial models do not require a transformation of the criterion variable.

The model given by Eq. 10-79 has only two predictor variables (i.e., X_1 and X_2) and is a second-order equation. In practice, a model may have more predictor variables and may be of higher order. In some cases the interaction terms (i.e., $X_i X_j$) may be omitted to decrease the number of coefficients that must be calibrated. However, if the interaction terms are omitted when they are actually significant, the goodness-of-fit statistics may suffer, unless the variation is explained by the other terms in the model; when this occurs, the coefficients may lack physical significance.

10.9.3 Fitting a Power Model

The bivariate power model of Eq. 10-78 can be calibrated by forming the following set of variables:

$$\hat{Z} = \ln \hat{Y} \tag{10-85a}$$

$$c = \ln b_0 \tag{10-85b}$$

$$W = \ln X \tag{10-85c}$$

These transformed variables form the following linear equation, which can be calibrated using bivariate linear regression analysis:

$$\hat{Z} = c + b_1 W \tag{10-86}$$

After values of c and b_1 are obtained, the coefficient b_0 can be determined using

$$b_0 = e^c \tag{10-87}$$

Base 10 logarithms can also be used; in such a case, Eqs. 10-85a and c will use the base 10 logarithm and Eq. 10-87 will use a base 10 rather than e.

The multivariate power model of Eq. 10-80 can be evaluated by making a logarithmic transformation of both the criterion and the predictor variables:

$$\hat{Z} = \ln \hat{Y} \tag{10-88}$$

$$c = \ln b_0 \tag{10-89}$$

$$W_i = \ln X_i, \qquad \text{for } i = 1, 2, \ldots, p \tag{10-90}$$

The resulting model has the form

$$\hat{Z} = c + \sum_{i=1}^{p} b_i W_i \tag{10-91}$$

The coefficients of Eq. 10-91 can be evaluated using a multiple regression analysis. The value of b_0 can be determined by making the transformation of Eq. 10-87. Again, a base 10 transformation can be used rather than a natural log transformation.

10.9.4 Goodness of Fit

Since most computer programs that perform multiple regression analyses include goodness-of-fit statistics as part of the output, it is important to recognize the meaning of these statistics. For nonlinear models in which the criterion variable $\hat{Y}$ is not transformed, the goodness-of-fit statistics are valid indicators of the reliability of the model; however, when the criterion variable is transformed, such as is necessary for the power model form, the principle of least squares is applied in the log–log space. As a result, the residuals that are used to compute the standard error of estimate, and therefore the correlation coefficient, are measured in the domain of the logarithm of $\hat{Y}$ and not the $\hat{Y}$ domain. Therefore, the goodness-of-fit statistics are not necessarily a reliable indicator of model reliability, especially since engineering decisions are made to predict Y and not $\log(Y)$. Therefore, when a model requires a transformation of the criterion variable $\hat{Y}$ in order to calibrate the coefficients, the goodness-of-fit statistics that are included with the multiple regression output should not be used as measures of reliability. Instead, values for the goodness-of-fit statistics should be recomputed using values of $\hat{Y}$ rather than $\log(\hat{Y})$. The correlation coefficient and the standard error of estimate should be computed using Eqs. 10-65 and 10-67, respectively. In summary, when an equation is calibrated with the criterion variable transformed, such as the log–log space, the least-squares concepts apply only to the transformed space, not to the measurement nontransformed space. The sum of the residuals based on the logarithms of Y will equal zero; however, the sum of the residuals of the transformed values of Y may not equal zero, and the sum of the squares of the errors may not be a minimum in the Y domain even though it is in the $\log(Y)$ domain. Furthermore, the assumption of a constant variance may also not be valid. Because these basic assumptions of regression are not valid in the X–Y space, many practitioners object to data transformations. However, in spite of these theoretical considerations, the transformations may provide reasonable estimates in the X–Y space.

10.9.5 Additional Model Forms

In addition to the polynomial and power models, other forms can provide good approximations to the underlying population relationship between two or more variables. An exponential model has the form

$$\hat{Y} = b_0 e^{b_1 X} \tag{10-92}$$

in which b_0 and b_1 are coefficients requiring values by fitting to data. Values for b_0 and b_1 can be obtained by linear bivariate regression after taking the logarithms of both sides of Eq. 10-92 as follows:

$$\ln \hat{Y} = \ln b_0 + b_1 X \tag{10-93}$$

In this case, $\hat{Y}$ is transformed; thus the correlation coefficient and standard error of estimate apply to the $\ln \hat{Y}$ and not to $\hat{Y}$. Also, the intercept coefficient obtained from the regression analysis for Eq. 10-93 must be transformed for use with Eq. 10-92.

A logarithmic curve can also be used to fit values of Y and X:

$$\hat{Y} = b_0 + b_1 \ln X \tag{10-94}$$

In this case, $\hat{Y}$ is not transformed, so the values of the correlation coefficient and the standard error of estimate are valid indicators of the goodness of fit in the $\hat{Y}$ space.

10.10 APPLICATIONS

10.10.1 One-predictor Polynomial of Sediment Yield versus Slope

A computer program was used to derive first-, second-, and third-order polynomials that regress sediment yield on watershed slope. The correlation matrix is given in Table 10-11. It is evident that the intercorrelation coefficients are very significant and should lead to irrational regression coefficients; furthermore, because of the high intercorrelation it is unlikely that the explained variance will improve significantly as the order of the equation increases. Table 10-12 provides a summary of the correlation coefficients, the standardized partial regression coefficients, and the partial regression coefficients. The second-

TABLE 10-11   Correlation Matrix: Sediment Yield (Y) versus Slope (S)

	S	S^2	S^3	Y
S	1.000	0.908	0.791	0.443
S^2		1.000	0.972	0.443
S^3			1.000	0.428
Y				1.000

TABLE 10-12   Summary of Polynomial Analyses for Sediment Yield versus Slope Models

Model	R	Intercept b_0	b_1	b_2	b_3	t_1	t_2	t_3
			Partial Regression Coefficients			Standardized Partial Regression Coefficients		
First-order	0.443	0.220	0.0263	—	—	0.443	—	—
Second-order	0.454	0.323	0.0136	2.31×10^{-4}	—	0.229	0.235	—
Third-order	0.499	0.042	0.0865	-0.0031	3.43×10^{-5}	1.456	-3.164	2.35

and third-order models do not provide a significant increase in the explained variation when compared with the linear model.

10.10.2 One-predictor Polynomial of Evaporation versus Temperature

The evaporation data were used to derive polynomials for the regression of evaporation on temperature. The correlation matrix, which is shown in Table 10-13, shows very high intercorrelation. Table 10-14 provides a summary of the regression analyses. Whereas the linear model explains approximately 34% of the variation in evaporation, the higher-order terms provide only a marginal increase in the explained variation. The quadratic model increases the explained variance by about 8%. The marginal improvement is also

TABLE 10-13   Correlation Matrix: Evaporation (E) versus Temperature (T)

	T	T^2	T^3	E
T	1.000	0.995	0.981	0.581
T^2		1.000	0.996	0.602
T^3			1.000	0.627
E				1.000

TABLE 10-14   Summary of Regression Analyses

Model	R	S_e[a]	t_1	t_2	t_3	b_0	b_1	b_2	b_3
First-order	0.581	0.0763	0.581	—	—	-0.114	0.00383	—	—
Second-order	0.648	0.0715	2.16	2.76	—	0.425	-0.0142	0.000143	—
Third-order	0.654	0.0711	3.11	-8.44	5.98	-0.239	0.0205	-0.000439	3.14×10^{-6}

[a]$S_y = 0.0936$ in./day.

evident from the small decrease in the standard error of estimate. In judging the rationality of a model, it is of value to examine the predicted values of evaporation for values of the predictor variable that are likely to be observed. The three models were used to estimate the daily evaporation depth for selected temperatures (Table 10-15). Because water freezes at 32°F, evaporation rates near zero are expected; only the linear model gives a rational estimate at 32°F. Estimates made using the second-order model decrease with increases in temperature for temperatures up to about 50°F. Both the second- and third-order models provide especially high rates at 100°F. In summary, the high intercorrelations lead to irrational estimates when the higher-order forms are used.

10.10.3 Single-predictor Power Model

A data set that consists of five observations will be used to illustrate the fitting of a power model that includes a single predictor variable. The data set is given in Table 10-16. The values were converted to natural logarithms and the coefficients of Eq. 10-86 were calibrated using least squares:

$$\ln \hat{Y} = -0.1 + 1.9 \ln X \qquad (10\text{-}95)$$

TABLE 10-15   Predicted Evaporation Rates (in./day)

	Temperature (°F)			
Model	32	50	75	100
First-order	0.009	0.078	0.173	0.269
Second-order	0.117	0.073	0.164	0.435
Third-order	0.070	0.081	0.154	0.561

TABLE 10-16   Data Base for Single-predictor Power Model

				Using Eq. 10-97		Using Eq. 10-96	
X	Y	$\ln X$	$\ln Y$	$\hat{Y}$	e	$\ln \hat{Y}$	e_e
2.718	7.389	1	2	−214.54	−221.929	6.048	−1.341
7.389	20.086	2	3	53.18	33.095	40.445	20.359
20.086	403.429	3	6	780.93	377.499	270.427	−133.002
54.598	2980.96	4	8	2759.03	−221.927	1807.96	−1173.00
148.413	8103.08	5	9	8136.00	32.923	12087.4	3984.31
					$\Sigma e = 0$		$\Sigma e_e \neq 0$

A correlation coefficient of 0.9851 was computed for Eq. 10-95. Equation 10-95 can be retransformed to the X–Y coordinate space:

$$\hat{Y} = 0.9048X^{1.9} \tag{10-96}$$

A linear analysis of the values of X and Y provides a correlation coefficient of 0.9975 and the following regression equation:

$$\hat{Y} = -370.3301 + 57.3164X \tag{10-97}$$

The residuals for Eqs. 10-96 and 10-97 are given in Table 10-16. Whereas the residuals for Eq. 10-97 have a sum of zero, the sum of the residuals for Eq. 10-96 is not equal to zero; however, the sum of the residuals for Eq. 10-95 will equal zero in the natural-log space.

The correlation coefficient and standard error of estimate that are computed for the model in the natural-log space are not valid indicators of the accuracy of the estimates in the domain of the nontransformed criterion variable. For example, the correlation coefficient of 0.9851 computed with Eq. 10-95 is not a measure of the accuracy of Eq. 10-96. Values of Y that were estimated with Eq. 10-96 are given in Table 10-16, and the resulting residuals are denoted as e_e. If the separation of variation concept is applied to Eq. 10-96, the components are

$$\sum_{i=1}^{n} (Y_i - \bar{Y})^2 = \sum_{i=1}^{n} (Y_i - \hat{Y}_i)^2 + \sum_{i=1}^{n} (\hat{Y}_i - \bar{Y})^2 \tag{10-98}$$

$$48{,}190{,}499.4 \neq 17{,}276{,}427.6 + 100{,}525{,}199.1 \tag{10-99}$$

The separation of variation is not valid for Eq. 10-96 because the model was calibrated by taking logarithms. The separation of variation does, however, apply to the logarithms. It is evident that, if the correlation coefficient were computed as the ratio of the explained variation to the total variation (Eq. 10-65), then from the values of Eq. 10-99 the correlation coefficient would be greater than 1.0, which is not rational. Thus the standard error of estimate, rather than the correlation coefficient, should be used as a measure of accuracy:

$$S_e = \left(\frac{17{,}276{,}427.6}{5 - 2} \right)^{0.5} = 2399.75 \tag{10-100}$$

The standard deviation of the criterion variable is 3470.97; therefore, the power model is not highly accurate, and the computed correlation for the log model (i.e., 0.9851) certainly does not reflect the accuracy of the model.

10.10.4 Multivariate Power Model

Sediment yield data will be used to illustrate a power model that includes more than one predictor variable. The small particle size index cannot be included as a predictor variable because some of the observations equal zero; thus the model regresses sediment yield (Y)

on three predictor variables: the precipitation/temperature ratio (P), the slope (S), and the soil index (I). The data are given in Table 10-17. The resulting regression equation in logarithm form is

$$\ln \hat{Y} = -4.563 - 1.123 \ln P = 0.8638 \ln S - 0.1089 \ln I \qquad (10\text{-}101)$$

Transforming Eq. 10-101 to the X–Y space yields

$$\hat{Y} = 0.01043 P^{-1.123} S^{0.8638} I^{-0.1089} \qquad (10\text{-}102)$$

TABLE 10-17   Sediment Yield Data[a]

P	S	I	Y
0.135	4.000	40.000	0.140
0.135	1.600	40.000	0.120
0.101	1.900	22.000	0.210
0.353	17.600	2.000	0.150
0.353	20.600	1.000	0.076
0.492	31.000	0.000	0.660
0.466	70.000	23.000	2.670
0.466	32.200	57.000	0.610
0.833	41.400	4.000	0.690
0.085	14.900	22.000	2.310
0.085	17.200	2.000	2.650
0.085	30.500	15.000	2.370
0.193	6.300	1.000	0.510
0.167	14.000	3.000	1.420
0.235	2.400	1.000	1.650
0.448	19.700	3.000	0.140
0.329	5.600	27.000	0.070
0.428	4.600	12.000	0.220
0.133	9.100	58.000	0.200

P	S	I	Y
0.149	1.600	28.000	0.180
0.266	1.200	15.000	0.020
0.324	4.000	48.000	0.040
0.133	19.100	44.000	0.990
0.133	9.800	32.000	0.250
0.133	18.300	19.000	2.200
0.356	13.500	58.000	0.020
0.536	24.700	62.000	0.030
0.155	2.200	27.000	0.036
0.168	3.500	24.000	0.160
0.673	27.900	22.000	0.370
0.725	31.200	10.000	0.210
0.275	21.000	64.000	0.350
0.150	14.700	37.000	0.640
0.150	24.400	29.000	1.550
1.140	15.200	11.000	0.090
1.428	14.300	1.000	0.170
1.126	24.700	44.000	0.170

[a]E.M. Flakman, "Predicting Sediment Yield in Western United States," *Journal of the Hydraulic Division,* ASCE, 98(HY12), 2073–2085, December 1972. Y = sediment yield, X_1 = climate variable, X_2 = watershed slope variable, and X_3 = soil aggregation variable.

A computer program provided a correlation coefficient of 0.725 and a standard error estimate of 1.019; however, these are for Eq. 10-101 and may not be accurate indications of the accuracy of Eq. 10-102. The standard error of estimate is expressed in logarithmic units. The standard error of estimate as measured in the X–Y space was computed as 0.595, which can be compared with a standard deviation of 0.817 for the 37 observations of sediment yield.

For a comparison, the following linear model was calibrated using the same three predictor variables:

$$\hat{Y}_{\text{linear}} = 0.8920 - 1.428P + 0.03664S - 0.01308I \qquad (10\text{-}103)$$

This resulted in a multiple correlation coefficient of 0.720 and a standard error of estimate of 0.592. Thus, for this case, there is no significant difference in the expected accuracy of the linear and power models.

10.10.5 Estimating Breakwater Costs

In Section 10.7.2, a bivariate model between the construction cost of a breakwater and length of the breakwater was developed. It was found to be inaccurate, with the suggestion given that other variables associated with cost were probably better determinants of the cost.

Table 10-18 includes the original measurements of length and cost along with a second predictor variable, the average depth of the breakwater. The correlation matrix is

	X_1	X_2	Y
X_1: Length	1.000	-0.229	0.176
X_2: Depth		1.000	0.905
Y: Cost			1.000

While length and depth are poorly correlated, depth and cost have a high correlation, with R^2 equal to 0.819.

A regression analysis yields the following equation:

$$\hat{Y} = -199.8 + 0.1904X_1 + 58.72X_2 \qquad (10\text{-}104)$$

The multiple correlation coefficient equals 0.987; thus $R^2 = 0.974$, which indicates that the two-predictor model provides accurate estimates of the cost. The standard error of estimate is $\$25.3 \times 10^3$, with $S_e/S_y = 0.174$.

The standardized partial regression coefficients are $t_1 = 0.405$ and $t_2 = 0.998$. While t_2 is below 1, it is still very large.

The intercept is also very large in magnitude. While none of the predicted costs is

TABLE 10-18 Data for Breakwater Cost Estimation

Length, X_1 (ft)	Depth, X_2 (ft)	Cost, Y ($\$ \times 10^3$)
450	6.3	230
625	9.6	490
670	3.1	150
730	10.1	540
810	5.4	280
880	3.5	120
1020	6.5	380
1050	2.9	170
1100	8.8	530
1175	3.9	290
1230	7.4	460
1300	3.3	230
1350	7.2	470
1510	4.6	360

negative, it is still possible because the intercept is negative. For example, at a length of 450 ft and depth of 2 ft, the estimated cost would be negative. Equation 10-104 should be used cautiously beyond the range of the data: $450 < X_1 < 1510$ ft and $2.9 < X_2 < 10.1$ feet.

It is interesting to compare the two-predictor multiple regression model with both one-predictor models. The length versus cost model was discussed in Section 10.7.2. It provided very poor accuracy, $R^2 = 0.031$. The depth versus cost model would explain 81.9% of the variation in cost (0.905^2). Since the multiple regression model explains 97.4%, adding the length to the model increases the explained variance by 15.5% (97.4 − 81.9); this is a significant increase, so the two-predictor model is preferable to either of the one-predictor models. Also, when the depth is added to the equation, the importance of the length increases. The t for the bivariate model is 0.176, and $t_1 = 0.405$ for the multiple regression model.

10.10.6 Trip Generation Model

For a regional-planning study, a transportation engineer collects data on traffic. The data of Table 10-19 give measurements of nonwork, home-based-trip-production rates for suburban, medium-density households (Y) as a function of the number of persons per household (X_1) and the number of vehicles per household (X_2). In this case, the predictor variables are integers.

The correlation matrix for the sample of 15 is given by

	X_1	X_2	Y
X_1	1.000	0.000	0.931
X_2		1.000	0.264
Y			1.000

The intercorrelation between the two predictor variables is zero because the data were collected on a grid. One advantage of having no intercorrelation is that the potential for irrational regression coefficients is minimized. The predictor–criterion correlation coefficient for the number of persons per household is large ($R^2 = 0.868$), while that for the number of vehicles per household is small ($R^2 = 0.070$). This may suggest that the bivariate equation $\hat{Y} = f(X_1)$ may be as accurate as the multiple regression model $\hat{Y} = f(X_1, X_2)$.

A multiple regression analysis produced the following equation:

$$\hat{Y} = -1.023 + 1.4467X_1 + 0.7100X_2 \tag{10-105}$$

The correlation coefficient is 0.968, which gives $R^2 = 0.937$. Since the intercorrelation between X_1 and X_2 is zero, the sum of the squares of the two predictor–criterion correlation coefficients equals the coefficient of multiple determination:

$$R^2 = R_{1Y}^2 + R_{2Y}^2 \tag{10-106a}$$

$$0.937 = (0.931)^2 + (0.264)^2 \tag{10-106b}$$

This equality holds only when there is no intercorrelation. It also shows that the multiple regression model explains 7% more variation than the bivariate model based solely on X_1.

The standard error of estimate is 0.6153, which gives $S_e/S_y = 0.271$. This value supports the conclusion suggested by the correlation coefficient that the model is highly accurate.

The standardized partial regression coefficients are $t_1 = 0.931$ and $t_2 = 0.264$. Since the intercorrelation coefficient is 0, the t values equal the predictor–criterion correlation coefficients. Again, the number of persons per household is shown to be the more important predictor.

The intercept of Eq. 10-105 is -1.023, which is irrational, since a value of 0 would be expected. If both X_1 and X_2 are 0, then Y should be 0. Furthermore, the magnitude of

TABLE 10-19   Regression Analysis

X_1	X_2	Y	$\hat{Y}$	e
1	0	0.9	0.42	−0.48
2	0	1.8	1.87	0.07
3	0	2.4	3.32	0.92
4	0	4.3	4.76	0.46
5+	0	6.6	6.21	−0.39
1	1	1.9	1.13	−0.77
2	1	2.7	2.58	−0.12
3	1	3.6	4.03	0.43
4	1	5.4	5.47	0.07
5+	1	7.7	6.92	−0.78
1	2+	2.1	1.84	−0.26
2	2+	3.1	3.29	0.19
3	2+	3.7	4.74	1.04
4	2+	5.9	6.18	0.28
5+	2+	8.3	7.63	−0.67

the intercept is large, that is, approximately 25% of the mean of Y and larger than the value of Y for $X_1 = 1$ and $X_2 = 0$. This suggests that another model structure may be needed to realistically represent the data.

The residuals are given in Table 10-19, and they also suggest that the linear model of Eq. 10-105 is an incorrect structure. There are very noticeable local biases. For each value of X_2 the residuals for $X_1 = 1$ and $X_1 = 5$ are negative; additionally, for each value of X_2 the residuals for $X_1 = 3$ and $X_1 = 4$ are positive. The trends for each value of X_2 suggest that a nonlinear structure would be more appropriate.

10.10.7 Estimation of the Reaeration Coefficient

To estimate the biochemical oxygen demand of a stream, the environmental engineer needs an estimate of the reaeration coefficient. The reaeration coefficient varies with the stream velocity (X_1, in meters per second), the water depth (X_2, in meters), and the water temperature (X_3, in °C). Table 10-20 gives a set of 21 measurements of the three predictor variables and estimated reaeration coefficients (Y). The correlation matrix is

	X_1	X_2	X_3	Y
X_1	1.000	0.839	0.139	0.436
X_2		1.000	0.240	0.456
X_3			1.000	0.621
Y				1.000

The intercorrelation is high ($|\mathbf{R}_{11}| = 0.276$) because of the high correlation between X_1 and X_2. The water temperature is not highly correlated with either velocity or depth. The regression analysis of the 21 sample values yields

$$\hat{Y} = 1.508 + 0.3315X_1 + 0.09731X_2 + 0.08672X_3 \qquad (10\text{-}107)$$

All the regression coefficients are rational in sign. Equation 10-107 results in a correlation coefficient of 0.715, which means that the equation explains 51.2% of the total variation. The standard error of estimate is 0.497 ($S_e/S_y = 0.758$), which suggests that the equation provides estimates of Y that are not highly accurate.

   Of the three predictor variables, the water temperature (X_3) would provide the highest R^2 for a bivariate model ($R^2 = 0.386$). The multiple regression model of Eq. 10-107 explains 51.2%. Thus the addition of X_1 and X_2 increases the explained variance by 12.6%. This means that more than one predictor variable should be used, but it does not necessarily mean that all three predictor variables are needed. It may be worthwhile developing two-predictor models: $\hat{Y} = f(X_1, X_3)$ and $\hat{Y} = f(X_2, X_3)$. Because X_1 and X_2 are highly correlated, both are probably not needed, so a two-predictor model may provide estimates that are just as accurate as the three-predictor model.

   The standardized partial regression coefficients are $t_1 = 0.299$, $t_2 = 0.070$, and $t_3 = 0.563$. These suggest that X_3 is most important, followed by X_1, and finally by X_2, which appears to be unimportant. This differs somewhat from the predictor–criterion correlation coefficients, which suggest that X_3 is most important, but that X_1 and X_2 are of equal importance. Since velocity is probably physically more important than depth, the t values probably give a more realistic assessment of the relative importance of the predictor variables than are given by the predictor–criterion correlation coefficients.

10.10.8 Estimating Slope Stability

Slope stability is very important to engineers who design with unrestrained slopes. The stability number is an index used by geotechnical engineers to assess the safety of a slope. Among others, the slope stability is a function of the angle of friction of the soil (X_1, in degrees) and the slope angle (X_2, in degrees). Twelve laboratory measurements

TABLE 10-20 Multiple Regression Analysis of the Stream Reaeration Coefficient (Y) on Stream Velocity (X_1, m/s), Water Depth (X_2, m), and Water Temperature (X_3, °C)

X_1	X_2	X_3	Y	$\hat{Y}$	e
1.4	1.24	10	2.16	2.69	0.80
0.6	0.87	9	2.50	2.57	0.07
2.2	1.87	13	2.54	3.55	1.01
1.5	0.99	11	2.60	3.06	0.46
0.5	0.85	14	2.71	2.97	0.26
1.6	0.97	8	3.00	2.83	−0.17
1.3	1.12	19	3.02	3.70	0.68
0.8	1.21	11	3.03	2.85	−0.18
1.6	1.64	14	3.27	3.41	0.14
2.3	1.91	6	3.42	2.98	−0.44
1.0	0.70	15	3.42	3.21	−0.21
1.9	1.62	12	3.45	3.34	−0.11
1.3	1.01	17	3.49	3.51	0.02
1.8	1.43	15	3.58	3.54	−0.04
1.7	1.05	15	3.61	3.47	−0.14
0.7	0.62	18	3.69	3.36	−0.33
2.2	1.58	12	3.85	3.43	−0.42
1.4	1.21	15	3.88	3.39	−0.49
1.8	2.07	20	3.99	4.04	0.05
1.5	1.59	16	4.51	3.55	−0.96
2.8	2.38	24	4.74	4.75	0.01

were made of the stability number for selected values of the angle of friction and the slope angle (see Table 10-21). The correlation matrix for the twelve tests is

	X_1	X_2	Y
X_1	1.000	0.520	0.415
X_2		1.000	-0.543
Y			1.000

The intercorrelation is moderate; the predictor–criterion correlation coefficients are not high, with the two values only representing 17.2% and 29.5% of the total variation in the stability number.

The least-squares regression equation is

$$\hat{Y} = 0.04234 + 0.002842X_1 - 0.005093X_2 \qquad (10\text{-}108)$$

TABLE 10-21 Multiple Regression Analysis of Slope Stability Values

X_1	X_2	Y	$\hat{Y}$	e
10	5	0.053	0.045	-0.008
10	10	0.005	0.020	0.015
20	5	0.081	0.074	-0.007
20	12	0.036	0.038	0.002
20	18	0.012	0.008	-0.004
30	6	0.102	0.097	-0.005
30	13	0.058	0.061	0.003
30	21	0.021	0.021	0.000
40	9	0.105	0.110	0.005
40	14	0.079	0.085	0.006
40	21	0.052	0.049	-0.003
40	26	0.027	0.024	-0.003

X_1 = friction angle (degrees), X_2 = slope angle (degrees), and Y = stability number.

The multiple correlation coefficient for Eq. 10-108 is 0.981, which means that the equation explains 96.2% of the variation in the measured values of the stability number. The standard error of estimate is 0.0072, which gives $S_e/S_y = 0.214$. The goodness-of-fit statistics suggest that Eq. 10-108 provides accurate estimates of the stability number.

Note that, in spite of the relatively low predictor–criterion correlation coefficients (i.e., 0.415 and −0.543), the total R^2 is very high. The intercorrelation actually enhances the reliability of the equation.

The goodness-of-fit statistics are revealing, but they do not give an entirely accurate assessment of the model. The standardized partial regression coefficients for X_1 and X_2 are $t_1 = 0.96$ and $t_2 = -1.04$. Because the latter is greater than 1, it is considered irrational and suggests that b_2 gives an unrealistic indication of the rate of change of Y with respect to X_2. The high accuracy indicated by the goodness-of-fit statistics suggests that the model is acceptable for estimation; however, the t_2 value suggests that the model may not be reliable upon extrapolation or for assessing model sensitivity.

The residuals (see Table 10-21) should be evaluated. Figure 10-12 shows a plot of the residuals as a function of the angle of friction (ordinate) and the slope angle (abscissa). While the sample size is probably too small to be very conclusive, there appears to be a region of positive residuals between friction angles of 9° and 15° and two regions of negative residuals, friction angles less than 6° and greater than 18°. This would suggest that the structure of the linear model may be incorrect.

The intercept also appears to be a problem. If both X_1 and X_2 are zero, then the stability number equals 0.042, which does not seem reasonable.

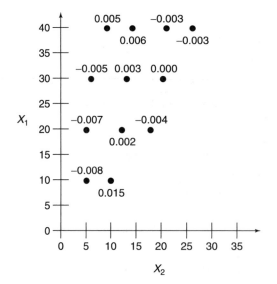

Figure 10-12　Residuals of a Multiple Regression Model versus Predictor Variables X_1 and X_2

10.11 PROBLEMS

10-1. Plot the following data, assess the degree of systematic variation, and assess the appropriateness of using a linear model:

X	0.7	1.3	2.8	5.1	7.3	8.9	11.5
Y	1.1	3.8	7.2	7.9	9.7	8.8	9.6

10-2. Plot the following data, and assess the quality of the data for prediction:

X	0.7	0.8	1.3	1.2	1.7	1.9	4.1
Y	1.0	1.8	1.2	2.1	1.0	2.8	5.7

10-3. Using the following values of X and Y, show that the total variation (TV) equals the sum of the explained variation (EV) and the unexplained variation (UV):

X	2	5	7	6	2
Y	1	3	5	7	9

Assume that the regression equation relating Y and X is given by

$$\hat{Y} = 4.5849 + 0.09434X$$

10-4. Using the following values of X and Y, show that the total variation (TV) equals the sum of the explained variation (EV) and the unexplained variation (UV). Assume that the regression equation relating Y and X is given by

$$\hat{Y} = 1.2 + 0.657X$$

Compute the correlation coefficient.

X	1	2	3	4	5	6
Y	2	2	3	5	4	5

10-5. For the following observations on X and Y, compute the correlation coefficient.

X	-3	-2	-1	0	1	2
Y	2	2	3	0	-2	-1

10-6. For the data given in Problem 10-3, compute the correlation coefficient using the standard normal transformation definition of Eq. 10-5.

10-7. Derive Eq. 10-5 from Eq. 10-4.

10-8. Given the following observations on X and Y, (a) construct a graph of X versus Y, and (b) compute the Pearson correlation coefficient.

X	1	2	3	4	5
Y	1	1	2	4	4

10-9. Given the following paired observations on variables X and Y and the values of the standardized variables Z_X and Z_Y, compute the Pearson correlation coefficient between each of the following pairs: (a) X and Y; (b) Z_X and Z_Y; (c) Z_X and X; (d) Z_X and Y; (e) X and $Y - 3$; and (f) X and $2Y$.

X	Y	Z_X	Z_Y
3	4	-1.5	-1.5
5	8	-0.5	0.5
6	7	0.0	0.0
7	6	0.5	-0.5
9	10	1.5	1.5

10-10. Using the data of Problem 10-2, fit the zero-intercept model of Eq. 10-9.

10-11. Using the following data, fit the zero-intercept model of Eq. 10-9.

Test 1	98	94	93	90	87	85	84
Test 2	96	94	91	85	93	86	90

10-12. Given the following four pairs of observations

X	1	1	2	2	?
Y	1	2	1	2	?

compute the correlation coefficient and the regression coefficients if (a) the fifth pair is ($Y = 8, X = 8$); (b) the fifth pair is ($Y = 2, X = 8$). (c) Plot both sets of points and draw the regression lines.

10-13. Show that Eq. 10-25 results in a larger sum of squares of the errors for the data of Table 10-2 than the value obtained with Eq. 10-24.

10-14. Given the following paired observations on Y and X, (a) graph Y versus X; (b) calculate the correlation coefficient; (c) determine the slope and intercept coefficients of the linear regression line; (d) show the regression line on the graph of part (a); (e) compute the predicted value of Y for each observed value of X; (f) calculate Σe and S_e. (g) What is the predicted value of Y for $X = 5$? $X = 10$?

X	3	5	6	7	9
Y	4	8	7	6	10

10-15. Compute the correlation coefficient for the data of Problem 10-2.

10-16. Compare correlation analysis and regression analysis as tools in decision making.

10-17. Using the following data set, regress (a) Y on X; (b) X on Y. (c) Compute the correlation coefficient for each regression. (d) Transform the regression of Y on X into an equation for computing X, and compare these coefficients with the coefficients for the regression of X on Y. (e) Why do the coefficients of parts (b) and (d) differ?

X	2	1	3	2
Y	1	2	4	5

10-18. The data shown are the results of a tensile test of a steel specimen, where Y is elongation in thousands of an inch that resulted when the tensile force was X thousands of pounds. Fit the data with a linear model and make a complete assessment of the results.

X	0.8	1.6	3.1	4.4	6.3	7.9	9.2
Y	2.8	4.9	6.5	8.1	8.8	9.1	8.9

10-19. The stream reaeration coefficient (Y) is related to the water temperature (T) in °C. Fit the following with a linear model and make a complete assessment of the results:

X	Y
14	2.89
23	4.20
17	4.17
11	3.69
20	3.78
15	3.56
18	3.35
11	2.69
25	3.58
19	3.06
13	2.41
24	3.30
16	3.05
23	4.18
21	4.08

10-20. Using the principle of least squares, derive the normal equations for the following model:

$$\hat{Y} = b_1X_1 + b_2X_2 + b_3X_3$$

10-21. The cost of producing power (P) in mills per kilowatt-hour is a function of the load factor (L) in percent and the cost of coal (C) in cents per million Btu. Perform a regression analysis to develop a regression equation for predicting P using the following data:

L	C	P
85	15	4.1
80	17	4.5
70	27	5.6
74	23	5.1
67	20	5.0
87	29	5.2
78	25	5.3
73	14	4.3
72	26	5.8
69	29	5.7
82	24	4.9
89	23	4.8

10-22. Visitation to lakes for recreation purposes varies directly with the population of the nearest city or town and inversely with the distance between the lake and the city. Develop a regression equation to predict visitation (V) in person-days per year, using as predictors the population (P) of the nearby town and the distance (D) in miles between the lake and the city. Use the data at the top of page 352.

D	P	V
10	10,500	22,000
22	27,500	98,000
31	18,000	24,000
18	9,000	33,000
28	31,000	41,000
12	34,000	140,000
21	22,500	78,000
13	19,000	110,000
33	12,000	13,000

10-23. Using the method of least squares, derive the normal equations for the model $\hat{Y} = b_1 X + b_2 X^2$. Evaluate the coefficients using the data of Problem 10-8.

10-24. Using the data of Problem 10-18, fit a linear model and a quadratic polynomial and make a complete assessment of each. Which model is best for prediction? Support your choice.

10-25. Using the data of Problem 10-19, fit a linear model and a quadratic polynomial and make a complete assessment of each. Which model is best for prediction? Support your choice.

10-26. Using the method of least squares, derive the normal equations for the model $\hat{Y} = ax^b$. Evaluate the coefficients using the data of Problem 10-8. (*Hint*: Convert the equation to linear form by making a logarithmic transformation.)

10-27. Using the data of Problem 10-18, fit a linear model and a power model. Make a complete assessment of each. Which model is best for prediction? Support your choice.

10-28. Using the data of Problem 10-19, fit a linear model and a power model. Make a complete assessment of each. Which model is best for prediction? Support your choice.

10-29. Using the data of Problem 10-21, fit a multiple power model. Make a complete assessment of the model.

10-30. Using the method of least squares, derive the normal equations for the model $\hat{Y} = b_0 + b_1 \sqrt{X}$. Evaluate the coefficients using the data of Problem 10-8.

Index